Electromagnetic Wave Control Techniques of Metasurfaces and Metamaterials

Jingda Wen
Beihang University, China

A volume in the Advances
in Chemical and Materials
Engineering (ACME) Book Series

Published in the United States of America by
IGI Global
Engineering Science Reference (an imprint of IGI Global)
701 E. Chocolate Avenue
Hershey PA, USA 17033
Tel: 717-533-8845
Fax: 717-533-8661
E-mail: cust@igi-global.com
Web site: http://www.igi-global.com

Library of Congress Cataloging-in-Publication Data

CIP Data in progress

British Cataloguing in Publication Data
A Cataloguing in Publication record for this book is available from the British Library.

All work contributed to this book is new, previously-unpublished material.
The views expressed in this book are those of the authors, but not necessarily of the publisher.

For electronic access to this publication, please contact: eresources@igi-global.com.

Advances in Chemical and Materials Engineering (ACME) Book Series

J. Paulo Davim
University of Aveiro, Portugal

ISSN:2327-5448
EISSN:2327-5456

MISSION

The cross disciplinary approach of chemical and materials engineering is rapidly growing as it applies to the study of educational, scientific and industrial research activities by solving complex chemical problems using computational techniques and statistical methods.

The **Advances in Chemical and Materials Engineering (ACME) Book Series** provides research on the recent advances throughout computational and statistical methods of analysis and modeling. This series brings together collaboration between chemists, engineers, statisticians, and computer scientists and offers a wealth of knowledge and useful tools to academics, practitioners, and professionals through high quality publications.

Coverage

- Materials to Renewable Energies
- Composites
- Ductility and Crack-Resistance
- Fatigue and Creep
- Thermo-Chemical Treatments
- Electrochemical and Corrosion
- Multifuncional and Smart Materials
- Heat Treatments
- Computational Methods
- Nanomaterials

IGI Global is currently accepting manuscripts for publication within this series. To submit a proposal for a volume in this series, please contact our Acquisition Editors at Acquisitions@igi-global.com or visit: http://www.igi-global.com/publish/.

The (ISSN) is published by IGI Global, 701 E. Chocolate Avenue, Hershey, PA 17033-1240, USA, www.igi-global.com. This series is composed of titles available for purchase individually; each title is edited to be contextually exclusive from any other title within the series. For pricing and ordering information please visit http://www.igi-global.com/book-series/advances-chemical-materials-engineering/73687. Postmaster: Send all address changes to above address.

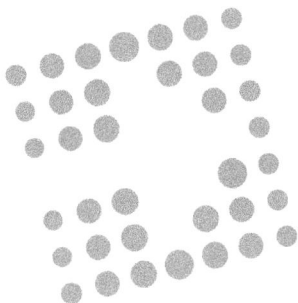

Titles in this Series

For a list of additional titles in this series, please visit:
http://www.igi-global.com/book-series/

Machining Polymer Matrix Composites Tools, Techniques, and Sustainability
Francisco Mata Cabrera (Universidad de Castilla-La Mancha, Spain) and Issam Hanafi (LSIA, ENSAH, Université Abdelmalek Essaâdi, Morocco)
Engineering Science Reference • copyright 2024 • 306pp • H/C (ISBN: 9781668499276) • US $275.00 (our price)

Biochemistry in the Space of the Highest Dimension
Gennadiy Vladimirovich Zhizhin (Russian Academy of Natural Sciences, Russia)
Engineering Science Reference • copyright 2024 • 326pp • H/C (ISBN: 9798369305881) • US $295.00 (our price)

Advancements in Renewable Energy and Green Hydrogen
Maria Simona Raboaca (National Research and Development Institute for Cryogenic and Isotopic Technologies, Râmnicu Vâlcea, Romania) Saheb Djohra (Renewable Energy Development Centre, Algeria) Omar Moussaoui (University Mohammed First, Oujda, Morocco) Traian Candin Mihaltan (Technical University of Cluj, Romania) and Mustapha Koussa (Renewable Energy Development Centre, Algeria)
Engineering Science Reference • copyright 2024 • 327pp • H/C (ISBN: 9798369310144) • US $260.00 (our price)

Production, Properties, and Applications of Engineered Cementitious Composites
S. Praveenkumar (PSG College of Technology, India) and J. Paulo Davim (University of Aveiro, Portugal)
Engineering Science Reference • copyright 2024 • 344pp • H/C (ISBN: 9781668481820) • US $250.00 (our price)

For an entire list of titles in this series, please visit:
http://www.igi-global.com/book-series/

IGI Global
PUBLISHER of TIMELY KNOWLEDGE

701 East Chocolate Avenue, Hershey, PA 17033, USA
Tel: 717-533-8845 x100 • Fax: 717-533-8661
E-Mail: cust@igi-global.com • www.igi-global.com

Table of Contents

Preface xii

Chapter 1
A Comprehensive Review for Beam Steering Technology: From Mechanical Prisms to Space-Time Metasurface 1
Yinpeng Wang, National University of Singapore, Singapore

Chapter 2
Metamaterial Antennas for Wireless Communication Systems 32
Smrity Dwivedi, IIT BHU, India

Chapter 3
Metamaterials and Metasurfaces for Sensor and Biosensor Applications 61
Maryam Ghodrati, Lorestan University, Iran
Ali Mir, Lorestan university, Iran
Jingda Wen, Tsinghua University, China

Chapter 4
Antenna Design for Beyond 5G: 5G and 6G Applications With Different Materials and Techniques Including Metamaterials and Metasurfaces 106
Smrity Dwivedi, IIT BHU, India

Chapter 5
Metamaterial and Metasurface for Wireless Power Transfer and Magneto-Inductive Waveguide 141
Thanh Son Pham, Vietnam Academy of Science and Technology, Vietnam
Xuan Thanh Pham, Hanoi University of Industry, Vietnam
Manh Kha Hoang, Hanoi University of Industry, Vietnam

Chapter 6
Planar Metamaterial Microwave Sensors for Characterization of Dielectric Materials 164
Man Seng Sim, Universiti Teknologi Malaysia, Malaysia
Kok Yeow You, Universiti Teknologi Malaysia, Malaysia
Stephanie Yen Nee Kew, Universiti Teknologi Malaysia, Malaysia
Raimi Dewan, Universiti Teknologi Malaysia, Malaysia
Fahmiruddin Esa, Universiti Tun Hussein Onn Malaysia, Malaysia
DiviyaDevi Paramasivam, Universiti Teknologi Malaysia, Malaysia
Fandi Hamid, Universiti Teknologi Malaysia, Malaysia

Chapter 7
Performance Analysis of Dual-Band Microstrip Patch Antenna With and Without Electromagnetic Band Gap Structures: A Comparative Study 207
Mehaboob Mujawar, Bearys Institute of Technology, India
Subuh Pramono, Universitas Sebelas Maret, Surakarta, Indonesia

Chapter 8
An Eight-Scallop Flower-Based Ultrathin and Bidirectional Multiple Passband FSS 225
Kanwar Preet Kaur, Department of Electronics and Communication Engineering, Charotar University of Science and Technology, India
Trushit Upadhyaya, Department of Electronics and Communication Engineering, Chandubhai S. Patel Institute of Technology, India
Upesh Patel, Department of Electronics and Communication Engineering, Chandubhai S. Patel Institute of Technology, India
Poonam Thanki, Department of Electronics and Communication Engineering, Chandubhai S. Patel Institute of Technology, India

Compilation of References 239

About the Contributors 275

Index 279

Detailed Table of Contents

Preface xii

Chapter 1
A Comprehensive Review for Beam Steering Technology: From Mechanical Prisms to Space-Time Metasurface 1
Yinpeng Wang, National University of Singapore, Singapore

Beam steering technology refers to technologies aiming at altering the propagation direction of the main lobe in radiation scenarios. Among the past decades, the beam steering technique has aroused people's attraction in various research directions. Simultaneously, multitudinous emerging realms like radar, biological imaging, wireless communication, and sensing are inseparable from beam steering. In this chapter, the authors have elaborated on the development of beam control technology from the classic mechanical prism to the latest spatiotemporal encoded metasurface. This review first covers geometric dependent methods, including mechanical control, MEMS control, liquid control, and optical control mechanisms, which are on the basis of Snell's law. After that, the phase control methods based on generalized Snell's law such as semiconductor components, electrically tunable materials, phase change materials, MEMS and spatiotemporal metasurfaces are discussed detailly. Finally, the authors summarize the methods of beam control and look forward to future possibilities.

Metamaterials and metasurfaces are two unique engineered materials which play a very important role in every field of research at almost every frequency range and band. Optical, microwave, terahertz metamaterials and metasurfaces now in demand due its special features and properties which can bend, absorb, propagate, and enhance, block the electromagnetic waves of any frequency. It has a special structure dimension which is related to wavelength of that wave which is passing through it. There are four types of metamaterials that have been defined according to permittivity and permeability values in different quadrants, as velocity is also different for these metamaterials and metasurfaces. In all applications area, design of antenna is the main issue on which researchers are working, and many types of antenna designs have been proposed and fabricated. For 5G, beyond 5G, and 6G applications, metamaterials are the best, chosen by designers apart from graphene, liquid crystal, and gold. In this chapter, the authors will cover all about metamaterials structures, properties and the main part is design of antenna for wireless applications.

Metamaterials are efficiently homogenizable arrangements of artificial structural components engineered to achieve beneficial and exotic electromagnetic (EM) properties not found in natural materials. Metasurfaces are the two-dimensional analogue of metamaterials consisting of single-layer or multi-layer stacks of planar structures. Both metamaterials and metasurfaces have great potential to be used in a wide range of applications, e.g., antennas, absorbers, slow light devices, photocatalysis, optical modulation devices, and ultra-sensitive sensors and biosensors. This chapter highlights the recent advances in sensing and biosensing technology based on metamaterials and metasurfaces as well as its main applications, focusing on the effect of material selection on device performance. This chapter offers a detailed and comprehensive analysis of optical sensors and biosensors based on metamaterials and metasurfaces for medical diagnoses, biomolecule detection like viruses, and bacteria, and disease diagnosis such as cancer.

Chapter 4

In the last few years, economic and social development has been greatly influenced by the advancements in the field of mobile communication and technology. 5G technology has emerged as an important of the future 2020 generation which is already in use. After the development of fifth generation technologies, researchers, scientist, and engineers are looking for wide bandwidth which should be improve wireless systems and devices to provide better services and fast experience. Also, the development of 5G wireless network technology is the response to the crucial factors that lead to this demand because of its ability to provide extremely fast internet speed, high bandwidth, high performance, reduced latency, and high reliability and better gain. Metamaterials (LHM) hold great promises for such devices and for many applications like medical, technologies, communications, defense, etc. Metamaterials have interesting properties, making them more amenable to transmit and receive in small quantities. In this chapter, the authors will review the recent design with advancement in 5G antennas for beyond 5G application and also we will give few design of antenna for 6G applications.

Chapter 5

Metamaterials, specifically in their two-dimensional form known as metasurfaces, have emerged as a focal point of extensive investigation within contemporary scientific discourse. This chapter serves to elucidate the intricacies of the theoretical underpinnings and the practical implementation of a metasurface architecture engineered to function within the frequency domain of megahertz (MHz). The metasurface under consideration exhibits considerable promise in the realm of wireless power transfer (WPT) systems, showcasing its potential to develop this domain. Furthermore, its intrinsic ability to propagate magneto-inductive waves (MIWs) imbues it with a multifaceted utility, enabling its application in diverse scenarios such as near-field information transmission and the development of structures that facilitate the simultaneous conveyance of both energy and information.

Metamaterials can be integrated into planar microwave sensors for field localization and enhancement as well as sensitivity improvement in permittivity-based sensing measurements. This chapter reviews metamaterial-based planar microwave sensors for characterizing dielectric materials. It begins by introducing planar microwave sensors. The subsequent section focuses on the sensing principles of planar microwave sensors loaded with metamaterial-inspired resonators, namely frequency shift, frequency splitting, and amplitude or phase variation. Furthermore, recent advances in metamaterial-integrated sensors are discussed, focusing on the types of samples under test, including solid samples in bulk and powder form, and liquid samples in fluid tubes or microfluidic channels. Furthermore, planar antennas loaded with metamaterial elements are also explored. To enhance understanding of frequency-variation sensors, a microstrip transmission line loaded with complementary split ring resonators is simulated. Finally, the design strategies of planar metamaterial-based sensors are reviewed.

This chapter investigates the optimization of dual-band MPA antennas in modern wireless communication systems. By integrating Electromagnetic Bandgap (EBG) structures, performance metrics such as bandwidth, gain, and radiation efficiency are enhanced. The chapter presents a thorough literature review, design methodology, and simulation results analyzing key metrics. Insights gained from this comparative study contribute to antenna design advancements. Additionally, the impact of antenna bending on wearable applications is explored, along with Specific Absorption Rate (SAR) analysis ensuring safety within regulatory limits.

Preface

Welcome to "Electromagnetic Wave Control Techniques of Metasurfaces and Metamaterials." In recent years, the fields of metasurfaces and metamaterials have undergone a transformative surge in attention and innovation, particularly in their applications within communication systems, smart IoT technologies, sensing devices, and beyond. The ability to control electromagnetic waves with precision and efficiency lies at the heart of these advancements, making metasurfaces and metamaterials indispensable in various domains.

This book is a collaborative effort aimed at providing researchers with a comprehensive resource to deepen their understanding and explore the latest developments in practical metamaterial and metasurface applications. It is structured into three distinct sections: elucidating the mechanisms underlying metasurface and metamaterial functionalities, showcasing their diverse applications, and delving into other pertinent topics surrounding these materials.

Our primary goal is to showcase the cutting-edge designs and techniques in wave control, with a special emphasis on applications such as metamaterial absorbers, metasurface converters, sensor technologies, MEMS integration, broadband metamaterials, tunable and reconfigurable designs, and intelligent metasurfaces. However, we welcome contributions that span theoretical insights, simulation methodologies, and fundamental mechanisms, ensuring a holistic exploration of these fascinating materials.

This book is designed to serve as a valuable resource for students, researchers, scientists, and industry professionals alike. Whether you're interested in microwave/terahertz/optical metamaterial absorbers, employing deep learning for optimized designs, developing metamaterial antennas for wireless communication, or exploring the frontier of near-zero index metamaterials, this compilation offers insights and methodologies to fuel your exploration.

We extend our gratitude to all contributors who have shared their expertise and perspectives, enriching this collective endeavor. We hope that "Electromagnetic Wave Control Techniques of Metasurfaces and Metamaterials" inspires new ideas, fosters collaborations, and contributes significantly to the ongoing advancements in this dynamic field.

Organization of the Book

Chapter 1: A Comprehensive Review for Beam Steering Technology: From Mechanical Prisms to Space-time Metasurface

This chapter delves into the evolution of beam steering technology, tracing its development from traditional mechanical prisms to the cutting-edge spatiotemporal encoded metasurfaces. Beginning with a review of geometric dependent methods such as mechanical, MEMS, liquid, and optical control mechanisms based on Snell's law, the discussion progresses to phase control methods utilizing semiconductor components, electrically tunable materials, phase change materials, MEMS, and spatiotemporal metasurfaces. The chapter provides detailed insights into various beam control methods and offers a glimpse into future possibilities in the field.

Chapter 2: Metamaterial Antennas for Wireless Communication Systems

Metamaterials and metasurfaces are explored in this chapter as essential components in wireless communication systems across different frequency ranges. The chapter covers the structural dimensions, unique properties, and design aspects of antennas for wireless applications. It highlights the significance of metamaterials in bending, absorbing, propagating, and enhancing electromagnetic waves, with a focus on their role in the design of antennas for 5G, beyond 5G, and 6G applications.

Chapter 3: Metamaterials and Metasurfaces for Sensor and Biosensor Applications

This chapter presents recent advances in sensing and biosensing technologies leveraging metamaterials and metasurfaces. It explores the utilization of these engineered materials in various applications such as medical diagnoses, biomolecule detection, and disease diagnosis. The chapter provides a comprehensive analysis of optical sensors and biosensors, emphasizing the impact of material selection on device performance and highlighting the potential for transformative applications in healthcare and biotechnology.

Chapter 4: Antenna Design for beyond 5G, 5G and 6G applications with different materials and techniques including Metamaterials and Metasurfaces

Focusing on the rapid advancements in mobile communication technology, this chapter reviews antenna designs tailored for beyond 5G, 5G, and 6G applications. It discusses the importance of wide bandwidth, improved wireless systems, and fast data transmission in meeting the demands of evolving communication networks. With a spotlight on antenna design strategies incorporating metamaterials and metasurfaces, the chapter offers insights into enhancing performance and reliability in next-generation wireless systems.

Chapter 5: Metamaterial and Metasurface for Wireless Power Transfer and Magneto-Inductive Waveguide

This chapter delves into the theoretical foundations and practical implementation of metasurfaces in wireless power transfer systems operating in the megahertz frequency domain. It explores the potential of metasurfaces to facilitate wireless power transfer and magneto-inductive wave propagation, enabling diverse applications such as near-field information transmission. The chapter highlights the multifaceted utility of metasurfaces in energy and information transmission scenarios.

Chapter 6: Planar Metamaterial Microwave Sensors for Characterization of Dielectric Materials

Offering a comprehensive overview of metamaterial-integrated planar microwave sensors, this chapter explores their role in characterizing dielectric materials. It discusses sensing principles, recent advances, and design strategies, focusing on enhancing sensitivity and localization in sensing measurements. The chapter also examines the integration of metamaterial elements into planar antennas, expanding the capabilities of microwave sensors for diverse applications.

Chapter 7: Performance Analysis of Dual-Band Microstrip Patch Antenna with and without Electromagnetic Band Gap Structures: A Comparative Study

This chapter investigates the optimization of dual-band microstrip patch antennas through the integration of electromagnetic bandgap (EBG) structures. It analyzes key performance metrics such as bandwidth, gain, and radiation efficiency, comparing antennas with and without EBG structures. Additionally, the chapter explores the impact of antenna bending on wearable applications and ensures compliance with safety regulations through Specific Absorption Rate (SAR) analysis.

Chapter 8: An Eight-Scallop Flower-based Ultrathin and Bidirectional Multiple Passband FSS

Presenting a novel ultrathin microwave passband frequency selective surface (FSS), this chapter discusses its design, fabrication, and performance evaluation. The FSS architecture, comprising concentric metallic eight-scallop flower (ESF) rings, demonstrates multiple passbands and stability across different polarization states. Fabricated using standard printed circuit board (PCB) techniques, the prototype's performance aligns closely with simulation results, showcasing its potential for practical applications.

Conclusion

As we conclude "Electromagnetic Wave Control Techniques of Metasurfaces and Metamaterials," we reflect on the immense strides made in understanding and harnessing these transformative materials. The collaborative effort of researchers, scientists, and industry professionals has culminated in a comprehensive resource that delves into the intricacies of metasurfaces and metamaterials, their mechanisms, applications, and theoretical underpinnings.

Throughout this book, we have explored a myriad of topics, from the fundamental principles of wave control to the practical implementations in communication systems, IoT technologies, sensing devices, and more. The diverse range of contributions has shed light on the versatility and potential of metasurfaces and metamaterials, showcasing their role as indispensable tools in modern engineering and scientific endeavors.

Chapter 1
A Comprehensive Review for Beam Steering Technology:
From Mechanical Prisms to Space-Time Metasurface

Yinpeng Wang
National University of Singapore, Singapore

ABSTRACT

Beam steering technology refers to technologies aiming at altering the propagation direction of the main lobe in radiation scenarios. Among the past decades, the beam steering technique has aroused people's attraction in various research directions. Simultaneously, multitudinous emerging realms like radar, biological imaging, wireless communication, and sensing are inseparable from beam steering. In this chapter, the authors have elaborated on the development of beam control technology from the classic mechanical prism to the latest spatiotemporal encoded metasurface. This review first covers geometric dependent methods, including mechanical control, MEMS control, liquid control, and optical control mechanisms, which are on the basis of Snell's law. After that, the phase control methods based on generalized Snell's law such as semiconductor components, electrically tunable materials, phase change materials, MEMS and spatiotemporal metasurfaces are discussed detailly. Finally, the authors summarize the methods of beam control and look forward to future possibilities.

DOI: 10.4018/979-8-3693-2599-5.ch001

INTRODUCTION

Beam steering represents an alluring technology, characterized by the manipulation of the main lobe's propagation direction within a specified radiation pattern. This capability facilitates the execution of intricate functions such as focusing, deflection, polarization conversion, and the generation of orbital angular momentum (OAM). Over recent years, this pervasive technology has found widespread applications in diverse domains, including radar (B. Li et al., 2023), biological imaging (Hernandez-Cardoso et al., 2017), wireless communication (Chen et al., 2022), and sensing (Shangguan et al., 2017). For instance, in radar systems, beam control enables directional scanning for the detection of unknown objects; in wireless communication, it is harnessed for antenna directional transmission and reception to mitigate losses and enhance the signal-to-noise ratio (SNR). Furthermore, within the field of sensors, beam control is leveraged to concentrate the beam on a specific area, thereby improving resolution.

The burgeoning interest in beam control is evidenced by recent research trends. A comprehensive analysis was conducted by querying the keyword 'beam steering' on Web of Science and utilizing analytical tools to tally papers included in the core collection of Web of Science spanning the years 1995 to 2022. As depicted in Figure 1 (a), a discernible surge in the number of published papers is evident, escalating from 10 papers per year in 1995 to 877 in 2022, underscoring the dynamic prospects within this field. Concurrently, a systematic collection of pertinent papers related to radio frequency, millimeter wave, terahertz, infrared, and visible light, categorized according to electromagnetic wave frequency bands, was undertaken. The outcomes, as portrayed in the pie chart in Figure 1 (b), reveal that traditional RF and millimeter wave beam steering continue to dominate half of the total papers. Simultaneously, the nascent investigations in terahertz and infrared band control exhibit rapid development, exemplifying the diversification of research within this domain.

From the perspective of regulation mechanism, beam control can be divided into two categories: geometry-based and phase-based control. Here, geometry-based control refers to utilize machinery, electricity, temperature, or light to change the relative position of the controller, thereby achieving direct control of the beam direction. This type of method includes mechanical control, micro-electromechanical control, liquid control, optical control, etc. In recent years, with the development of phased array technology, the implementation of phase-controlled beamforming has become more popular. In such approaches, by accurately adjusting the phase of each subunit, the equivalent beam direction can be precisely determined. Phase tuning can be achieved through various methods such as semiconductor components, electrically tunable materials, phase change materials, MEMS, spatiotemporal metasurfaces, etc. In the following parts of this chapter, we will first introduce the mechanisms and examples of geometry-based approaches, and then focus on phase-based methods. Lastly, we will elaborate the conclusion and prospects for future research directions at the end of this chapter.

Figure 1. Search results of "beam steering" on web of science: (a) Number of articles published annually from 1995 to 2022, (b) Proportion of articles published in different frequency bands

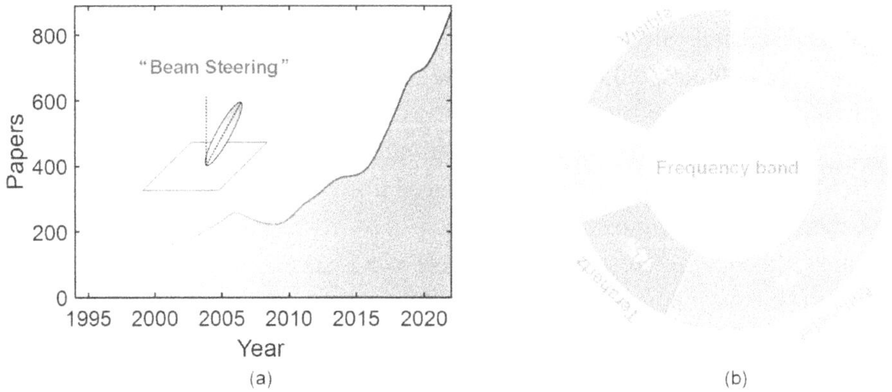

GEOMETRY BASED BEAM STEERING

Geometry based control is the most straightforward beam manipulation method with the most direct principle. Assuming the wave source is stationary, the incident angle can be altered by changing the geometric position of the refracting surface (or reflecting surface), thereby obtaining the desired refracted (or reflected) light. As shown in Figure 2 (a-b), the incident light and refracted light (or reflected light) satisfy the Snell's law:

$$n_1 \sin\theta_1 = n_2 \sin\theta_2 \tag{1}$$

where n_1 and n_2 represent the refractive index of the medium in which the incident and refracted light are located (for reflection cases, $n_2 = n_1$). The incidence angle θ_1 can be quantitatively adjusted using machinery, electricity, or light, in order to acquire the required exit direction θ_2. Next, we will explore different mechanisms of geometry-based beam steering.

Mechanical Controlled

Mechanical-based control uses a motor to realize the translation or rotation of the refractive/reflective surface to change the incident angle, thereby directly changing the direction of the outgoing beam. This classic method is already quite mature, which has a wide adjustment range. Beam manipulation using reflected light is typically achieved using mirrors. In 2010, Duma *et al.* (Duma & Rolland, 2010) introduced a biaxial polygonal mirror (PM) to achieve spatial scanning. The design consists of two axes, each connected to a reflector. The mirror on the main axis is used for lateral control of two-dimensional plane beams, while the other mirror is used to control beam pitch. This design enhances the range of beam adjustment.

Employing refraction to control the direction of the beam usually requires a prism or lens. For example, Li *et al.* (A. Li et al., 2016) achieved the manipulation of the beam in a two-dimensional plane using a Risley prism in 2016. As displayed in Figure 2 (a), the Risley system consists of two wedge-shaped prisms placed side by side, each of which can be rotated by an independent motor. Due to the angle difference between the two lenses, the beam can be deflected twice, thereby increasing the adjustment range. Figure 2 (b) shows another method of using mechanical scanning to control the direction of refractive beams by lens, which was proposed by Alonso del Pino *et al.* (Alonso-delPino et al., 2019) in 2019. The hemispherical silicon lens is located above the waveguide, and the position of the waveguide can be controlled by a piezoelectric motor. By changing the relative position of the waveguide and lens, beam steering can be achieved from -13 ° to 17 ° at a working frequency of 550 GHz.

Although the mechanical control principle is simple and the cost is fairly low, due to the limitation of its scanning speed by the electric motor, the beam control rate is not sufficient for high-speed applications. On the other hand, the bulky size of driving devices also hinders the integration of the control component.

Figure 2. Two classical geometry-based beam steering approaches: (a) Risley prism, and (b) translational lens

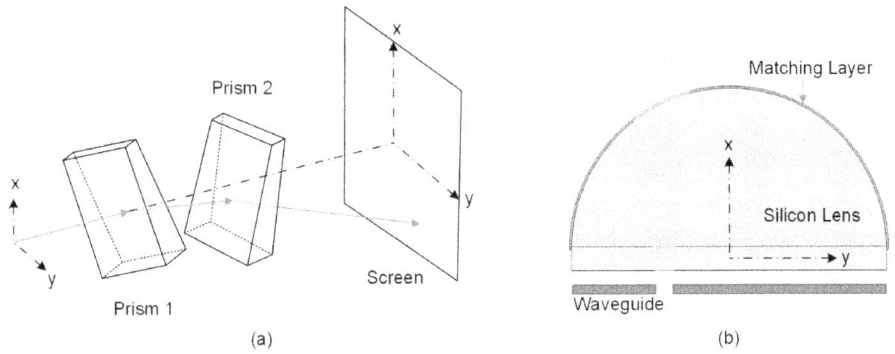

MEMS Controlled

Micro electromechanical systems (MEMS) integrate micro machinery and electronic circuits, greatly reducing the volume of traditional mechanical beam control devices and improving response speed. The basic principle of beam control in MEMS is to use physical mechanisms such as electrostatic/electromagnetic/piezoelectric or electrothermal actuation to drive the rotation of the reflect mirror, thereby enabling the reflected light to deflect.

Electrostatic Actuator

The actuator driven by static electricity utilizes the electrostatic attraction between the capacitor plates to drive the plate to rotate, thereby driving the mirror to rotate. Here, the driving force F_l can be obtained from the derivative of the total energy storage to the direction of rotation l:

$$F_l = \frac{\partial}{\partial l}\left(\frac{1}{2}C_t U^2\right) \tag{2}$$

where C_t is the total capacity and U is the voltage applied to the flat plates. The advantages of electrostatic actuators are their simple structure and fast response speed. Hofman *et al.* (Hofmann et al., 2012) proposed a beam controller based on electrostatic actuation in 2012. As shown in Figure 3 (a), the controller consists of stacked vertical combs. By changing the voltage of the upper and lower plates of the fixed comb, the tilt direction of the reflecting mirror can be altered. In fact, they ultimately achieved a one-dimensional scanning angle of about 86°.

Electromagnetic Actuator

The electromagnetic actuator employs the Ampere force, generated by a conducting wire in a magnetic field, as the driving force. The Ampere force F can be calculated as

$$F = BIl \sin\theta \tag{3}$$

where l and B are the length and magnetic field while I and θ are the currency and the angle between B and I. Electromagnetic drive avoids the drawbacks of electrostatic pull-in effect, resulting in a more flexible controlling process. For beam steering, Li *et al.* (F. Li et al., 2017) proposed a beam scanner based on an electromagnetic actuator in 2017. As shown in Figure 3 (b), the bottom of the reflector surface is coated with magnetic material nickel, and its lower side is two sets of coils. Under the driving currents of 21 mA and 38 mA, the author achieved deflection angles of ± 4.5 ° and ± 7.6 ° for two axes.

Figure 3. Beam steering mechanisms based on micro electromechanical systems: (a) Electrostatic actuator, (b) Electromagnetic actuator, (c) Piezoelectric actuator, and (d) Electrothermal actuator

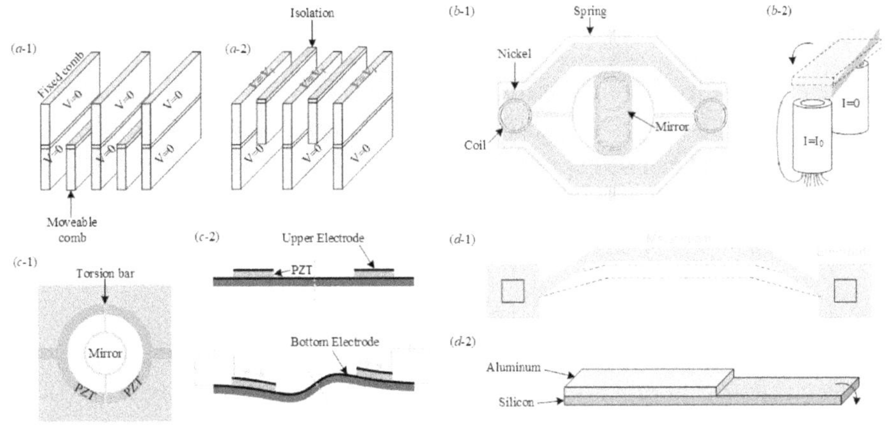

Piezoelectric Actuator

Piezoelectric actuators rely on the inverse piezoelectric effect: applying a suitable voltage to them can cause required deformation. As exhibited in Figure 3 (c), the classic one fixed end double-layer cantilever beam is composed of a piezoelectric layer and an ordinary elastic layer, and the quantitative relationship between the bending center angle θ and the applied voltage is:

$$\theta = \frac{2d_{31}(t_p + t_e)A_pE_pA_eE_3L}{4(E_pI_p + E_eI_e)(A_pE_p + A_eE_e) + A_pE_pA_eE_e(t_p + t_e)^2} \tag{4}$$

where A_p/A_e, E_p/E_e, T_p/t_e are the cross-sectional area, Young's modulus, and thickness of the piezoelectric/elastic layer, respectively. Besides, L, E_3 and d_{31} represents the length, longitudinal electric field, and piezoelectric coefficient. Piezoelectric actuators typically have a fast response speed, thus could be used to tune the beam deflection direction at real time. Ye *et al.* (Ye et al., 2017) proposed a beam scanner based on piezoelectric material PZT in 2017. They applied 5V AC voltage to PZT ceramics and amplified the amplitude using a lever structure. By elaborate optimize the parameters, they ultimately achieved a two-axis scanning range of 41.9°×40.3° with a power consumption of only 16 mW.

Electrothermal Actuator

As presented in Figure 3 (d), electrothermal actuators usually adopt a thermal bimorph structure, which converts the temperature changes caused by Joule heat into the displacement of the mechanical beam. Electrothermal actuators are composed of two materials with different thermal expansion coefficients (α_1, α_2) and a common length L. With the temperature difference ΔT, the center angle of the bending beam with one fixed end can be expressed as

$$\theta = \frac{6w_1w_2E_1E_2t_1t_2(t_1 + t_2)(\alpha_1 - \alpha_2)\text{Ä}TL}{(w_1E_1t_1^2)^2 + (w_2E_2t_2^2)^2 + 2w_1w_2E_1E_2t_1t_2(2t_1^2 + 3t_1t_2 + 2t_2^2)} \tag{5}$$

where w_i, T_i, E_i, α_i represents the width, thickness, Young's modulus, and thermal expansion coefficient of two beams with different materials. Wang *et al.* (H. Wang et al., 2018) proposed a beam scanner based on electrothermal actuators in 2018. The control rod is composed of an Al-SiO_2-Al structure, and the displacement of the top of the wafer can be controlled by controlling temperature, thereby changing the deflection angle of the reflector. Experiments have validated that by applying a voltage of 4.5V, the device can achieve an angle control range of $\pm$ 8.5 ° on both axes.

Although MEMS devices have advantages such as fast respond speed and low power consumption compared to traditional mechanical scanning, their working angle range is usually limited, and their durability and impact resistance are also concerning.

Liquid Controlled

Liquid based beam control method is another emerging method, which has the advantages of wide bandwidth, low power consumption, and simple structure. Due to the avoidance of mechanical structures, its reliability has also been improved. The mechanism of using liquids to control light deflection varies, but they all exploit liquid surfaces to achieve refractive light deflection. Generally speaking, these methods (Y. Cheng et al., 2021) can be divided into dielectric electrophoresis effect, electrowetting effect, hydraulic control, etc.

Dielectric Electrophoresis Effect

The principle of beam steering based on dielectric electrophoresis effect is illustrated in Figure 4 (a) to introduce two liquids with different dielectric constants to generate dielectric force when imposing electric fields, thereby causing deformation of the liquid surface. Among them, the force density at the liquid level can be defined as

$$F = \frac{\varepsilon_0}{2}(\varepsilon_1 - \varepsilon_2)\ \nabla\ (\mathbf{E} \bullet \mathbf{E}) \tag{6}$$

where ε_0, ε_1 and ε_2 are the vacuum dielectric constant and the relative dielectric constant of the two liquids. **E** is the electric field at the interface. Lin *et al.* (Lin et al., 2009) achieved broadband beam control using the principle of dielectrophoresis in 2009. The equipment they adopted is shown in Figure 3, which is consisting of wedge-shaped indium tin oxide (ITO) glass plates, optical fluid (SL-5267), and deionized water. The author achieved a light deflection angle of approximately 0.87° at a 180 V electrode voltage.

Electrowetting Effect

The electrowetting effect is another liquid based beam control method, mainly supported by the phenomenon of electrocapillarity. Figure 4 (b) reveals a micro container with electrode plates placed on both sides. When not electrified, the liquid surface bends into an arc on both sides due to surface tension. After imposing different voltages to the left and right plates, the contact angle between the liquid and the wall alters, causing the liquid level to tilt. Through theoretical calculation, the contact angle at voltage V is θ_V.

$$\cos\theta_V = \cos\theta_0 + \frac{\varepsilon V^2}{2d\gamma} \tag{7}$$

where ε, θ_0, d and γ are the permittivity, initial contact angle, the thickness of the dielectric layer and the surface tension coefficients. As early as 2007, Smith *et al.* (Smith et al., 2006) used an electrowetting prism to achieve beam control. They used the liquid surface of a mixture of water/glycerol/potassium chloride solution as the refractive surface of the beam and achieved continuous beam control of ± 7° by controlling the voltage of the two plates. Sometimes, in order to reduce the volatilization of conductive solutions, a layer of nonconductive and nonvolatile liquid can be added on its top. In 2011, Cheng *et al.* (J. Cheng & Chen, 2011) also utilized electrowetting technology to actively control beam steering. They used the interface between potassium chloride solution and silicone oil as the beam refractive surface and realized the deflection and guiding of the beam at an incidence angle of 0-15° by controlling the voltage of the two electrodes.

Hydraulic Control

In addition to the above two principles, there is also a mechanism based on hydraulic control. By encapsulating the liquid in a sealed film and modify the pressure distribution inside the film, the geometric shape of the film can be changed, thereby deflecting the beam. Based on this principle, Shaw *et al.* (Shaw, 2007) achieved a beam deflection of ±4.4° in 2007 by controlling the deformation of the liquid lens using the displacement of 112 screws.

Although liquid control has a few advantages, the drawbacks are also obvious. Generally speaking, driving a liquid level deviation requires a high driving voltage (hundreds of Voltes). Even so, its adjustable deflection angle is still small, and the steering speed is also quite slow, making it difficult to achieve the required accuracy.

Figure 4. Liquid controlled beam steering methods: (a) Dielectric electrophoresis effect, and (b) electrowetting effect.

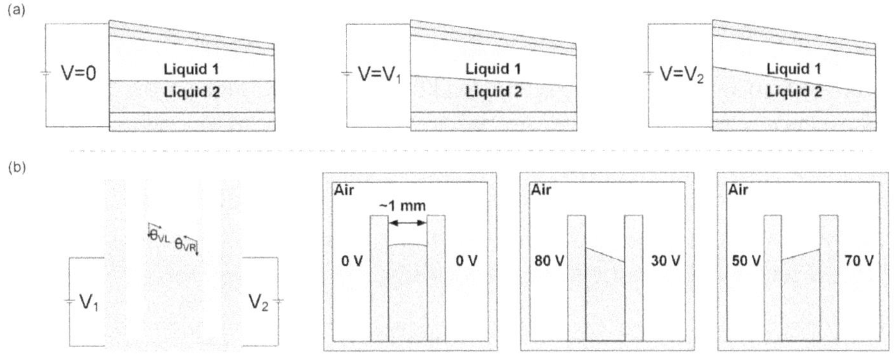

Optical Control

The aforementioned methods almost invariably use electrical control to achieve geometric state transforms. In fact, similar functions can also be realized through optical control. Taking liquid crystal (LC) based polymers as an example, induced phase transitions occur under specific illuminating conditions, resulting in macroscopic deformation. Typically, polymer materials are composed of LC monomers, LC crosslinkers, and photoinitiators. When illuminated by a beam of light, the corresponding portion of the polymer transforms from a liquid crystal phase to an amorphous phase. Due to the different expansion coefficients of the two phases, geometric bending of the polymer can occur. The use of geometric bending can drive reflective or refractive devices, thereby achieving beam steering.

In 2017, Nocentini *et al.* (Nocentini et al., 2017) utilized liquid crystal elastomer fibers to achieve 360° beam rotation in a plane. The optical fiber is manufactured by pulling a wire from LC monomer mixture using a bar under UV lamp curing. During the beam modulation process, green laser is used to irradiate the fiber horizontally, and the rotation angle is adjusted by controlling the laser power, thereby controlling the reflected light. In 2017, Li *et al.* (C. Li et al., 2017) utilized azo cholesterol type liquid crystal polymer (azo ChLCP) to achieve over 90% reflected light deflection in the broadband visible and near-infrared bands with an angle tuning range of 54°. As shown in Figure 5 (a), they used a purple light pump to make the flat plate undergo photomechanical bending, and a green light pump to restore the original structure in just 22 seconds. In addition, the plane deflection angle can be fixed by turning off the pump source during the deformation process, thereby flexibly adjusting the pointing angle.

Apart from reflective beam control, researchers have also realized beam steering for transmission lights. In 2023, Zhuang *et al.* (Zhuang et al., 2023) proposed a phase discontinuity metasurface with C-shaped split rings based on liquid crystal elastomers to control terahertz beams. Displayed in Figure 5 (b), the proposed metasurface uses a linearly focused infrared beam as the pump source, ultimately achieving a 22° angle deflection for transmitted light at 0.68 THz. By controlling the power of the infrared pump source, the deflection angle can be flexibly adjusted. After removing the pump source, the bended plate can return to its original state within 5 seconds, emerging strong reusability.

Figure 5. Optical controlled beam steering: (a) Azo cholesterol type liquid crystal polymer and (b) liquid crystal elastomers

(a) (b)

Although optical deformation can achieve large deflection angles, its drawbacks cannot be ignored. The corresponding time for this type of regulation method is usually on the order of seconds, making it arduous to apply to high-speed systems.

PHASE BASED BEAM STEERING

Phase control is another beam manipulation category which does not use mechanical components to bend the light. Based on the interference principle, it utilizes the carefully designed phase difference between each unit to achieve the equivalent beam propagation direction. The theoretical basis for phase-based beam manipulation method is the Generalized Snell's law exhibited in Figure 6, which is first brought by Yu *et al.* (Yu et al., 2011) in 2011.

Figure 6. The principle of generalized Snell's law

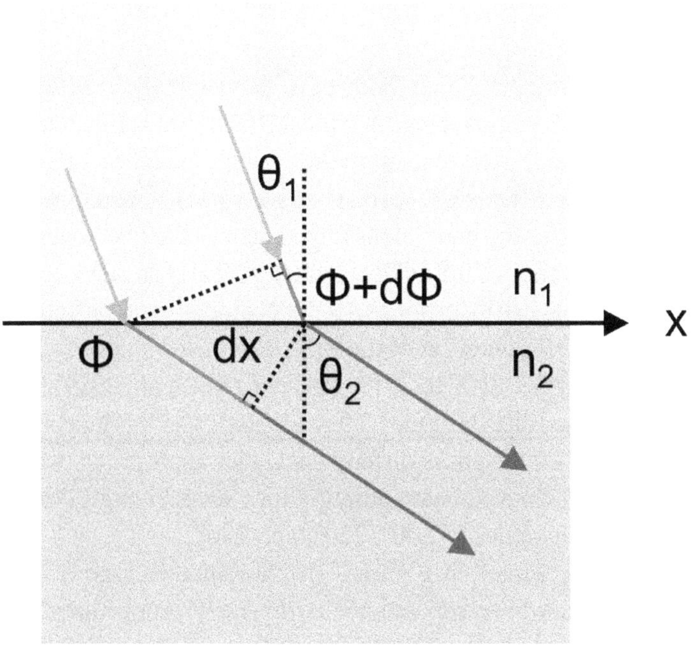

$$n_1 sin\theta_1 - n_2 \sin\theta_2 = \frac{\lambda_0}{2\pi}\frac{d\Phi}{dx} \quad (8)$$

where n_1 and n_2 represent the refractive index of the medium in which the incident and refracted light are located (for reflection cases, $n_2 = n_1$). Unlike geometry-based methods which alter the incidence angle θ_1 to acquire the desired angle θ_2, people change the phase gradient $\frac{d\Phi}{dx}$ to tune θ_2. Here, we will elaborate several mechanizations of phase-based beam steering.

Semiconductor Components

Semiconductor devices, such as diodes and transistors, have been widely used in the integrated circuit industry, laying a solid foundation for the development of modern society. By applying a bias voltage to semiconductor components, their equivalent circuit can be changed, thereby changing their electromagnetic parameters. Researchers can achieve regular phase differences while ensuring consistent reflection amplitudes by meticulously adjusting these parameters. According to the generalized Snell's law, this will cause the refracted/reflected light to deflect towards a specific angle.

Diode

PIN diodes and varactor diodes both have been widely used in configurable metasurface to realize beam steering. A typical PIN diode is illustrated in Figure 7 (a). In conduction state, it is equivalent to the series connection of an inductor and a resistor. While for cut-off state, it corresponds to series connection of an inductor of a capacitor. With different equivalent impedance at the two states, it is feasible to tune the phase difference of the adjacent units. As early as 2013, Sabapathy *et al.* (Sabapathy et al., 2013) have incorporated RF PIN diodes to reconfigurable antenna. By adopting ON and Off condition of the PIN, they have achieved a beam steering angle of 0° and ±30° at 5.8 GHz. In 2019, Ma *et al.* (Ma et al., 2019) brought a 2-bit digital metasurface based on PIN diodes. They adopted two PINs independently coded by a FPGA to manipulate the phase difference. Under appropriate two-dimensional coding matrices, they achieved beam pointing for elevation angles of 20°, 40°, and 60°, as well as azimuth angles of 200°, 220°, and 240°.

Varactor diodes, presented in Figure 7 (b), are another type of tunable active semiconductor device, whose capacitance is inversely proportional to the square root of the reverse bias voltage. Imposing a different voltage will definitely alter the equivalent electromagnetic parameters, hence changing the phase difference. In 2012, Zainud-Deen *et al.* (Zainud- Deen et al., 2012) designed a 21×21 reflection array printed on a substrate containing a varactor diode. The author adjusted the bias voltage to change the capacitance so that the phase value of each unit can be the design value. They achieved a beam deflection of 70 ° at a frequency of 13GHz. In 2018, Costanzo *et al.* (Costanzo et al., 2018) devised a double-layer active reflection array, with each unit consisting of two stacked rectangular plates and a varactor diode loaded inside. By controlling voltage bias, they achieved a phase range of up to 318 ° and a steering angle range of around ± 40° at 11 GHz.

It is worth noting that although classical diodes are often used as tuning devices in the microwave band, they are strait to control electromagnetic waves in higher frequency bands due to the constraints of response rate and parasitic effects. Schottky gate is a metal semiconductor interface with rectification characteristics. In recent years, due to the higher operating frequency of Schottky based semiconductor devices, they have been used for beam control in millimeter wave and terahertz bands. In 2014, Karl *et al.* (Karl et al., 2014) constructed a tunable active metasurface based on Schottky gate structure. Like the classic diodes mentioned earlier, they also use voltage bias to adjust resonance characteristics. The unit structure proposed by the author consists of a metal open ring, an n-doped epilayer, and an intrinsic gallium arsenide substrate. By optimizing the structure, the author achieved a light deflection of 36.1° at 400 GHz.

Transistors

Researchers have also introduced transistor technology into the field of beam control. By altering the ON/OFF state of the transistor, its equivalent impedance can be modified, thereby transforming the electromagnetic response. For example, Complementary metal–oxide–semiconductor transistor (CMOS) can be approximated as a resistor in the ON state, while a capacitor in the OFF state. In 2020, Venkatesh *et al.*(Venkatesh et al., 2020) manufactured an 8-bit programmable digital metasurface. Exhibited in Figure 7 (c), the structural unit of this design consists of a top copper, a silicon dioxide layer, a lossy silicon substrate, and a low resistance silicon substrate, with eight switches evenly arranged around the circumference. By controlling the ON and OFF combination of the switch, the author achieved an angle deviation of ± 30 ° at 300GHz.

In addition to COMS, there have also been new solutions in recent years that use non silicon processes to achieve beam manipulation. Recently, high electron mobility transistors (HEMT) have gradually attracted attention. This device, elaborated in Figure 7 (d), has a large dynamic carrier density range, high electron drift velocity, and small parasitic capacitance, enabling it suitable for the terahertz band. By applying a bias voltage, the carrier concentration can be changed, thereby regulating amplitude and phase information. In 2023, Lan *et al.* (Lan et al., 2023) described a HEMT based on gallium nitride technology that utilizes different bias voltages to achieve 1-bit digital encoding. Their proposed structural units are made on SiC substrates covered with GaN, and heterojunctions are formed by AlGaN/GaN. After exquisitely structural design, the author ultimately achieved a deflection angle range of 20-60° at 340 GHz, opening possibilities for high frequency beam steering.

Figure 7. Semiconductor components for beam steering: (a) PIN diodes, (b) Varactor diodes, (c) complementary metal–oxide–semiconductor transistor, and (d) high electron mobility transistors

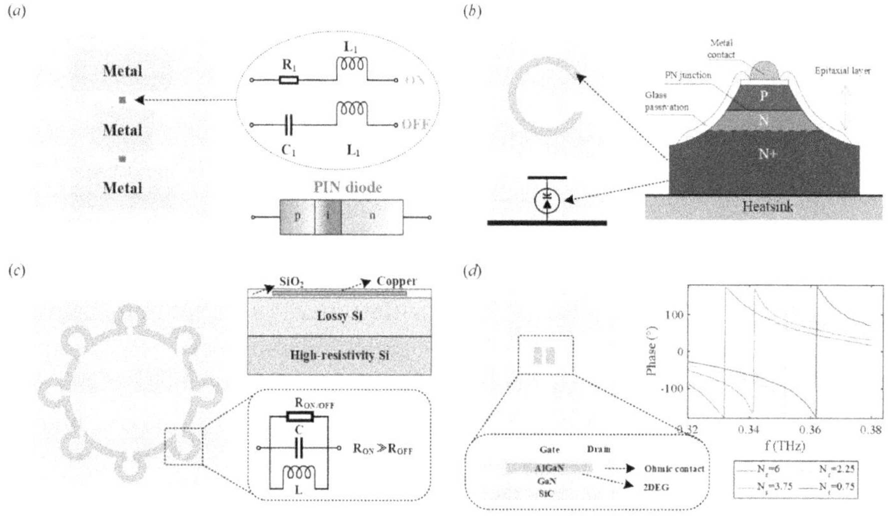

To sum up, beam control schemes based on semiconductor active devices have already flourished in the microwave band, and research in the millimeter wave or terahertz band is also rapidly developing. However, how to further improve the operating frequency and response speed of devices may become valuable future research findings.

Electrically Tunable Material

Electrically tunable materials are materials whose optical properties can be modified by applying an electric field. Although the modulation mechanism of electric field on different materials is sophisticated, the core idea is still achieving phase difference under different biases voltage. For the realm of beam manipulation, commonly adopted electrical tuning materials include liquid crystals and graphene, which will be introduced below:

Liquid Crystal

Liquid crystal is a tunable material with mature industrial technology, which has a wide dynamic range, high efficiency, relatively low power consumption, and low cost. The principle of using liquid crystal as beam control is shown in Figure

8 (a). The metal back panel is filled with a positive nematic liquid crystal, whose molecular shape is like a rod. The upper and lower electrodes are coated with orientation agents, causing the liquid crystal molecules between the electrode plates to tend to align parallel to the electrode plates due to van der Waals forces. After applying sufficient voltage, under the action of an electrostatic field, liquid crystal molecules tend to align parallel to the direction of the electric field. Liquid crystals have different dielectric constants in two directions:

$$\varepsilon_p = 1 + \frac{NFh}{\varepsilon_0}\left[\bar{\alpha} + \frac{2}{3}\Delta\alpha S + \frac{Fp^2}{3k_BT}\left\{1 - (1 - 3\cos^2\beta)S\right\}\right] \tag{9}$$

$$\varepsilon_v = 1 + \frac{NFh}{\varepsilon_0}\left[\bar{\alpha} - \frac{1}{3}\Delta\alpha S + \frac{Fp^2}{3k_BT}\left\{1 + \frac{1}{2}(1 - 3\cos^2\beta)S\right\}\right] \tag{10}$$

where β is the angle between the dipole moment and the molecular axis and S is the ordered parameter; ε_0 is the vacuum dielectric constant and α is the electron molecule polarizability. NF, h, and p are parameters related to LC molecules while k_B and T are Boltzmann constant and temperatures, respectively. By changing the bias voltage, the ordered parameter S can be regulated, so that the dielectric constant of the liquid crystal can also be conveniently dominated in ε_p and ε_v, ultimately influencing the resonant frequency and phase response.

There have been quite a few works about employing LC to manipulate beams. For example, in 2018, Komar *et al.* (Komar et al., 2018) achieved a deflection angle of up to 12° for a 745 nm transmission red light beam using a silicon nano dielectric metasurface soaked in liquid crystals. Besides, in 2020, Wu *et al.* (J. Wu et al., 2020) constructed a liquid crystal-based 1-bit reconfigurable metasurface driven by FPGA, achieving a reflected light deflection of 31.5° at 672 GHz. The proposed metasurface is a metal-insulator-metal configuration, composed of silica substrate, a metal ground, liquid crystal and metal pattern. In 2023, Fu *et al.* (Fu et al., 2023) utilized digitally encoded liquid crystal metasurfaces to achieve two-dimensional beam control at two frequency points, 94 and 220 GHz. Their proposed structure consists of FR4 substrate, feed, polyimide, liquid crystal, polyimide, metal superatoms, and top quartz. By controlling the encoding sequence, a scanning range of 20°-60° and 25°-60° in both directions of x and y can be achieved, and the simulation results coincide well with experimental testing.

Although liquid crystal materials have a complete industry chain, with the increasing demand for precision equipment, they have gradually exposed many shortcomings. Firstly, the corresponding responding time of LC devices is usually tens of milliseconds, making it difficult to apply to high-speed and real-time scenarios.

In addition, the size of LC devices is too large, hindering the miniaturization and integration. Consequently, how to overcome the related drawbacks is a worthwhile research direction in the future.

Figure 8. Electrical tunable materials for beam steering: (a) Liquid crystal and (b) graphene

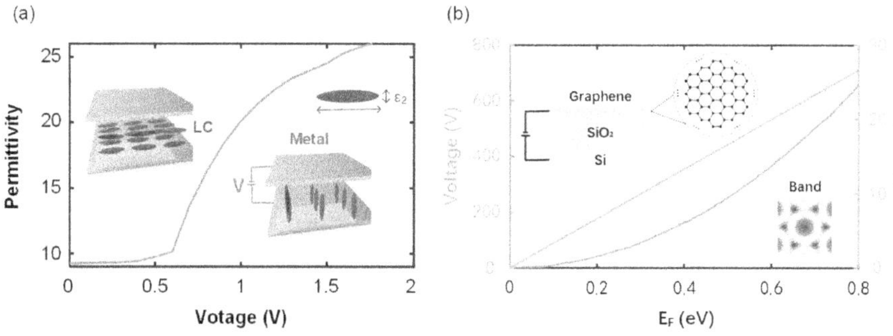

Graphene

Graphene is composed of a single layer of carbon atoms, with a hexagonal lattice shape. As an emerging material, its high carrier mobility, high robustness, and flexible tunability have led to its widespread application in the terahertz and optical frequency bands. The effective relative dielectric constant of graphene ε_{eff} can be written as:

$$\varepsilon_{eff} = 1 + \frac{i\sigma_s(E_F)}{\varepsilon_0 \omega d} \quad (11)$$

where ω is the angular frequency and d is the thickness of graphene. ε_0 is the vacuum dielectric constant and σ_s is the surface conductivity, which can be determined by the Fermi level E_F.

$$\sigma_s(E_F) = \frac{ie^2 E_F}{\pi \hbar^2(\omega + i/\tau)} \quad (12)$$

where e is the electronic charge, and $\hbar$ is the reduced Planck constant; ω is the angular frequency, and τ is the relaxation time. Applying external bias voltage V_g can cause the Fermi level to shift up and down:

$$E_F = \hbar v_f \sqrt{\pi \varepsilon V_g / ed} \quad (13)$$

It is feasible to change the electromagnetic parameters of graphene, thereby adjusting the resonant frequency and phase information, and achieving beam manipulation function. In 2019, Hosseinejad *et al.* (Hosseininejad et al., 2019) proposed a reconfigurable metasurface based on graphene to operate in the terahertz band, achieving three-dimensional beam pointing at a frequency of 2THz. The metasurface they designed consists of a gold substrate, a silicon substrate, graphene 1, high-density polyethylene, and graphene 2. By independently controlling the bias voltage of two layers of graphene, 2-bit independent encoding can be achieved. In 2021, Ai *et al.* (Ai et al., 2021) designed a graphene based metasurface that achieved a reflection angle of up to 42 ° at a frequency of 12.32THz. The proposed metasurface comprises a metal layer at the bottom, a dielectric layer in the middle, and a graphene layer at the top. By designing the geometric dimensions of graphene ribbons and controlling the bias voltage, they achieved dynamic regulation of single, double, and triple beams. In 2023, Zheng *et al.* (Zheng et al., 2023) designed a graphene reconfigurable antenna operating in the millimeter wave frequency band, achieving 30° single beam deflection, dual beam/four beam generation, and radar cross section reduction. Their proposed metasurface contains four layers: graphene layer, polyvinyl chloride (PVC), Taconic TLY-5 substrate, and copper layer. The developed control circuits are more concise and practical.

Graphene materials have a satisfied tuning range in the terahertz to mid infrared band, while the absorption loss is large when the wavelength is further shortened. In addition, it is hard to process single-layer graphene materials, which increases the cost of application. Nevertheless, with the development of industry, graphene-based beam control schemes have considerable prospects.

Phase Change Material

Phase change materials are a special type of material that can undergo phase change under external stimuli, thereby altering its physical properties. In recent years, the tunable nature of phase change materials has enabled them widely adopted in scientific research. In the field of beam control, due to the significant phase difference before and after phase transition, it is easy to achieve a wide tuning range. Common phase change materials include vanadium dioxide and sulfur-based compound materials (such as GST).

Vanadium Dioxide

Vanadium dioxide is a common phase change material with the advantages of fast response and large modulation depth. As displayed in Figure 9 (a), vanadium dioxide has two crystal forms that can experience transformation under physical fields such as temperature, electric field, light, and pressure at specific thresholds. Taking temperature as an example, as shown in Figure x (b), when T<341K, it is a monoclinic crystal and exhibits an insulating state. When T>341K or above, it is a tetragonal crystal and presents a metallic state. According to the Druid model, the relative dielectric constant of vanadium dioxide can be expressed as

$$\varepsilon(\omega) = \varepsilon_\infty - \frac{\omega_p^2 \frac{\sigma}{\sigma_0}}{\omega^2 + i \cdot \omega \cdot \omega_d} \tag{14}$$

Among them, the high-frequency dielectric constant ε_∞ is 12, and plasma frequency ω_p is $1.4 \times 10^{15}\,s^{-1}$. The damping frequency ω_d equals $5.75 \times 10^{13}\,s^{-1}$ and the ideal conductivity $\sigma_0 = 3 \times 10^5\,S/m$. During the phase transition process, the conductivity of vanadium dioxide can be increased by 2-5 orders of magnitude, which directly leads to a change in the dielectric constant. As a result, the amplitude and phase curves shift, providing the possibility for beam steering.

In recent years, there have been series of studies using vanadium dioxide as a beam control solution. For example, in 2016, Hashemi *et al.* (Hashemi et al., 2016) designed and manufactured a transmission metasurface based on vanadium dioxide, which achieved beam deflection of up to 44 ° in both horizontal and vertical directions for 100GHz signals. Their structure consists of a silicon substrate, vanadium dioxide, silicon dioxide, heating electrodes, and gold at the top. They use heating electrodes to control the temperature of each unit, thereby controlling the phase gradient between adjacent units. In addition, Takase *et al.* (Takase & Takahara, 2021) designed an all dielectric metasurface that operates at a wavelength of 1550nm, comprising of a columnar silicon layer, a vanadium dioxide layer, and an alumina substrate. For the insulation phase and the metal phase, the author achieved a deflection of the transmitted beam of 24.4°. Compared to plasma metasurfaces with metal-dielectric-metal structures, pure dielectric structures have less transmission loss and can be applied to higher frequencies. In addition to regulating the transmitted light, researchers have also attempted to tune the reflected light. In 2022, Yang *et al.* (Yang et al., 2022) proposed a programmable metasurface approach based on VO_2, which operates at 218GHz. For 1-bit metasurfaces, a carefully designed encoding sequence can achieve wide-angle scanning from −60° to 60 °. For two-bit encoding, they use two different units, combined with two different crystal states to represent four different encodings, with similar reflection amplitudes and 90 °

phase differences between them. Compared to the 1-bit case, the 2-bit metasurface has a higher degree of freedom in beam adjustment.

Figure 9. Phase change material for beam steering: (a) Vanadium dioxide and (b) germanium-antimony-tellurium

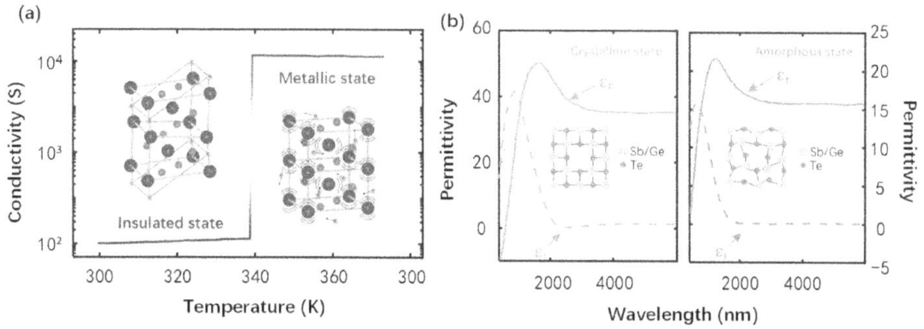

It is worth noting that almost all current research work adopts the method of thermal triggered phase transition for tuning, whose response time is usually on the order of milliseconds, which is not applicable for real-time scenarios. When activating by electric field stimulation, the speed of this transition significantly increases, making it a future research direction.

GST

Germanium-antimony-tellurium (GST) is also a widely used phase change material for beam tuning, with advantages such as fast speed and low power consumption. Demonstrated in Figure 9 (b), there are two states in solid-state GST, which can convert crystalline to amorphous states under thermal, electrical, or light stimulation. Taking temperature as an example, when the temperature of amorphous GST exceeds the crystallization temperature (433K), the amorphous state will transform into a steady-state hexagonal structure. It is worth noting that once the phase transition process is completed, even if the external excitation is removed and returned to the room temperature environment, GST can still maintain its post phase transition state for a long time, exhibiting non volatility. During the phase transition, the electromagnetic parameters (such as the complex dielectric constant) of GST will undergo significant changes, as shown in Figure x. In addition, by controlling the ratio of crystalline to amorphous states, the equivalent dielectric constant of the mixed state can be obtained using the Lorentz Lorenz formula

$$\frac{\varepsilon^{eff} - 1}{\varepsilon^{eff} + 2} = \alpha\frac{\varepsilon_c - 1}{\varepsilon_c + 2} + (1 - \alpha)\frac{\varepsilon_a - 1}{\varepsilon_a + 2} \quad (15)$$

where ε^{eff} represents the equivalent dielectric constant of the mixed state while ε_c and ε_a represents the dielectric constants of crystalline and amorphous states, respectively. The tunable parameter α represents the molar fraction of crystalline states. By utilizing the phase transition of GST, metasurface can be designed to achieve beam steering.

In 2018, Galaretta *et al.* (De Galarreta et al., 2018) reported a reconfigurable phase transition metasurface device based on GST, which achieved anomalous reflection of approximately 33.6 ° at 1550nm. The designed structure is a classic MIM metasurface, where the aluminum-silica-aluminum layer is located on the silicon substrate, and GST is filled between the aluminum and silica. For transmission scenarios, in 2022, Nemati *et al.* (Nemati et al., 2022) proposed and manufactured a GST based metamaterial that can achieve 15 ° deflection of the transmitted beam at 1550nm. The metasurface does not contain a metal structure so that resulting in lower losses. In addition, by accurately changing the temperature of GST nanorods, the ratio of abnormal transmission light to normal transmission light intensity can be controlled. GST phase change materials can also be applied to the mid-infrared band, for example, in 2021, Song *et al.* (Song et al., 2021) constructed a catenary based metasurface, achieving the beam deflection within a wide mid infrared range between 8 μm-9.5 μm. The proposed structure includes a gold layer at the bottom, GST, magnesium fluoride at the middle, and a gold catenary at the top. In the crystalline state, only specular reflection appears, but when it transitions to the amorphous state, an anomalous reflection light of about 20° occurs.

The drawbacks of GST materials are their low crystalline resistivity and crystallization temperature, as well as poor thermal stability. In order to maximize its merits and potential, researchers have been attempting to modify GST materials to further improve their performance, while also continuously exploring various types of new phase change materials.

MEMS

Apart from directly drive the reflector to achieve the purpose of light deflecting, MEMS devices can also control the beam by phase arrays. A classic design is presented in Figure 10 (a), which places a metal sheet below the metasurface and uses MEMS to drive its movement. When the metal sheet is close to the metasurface, it can provide a significant additional phase difference, resulting in anomalous reflected beams. However, when the metal sheet is far from the metasurface, no additional phase difference exists, so that only normal reflected beams are generated. Based on this mechanism, a few works have been proposed:

In 2014, Megens *et al.* (Megens et al., 2014) described a 32×32 MEMS based optical metasurfaces within a response time of 4 μs. The proposed structure adopts electrostatic drive, and each structural unit can be independently addressed. By periodically controlling the position of the reflecting surface, an adjustable phase difference can be formed. The author ultimately achieved a beam control range of approximately 2° at 1550nm. Besides, in 2018, Arbabi *et al.* (Arbabi et al., 2018) introduced MEMS technology into metasurfaces and constructed a lens with adjustable focal length. The structure they designed consists of a metasurface and a MEMS sealed cavity. Applying voltage to the two plates of the sealed cavity can alter the spacing, thereby changing the phase different. In fact, the author achieved focus lengths ranging from 565 to 629 μm at 915nm. In 2021, Meng *et al.* (Meng et al., 2021) combined thin film piezoelectric MEMS with optical metasurfaces of gap surface plasmons (GSPs) to achieve polarization independent beam steering for two-dimensional reflected wave. The structure they designed consists of silicon dioxide at the top, gold squares with different side lengths, air gaps, and gold plates at the bottom. By utilizing the inverse piezoelectric effect to control the air gap between the MEMS mirror and the metasurface, an anomalous reflected beam of 10.5 ° can be generated with normally incident. Through this principle, the author also achieved planar focusing control.

At present, beams steering based on MEMS phase modulation is still developing fast. Existing works generally demonstrate relatively small steering angle. Therefore, developing MEMS phase metasurfaces with a large angle control range with new actuation mechanism could be a future topic.

STC Metasurface

The usage of phase gradient to achieve beam deflection usually only involves spatial information encoding, while without utilizing the temporal dimension. In recent years, digital space time coded (STC) metasurfaces have gradually attracted people's attention. Displayed in Figure 10 (b), The STC metasurface utilizes FPGA to send a time-domain periodic pulse sequence to each subunit, significantly expanding the information content of the metasurface, enabling low-bit metasurfaces to achieve the effect of high-bit ones. Assuming that for an $N \times M$ metasurface is encoded with a sequence of time length L, the reflectivity of a certain unit with spatial coordinates (p, q) at the nth time step l can be expressed as

$$\Gamma_{pq}^{l} = A_{pq}^{n} e^{j\varphi_{pq}^{n}} \tag{16}$$

where A and φ are amplitude and phase. The knowledge of Fourier transform can be introduced to derive the equivalent reflection coefficient a_{pq}^{m} for m-th harmonics:

$$a_{pq}^{m} = \sum_{l=1}^{L} \frac{\Gamma_{pq}^{l}}{L} \mathrm{sinc}\left(\frac{\pi m}{L}\right) e^{\frac{-j\pi m(2l-1)}{L}} \tag{17}$$

It can be observed that there is an additional phase in the harmonic term, which can be used to achieve beam direction control through phase gradient metasurface. By applying Fourier series expansion, we can acquire the far-field scattering pattern of the m-th harmonic wave F_{m}:

$$F_{m}(\theta, \varphi) = \sum_{q=1}^{N} \sum_{p=1}^{M} E_{pq}(\theta, \varphi)\, a_{pq}^{m}\, e^{j\frac{2\pi}{\lambda_c}[(p-1)d_x \sin\theta\cos\varphi + (q-1)d_y \sin\theta\sin\varphi]} \tag{18}$$

where E_{pq} indicates the far-field scattering pattern of the unit cell. Based on the aforementioned theory, a few works have been presented. In 2018, Zhang *et al.* (Zhang et al., 2018) first utilized STC metasurface to achieve harmonic beam control. The metasurface they constructed is an 8×8 digital phase arrays consisting of PIN diodes. The author used FPGA to generate 8-bit signal series with a time length of 2 μs. They have achieved deflections of $\pm$ 15, $\pm$ 30, and $\pm$ 45° for $\pm$ 1, $\pm$ 2, and $\pm$ 3 harmonic at a center frequency of 9.8GHz. Based on similar principles, Wu *et al.* (G.-B. Wu et al., 2022) proposed a sideband free spatiotemporal encoded metasurface antenna in 2022, which realized the radiation of guided waves inside the waveguide into free space. The proposed metasurface unit is still composed of PIN diodes driven by FPGA. By applying different coding sequences, they attained the scanning of the first harmonic with a center frequency of 27GHz from -50° to 30°. In addition, through spatial amplitude modulation, they also achieved continuous scanning of the fundamental frequency beam at -20°-50°. Moreover, the author also fulfilled simultaneous control of diverse beams by superposing basic signals, providing the possibility for multi-objective regulation. The above two methods use the conduction and cutoff of PIN diodes to represent ON and OFF. Dai *et al.* (Dai et al., 2018) reported a digital STC metasurface working at 3.7GHz, which utilizes different bias voltage combinations of varactor diodes to actualize phase difference regulation. Through theoretical derivation they proved that applying distinct phase differences can bring about harmonic intensity control, while introducing time delay can achieve wavefront shaping. The author utilized different coding sequences and bias voltage combinations to attain harmonic beam deflection of 0-30° with a 0-10dB attenuation amplitude. Besides, Wang *et al.* (S. R. Wang et al., 2022) proffered a multi-partition asynchronous space-time-coding digital metasurface based on varactor diodes. Their design allows metasurfaces to have different encoding sequences in different spatial regions, thereby effecting frequency division multiplexing. The authors have designed communication experiments and completed directional transmission of specified images.

Figure 10. Beam steering technology based on (a) MEMS and (b) STC metasurface

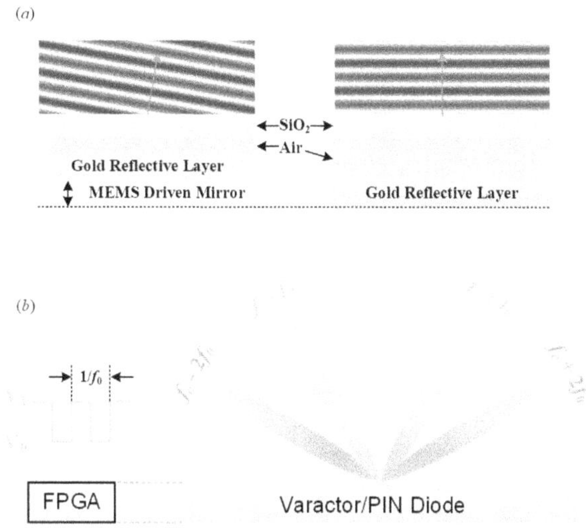

Although spatiotemporal encoding metasurface greatly expand the freedom of beam manipulation, they still have many limitations. Due to the operating frequency of the diodes, the metasurface proposed today can only be applied in the microwave band. In future research, expanding the frequency to higher bands poses high significance. On the other hand, generating a sophisticated coding sequence is also a challenge task. However, deep learning technique, which has already been adopted in forward and inverse computational electromagnetics (Y. Li et al., 2020; Qi et al., 2020; Ren et al., 2022) and other physics problems (Y. Wang et al., 2021, 2021, 2022; Y. Wang & Ren, 2022, 2023), may supply the possibility in designing the coding sequence.

CONCLUSION

In this chapter, an exhaustive review of beam manipulation techniques has been undertaken, delineating two primary categories: geometric-dependent and phase-dependent methodologies. The former encompasses mechanical control, MEMS control, liquid control, and optical control, all of which induce alterations in incident angles through geometric deformation, consequently influencing the exit angles. Nevertheless, these approaches necessitate mechanical components, rendering them susceptible to damage. In contrast, phase-based metasurfaces have emerged as transformative, capitalizing on the generalized Snell's law to manipulate light beams through artificially generated phases. This class of techniques encompasses active semiconductor devices suitable for microwave frequency bands, phase change materials, and electrically tunable materials applicable to terahertz and infrared bands, as well as MEMS systems spanning a broad frequency spectrum. It is imperative to acknowledge that prevalent outcomes are often reliant on numerical simulations, warranting empirical verification of accuracy and expeditiousness in practical experiments. Moreover, the incorporation of time-domain information into space-time encoding metasurfaces substantially augments the degree of beam control. However, the present constraints of regulatory equipment confine the applicability of these methodologies solely to the microwave frequency band, necessitating their integration into nascent tunable devices for broader utility.

As delineated previously, beam manipulation technology exhibits extensive applications across diverse domains, including communication, radar, sensing, and biomedical imaging, thereby assuming heightened significance in contemporary research endeavors. The author posits that reconfigurable metasurfaces, particularly when integrated with deep learning techniques, represent a compelling solution. Nevertheless, the intricacies associated with attaining two-dimensional pixel-level encoding state control pose a considerable challenge. In this context, on-chip methodologies rooted in microelectronic integrated manufacturing emerge as a promising avenue for addressing these challenges.

REFERENCE

Ai, H., Kang, Q., Wang, W., Guo, K., & Guo, Z. (2021). Multi-Beam Steering for 6G Communications Based on Graphene Metasurfaces. *Sensors (Basel)*, 21(14), 4784. 10.3390/s2114478434300521

Alonso-delPino, M., Jung-Kubiak, C., Reck, T., Llombart, N., & Chattopadhyay, G. (2019). Beam Scanning of Silicon Lens Antennas Using Integrated Piezomotors at Submillimeter Wavelengths. *IEEE Transactions on Terahertz Science and Technology*, 9(1), 47–54. 10.1109/TTHZ.2018.2881930

Arbabi, E., Arbabi, A., Kamali, S. M., Horie, Y., Faraji-Dana, M., & Faraon, A. (2018). MEMS-tunable dielectric metasurface lens. *Nature Communications*, 9(1), 812. 10.1038/s41467-018-03155-629476147

Cheng, J., & Chen, C.-L. (2011). Adaptive beam tracking and steering via electrowetting-controlled liquid prism. *Applied Physics Letters*, 99(19), 191108. 10.1063/1.3660578

Cheng, Y., Cao, J., & Hao, Q. (2021). Optical beam steering using liquid-based devices. *Optics and Lasers in Engineering*, 146, 106700. 10.1016/j.optlaseng.2021.106700

Costanzo, S., Venneri, F., Raffo, A., & Di Massa, G. (2018). Dual-Layer Single-Varactor Driven Reflectarray Cell for Broad-Band Beam-Steering and Frequency Tunable Applications. *IEEE Access : Practical Innovations, Open Solutions*, 6, 71793–71800. 10.1109/ACCESS.2018.2882093

Dai, J. Y., Zhao, J., Cheng, Q., & Cui, T. J. (2018). Independent control of harmonic amplitudes and phases via a time-domain digital coding metasurface. *Light, Science & Applications*, 7(1), 90. 10.1038/s41377-018-0092-z30479756

De Galarreta, C. R., Alexeev, A. M., Au, Y., Lopez-Garcia, M., Klemm, M., Cryan, M., Bertolotti, J., & Wright, C. D. (2018). Nonvolatile Reconfigurable Phase-Change Metadevices for Beam Steering in the Near Infrared. *Advanced Functional Materials*, 28(10), 1704993. 10.1002/adfm.201704993

Duma, V.-F., & Rolland, J. P. (2010). Mechanical Constraints and Design Considerations for Polygon Scanners. In Pisla, D., Ceccarelli, M., Husty, M., & Corves, B. (Eds.), *New Trends in Mechanism Science* (pp. 475–483). Springer Netherlands. 10.1007/978-90-481-9689-0_55

Fu, Y., Fu, X., Yang, S., Peng, S., Wang, P., Liu, Y., Yang, J., Wu, J., & Cui, T. J. (2023). Two-dimensional terahertz beam manipulations based on liquid-crystal-assisted programmable metasurface. *Applied Physics Letters*, 123(11), 111703. 10.1063/5.0167812

Hashemi, M. R. M., Yang, S.-H., Wang, T., Sepúlveda, N., & Jarrahi, M. (2016). Electronically-Controlled Beam-Steering through Vanadium Dioxide Metasurfaces. *Scientific Reports*, 6(1), 35439. 10.1038/srep3543927739471

Hernandez-Cardoso, G. G., Rojas-Landeros, S. C., Alfaro-Gomez, M., Hernandez-Serrano, A. I., Salas-Gutierrez, I., Lemus-Bedolla, E., Castillo-Guzman, A. R., Lopez-Lemus, H. L., & Castro-Camus, E. (2017). Terahertz imaging for early screening of diabetic foot syndrome: A proof of concept. *Scientific Reports*, 7(1), 42124. 10.1038/srep4212428165050

Hofmann, U., Janes, J., & Quenzer, H.-J. (2012). High-Q MEMS Resonators for Laser Beam Scanning Displays. *Micromachines*, 3(2), 509–528. 10.3390/mi3020509

Hosseininejad, S. E., Rouhi, K., Neshat, M., Faraji-Dana, R., Cabellos-Aparicio, A., Abadal, S., & Alarcón, E. (2019). Reprogrammable Graphene-based Metasurface Mirror with Adaptive Focal Point for THz Imaging. *Scientific Reports*, 9(1), 2868. 10.1038/s41598-019-39266-330814570

Karl, N., Reichel, K., Chen, H.-T., Taylor, A. J., Brener, I., Benz, A., Reno, J. L., Mendis, R., & Mittleman, D. M. (2014). An electrically driven terahertz metamaterial diffractive modulator with more than 20 dB of dynamic range. *Applied Physics Letters*, 104(9), 091115. 10.1063/1.4867276

Komar, A., Paniagua-Domínguez, R., Miroshnichenko, A., Yu, Y. F., Kivshar, Y. S., Kuznetsov, A. I., & Neshev, D. (2018). Dynamic Beam Switching by Liquid Crystal Tunable Dielectric Metasurfaces. *ACS Photonics*, 5(5), 1742–1748. 10.1021/acsphotonics.7b01343

Lan, F., Wang, L., Zeng, H., Liang, S., Song, T., Liu, W., Mazumder, P., Yang, Z., Zhang, Y., & Mittleman, D. M. (2023). Real-time programmable metasurface for terahertz multifunctional wave front engineering. *Light, Science & Applications*, 12(1), 191. 10.1038/s41377-023-01228-w37550383

Li, A., Sun, W., Yi, W., & Zuo, Q. (2016). Investigation of beam steering performances in rotation Risley-prism scanner. *Optics Express*, 24(12), 12840. 10.1364/OE.24.01284027410303

Li, B., Lin, Q., & Li, M. (2023). Frequency–angular resolving LiDAR using chip-scale acousto-optic beam steering. *Nature*, 620(7973), 316–322. 10.1038/s41586-023-06201-637380781

Li, C., Chen, C., Yu, C., Jau, H., Lv, J., Qing, X., Lin, C., Cheng, C., Wang, C., Wei, J., Yu, Y., & Lin, T. (2017). Arbitrary Beam Steering Enabled by Photomechanically Bendable Cholesteric Liquid Crystal Polymers. *Advanced Optical Materials*, 5(4), 1600824. 10.1002/adom.201600824

Li, F., Zhou, P., Wang, T., He, J., Yu, H., & Shen, W. (2017). A Large-Size MEMS Scanning Mirror for Speckle Reduction Application. *Micromachines*, 8(5), 140. 10.3390/mi8050140

Li, Y., Wang, Y., Qi, S., Ren, Q., Kang, L., Campbell, S. D., Werner, P. L., & Werner, D. H. (2020). Predicting Scattering From Complex Nano-Structures via Deep Learning. *IEEE Access : Practical Innovations, Open Solutions*, 8, 139983–139993. 10.1109/ACCESS.2020.3012132

Lin, Y.-J., Chen, K.-M., & Wu, S.-T. (2009). Broadband and polarization-independent beam steering using dielectrophoresis-tilted prism. *Optics Express*, 17(10), 8651. 10.1364/OE.17.00865119434198

Ma, Q., Bai, G. D., Jing, H. B., Yang, C., Li, L., & Cui, T. J. (2019). Smart metasurface with self-adaptively reprogrammable functions. *Light, Science & Applications*, 8(1), 98. 10.1038/s41377-019-0205-331700618

Megens, M., Yoo, B.-W., Chan, T., Yang, W., Sun, T., Chang-Hasnain, C. J., Wu, M. C., & Horsley, D. A. (2014). *High-speed 32×32 MEMS optical phased array* (W. Piyawattanametha & Y.-H. Park, Eds.; p. 89770H). 10.1117/12.2044197

Meng, C., Thrane, P. C. V., Ding, F., Gjessing, J., Thomaschewski, M., Wu, C., Dirdal, C., & Bozhevolnyi, S. I. (2021). Dynamic piezoelectric MEMS-based optical metasurfaces. *Science Advances*, 7(26), eabg5639. 10.1126/sciadv.abg563934162551

Nemati, A., Yuan, G., Deng, J., Huang, A., Wang, W., Toh, Y. T., Teng, J., & Wang, Q. (2022). Controllable Polarization-Insensitive and Large-Angle Beam Switching with Phase-Change Metasurfaces. *Advanced Optical Materials*, 10(5), 2101847. 10.1002/adom.202101847

Nocentini, S., Martella, D., Wiersma, D. S., & Parmeggiani, C. (2017). Beam steering by liquid crystal elastomer fibres. *Soft Matter*, 13(45), 8590–8596. 10.1039/C7SM02063E29105720

Qi, S., Wang, Y., Li, Y., Wu, X., Ren, Q., & Ren, Y. (2020). Two-Dimensional Electromagnetic Solver Based on Deep Learning Technique. *IEEE Journal on Multiscale and Multiphysics Computational Techniques*, 5, 83–88. 10.1109/JMMCT.2020.2995811

Ren, Q., Wang, Y., Li, Y., & Qi, S. (2022). *Sophisticated Electromagnetic Forward Scattering Solver via Deep Learning*. Springer Singapore. 10.1007/978-981-16-6261-4

Sabapathy, T., Jamlos, M. F. B., Ahmad, R. B., Jusoh, M., Jais, M. I., & Kamarudin, M. R. (2013). ELECTRONICALLY RECONFIGURABLE BEAM STEERING ANTENNA USING EMBEDDED RF PIN BASED PARASITIC ARRAYS (ERPPA). *Electromagnetic Waves*, 140, 241–261. 10.2528/PIER13042906

Shangguan, M., Xia, H., Wang, C., Qiu, J., Lin, S., Dou, X., Zhang, Q., & Pan, J.-W. (2017). Dual-frequency Doppler lidar for wind detection with a superconducting nanowire single-photon detector. *Optics Letters*, 42(18), 3541. 10.1364/OL.42.00354128914897

Shaw, D. (2007). Design and analysis of an asymmetrical liquid-filled lens. *Optical Engineering (Redondo Beach, Calif.)*, 46(12), 123002. 10.1117/1.2821426

Smith, N. R., Abeysinghe, D. C., Haus, J. W., & Heikenfeld, J. (2006). Agile wide-angle beam steering with electrowetting microprisms. *Optics Express*, 14(14), 6557. 10.1364/OE.14.00655719516833

Song, R., Deng, Q., Zhou, S., & Pu, M. (2021). Catenary-based phase change metasurfaces for mid-infrared switchable wavefront control. *Optics Express*, 29(15), 23006. 10.1364/OE.43484434614576

Takase, H., & Takahara, J. (2021). Switchable wavefront control using an all-dielectric metasurface mediated by VO_2. *Applied Physics Express*, 14(3), 032007. 10.35848/1882-0786/abdd13

Venkatesh, S., Lu, X., Saeidi, H., & Sengupta, K. (2020). A high-speed programmable and scalable terahertz holographic metasurface based on tiled CMOS chips. *Nature Electronics*, 3(12), 785–793. 10.1038/s41928-020-00497-2

Wang, H., Zhou, L., Zhang, X., & Xie, H. (2018). Thermal Reliability Study of an Electrothermal MEMS Mirror. *IEEE Transactions on Device and Materials Reliability*, 18(3), 422–428. 10.1109/TDMR.2018.2860286

Wang, S. R., Chen, M. Z., Ke, J. C., Cheng, Q., & Cui, T. J. (2022). Asynchronous Space-Time-Coding Digital Metasurface. *Advancement of Science*, 9(24), 2200106. 10.1002/advs.20220010635751468

Wang, Y., Gao, H., & Ren, Q. (2022). Differential Operator Approximation Based Tightly Coupled Multiphysics Solver Using Cascaded Fourier Network. *Advanced Theory and Simulations*, 5(11), 2200409. 10.1002/adts.202200409

Wang, Y., & Ren, Q. (2022). A versatile inversion approach for space/temperature/time-related thermal conductivity via deep learning. *International Journal of Heat and Mass Transfer*, 186, 122444. 10.1016/j.ijheatmasstransfer.2021.122444

Wang, Y., & Ren, Q. (2023). *Deep Learning-Based Forward Modeling and Inversion Techniques for Computational Physics Problems* (1st ed.). CRC Press., 10.1201/9781003397830

Wang, Y., Zhou, J., Ren, Q., Li, Y., & Su, D. (2021). 3-D Steady Heat Conduction Solver via Deep Learning. *IEEE Journal on Multiscale and Multiphysics Computational Techniques*, 6, 100–108. 10.1109/JMMCT.2021.3106539

Wu, G.-B., Dai, J. Y., Cheng, Q., Cui, T. J., & Chan, C. H. (2022). Sideband-free space–time-coding metasurface antennas. *Nature Electronics*, 5(11), 808–819. 10.1038/s41928-022-00857-0

Wu, J., Shen, Z., Ge, S., Chen, B., Shen, Z., Wang, T., Zhang, C., Hu, W., Fan, K., Padilla, W., Lu, Y., Jin, B., Chen, J., & Wu, P. (2020). Liquid crystal programmable metasurface for terahertz beam steering. *Applied Physics Letters*, 116(13), 131104. 10.1063/1.5144858

Yang, D., Wang, W., Lv, E., Wang, H., Liu, B., Hou, Y., & Chen, J. (2022). Programmable VO2 metasurface for terahertz wave beam steering. *iScience*, 25(8), 104824. 10.1016/j.isci.2022.10482435992076

Ye, L., Zhang, G., & You, Z. (2017). 5 V Compatible Two-Axis PZT Driven MEMS Scanning Mirror with Mechanical Leverage Structure for Miniature LiDAR Application. *Sensors (Basel)*, 17(3), 521. 10.3390/s1703052128273880

Yu, N., Genevet, P., Kats, M. A., Aieta, F., Tetienne, J.-P., Capasso, F., & Gaburro, Z. (2011). Light Propagation with Phase Discontinuities: Generalized Laws of Reflection and Refraction. *Science*, 334(6054), 333–337. 10.1126/science.121071321885733

Zainud- Deen., S. H., Gaber, Shaymaa. M., & Awadalla, K. H. (2012). Beam steering reflectarray using varactor diodes. *2012 Japan-Egypt Conference on Electronics, Communications and Computers*, (pp. 178–181). IEEE. 10.1109/JEC-ECC.2012.6186979

Zhang, L., Chen, X. Q., Liu, S., Zhang, Q., Zhao, J., Dai, J. Y., Bai, G. D., Wan, X., Cheng, Q., Castaldi, G., Galdi, V., & Cui, T. J. (2018). Space-time-coding digital metasurfaces. *Nature Communications*, 9(1), 4334. 10.1038/s41467-018-06802-030337522

Zheng, B., Rao, X., Shan, Y., Yu, C., Zhang, J., & Li, N. (2023). Multiple-Beam Steering Using Graphene-Based Coding Metasurfaces. *Micromachines*, 14(5), 1018. 10.3390/mi1405101837241641

Zhuang, X., Zhang, W., Wang, K., Gu, Y., An, Y., Zhang, X., Gu, J., Luo, D., Han, J., & Zhang, W. (2023). Active terahertz beam steering based on mechanical deformation of liquid crystal elastomer metasurface. *Light, Science & Applications*, 12(1), 14. 10.1038/s41377-022-01046-636596761

Chapter 2
Metamaterial Antennas for Wireless Communication Systems

Smrity Dwivedi
IIT BHU, India

ABSTRACT

Metamaterials and metasurfaces are two unique engineered materials which play a very important role in every field of research at almost every frequency range and band. Optical, microwave, terahertz metamaterials and metasurfaces now in demand due its special features and properties which can bend, absorb, propagate, and enhance, block the electromagnetic waves of any frequency. It has a special structure dimension which is related to wavelength of that wave which is passing through it. There are four types of metamaterials that have been defined according to permittivity and permeability values in different quadrants, as velocity is also different for these metamaterials and metasurfaces. In all applications area, design of antenna is the main issue on which researchers are working, and many types of antenna designs have been proposed and fabricated. For 5G, beyond 5G, and 6G applications, metamaterials are the best, chosen by designers apart from graphene, liquid crystal, and gold. In this chapter, the authors will cover all about metamaterials structures, properties and the main part is design of antenna for wireless applications.

DOI: 10.4018/979-8-3693-2599-5.ch002

INTRODUCTION

In todays era, the innovation to 5G wireless technology enabled IoT devices at a millimetre-wave band will revolutionize the electronics and telecommunication field (John Colaco et al. 2021). This and technologies of wireless communications is one of the biggest human achievements in terms of 5G and IoT technology. Now, 5G technology and IoT technology thinking will become reality for smart devices (A. Helena et al. 2020). It could be a big dream come true and a game-changer for humans particularly in the field of education, health, agriculture, autonomous vehicles, industry and many other related applications. Hence, in this proposed chapter, author has elaborated a design and analysis with specifications of metamaterials based antenna for 5G and beyond technologies for various 5G enabled advanced IoT Devices. IoT-based smart devices are usually expanding their support for Internet access (broadband) beyond devices, such as smart/mobile phones, laptops, and so on. The IoT based smart devices have been amalgamated with cutting-edge technology to manage and communicate smoothly using 5G and beyond wireless technologies. 5G wireless technologies grasps an important role in setting up the platform for real-time communications of IoT devices due to its potential to offer greater bandwidth with millimetre bands compared with 4G or 3G wireless technology previously. 5G wireless network technology combines the spectrum and access networks to meet the customer's capacity and coverage needs to fulfill the demand (G. GSMA 2019). The internet of everything specifies the synchronisation of various smart electronic devices such as tablets, smartphones, laptops, multiple machines such as smart vehicles equipped with sensors with IoT communication, and wireless or wired connection of consumer appliances connected through the internet (G. GSMA 2019) in present era. 5G technology will provide massive support for IoT devices as a useful part of the human digital world as the digital population is growing more than the human population as per need. The latest 5G enabled IoT smart device's trends are extending to sensor-based IoT competencies to actuators, robots, and drones for distributed synchronization system. The biggest and crucial challenge facing 5G enabled IoT devices is cyber security as hackers proliferate and infiltrate and target the server's proxy. Hence secured communication of a machine to a machine has a dynamic role in emerging IoT smart devices (J W. Ejaz et al. 2016). During the COVID-19 epidemic, this advancement would also allow real-time video and audio quality data for patient data analysis and wireless uninterrupted communication of detected health parameters. In this chapter, few designs have been given for antennas for effective multi-band operation useful for 5G based IoT devices and applications. In this chapter, antenna loaded the metamaterial SRR structure on the microstrip patch to boost the performance has discussed. The performance is then analysed in comparison with and without loading metamaterial SRR structure.

PROPERTIES OF METAMATERIAL

As already discussed about the material above, for many decades, (C. L. Holloway et al. 2009; V. G. Veselago 1968), metamaterial has caught a wide attention of so many researcher. As it is artificial and synthetic materials which is used to achieve important properties that is actually not found in nature. So, it can be also said that metamaterial is artificial dielectric in words of electromagnetics. As there few selected metamaterials structures which can affect electromagnetic radiation as well as sound, which is also not available in bulk materials that are exist. As it is mentioned, that metamaterials found with two major categories, resonant and non-resonant types materials, that depends basically on the oscillation of waves and its periodicity of structures which is decided. As per resonant types of metamaterials, they have well defined specific permittivity and permeability, which can be further divided into sub categories as given. These are double negative (DNG), negative permittivity (ENG), negative permeability (MNG) and negative refractive index types which are frequently used resonant type metamaterials, in other type, there are two more known as, anisotropic and hyperbolic that are the non-resonant type metamaterials, which can be used with important specified bands for electromagnetic applications. In research era, according to metamaterials, their structures are developed more commonly and that is used in waveguides, design of antennas, filters design and many other major applications. Left handed metmaterials (LHM), are with negative permittivity and negative permeability, known as DNG materials, which are used for design of coplanar waveguides, all types of filters (low pass, high pass) and microstrip lines and Antennas (D. R. Smith et al. 2000; C. L. Holloway et al. 2003). This type LHM based design of microstrip antennas have high efficiency, high gain, low profile and wide bandwidth, which is frequently used in wide band applications areas. Microstrip antennas which is previously designed having radiation not only from patch, but also from substrate material which supports patch and ground plane both, that is basically called as surface waves, which further creates radiation losses. So, this antenna has less bandwidth, low efficiency, poor directivity and higher losses (E. Shamonina et al. 2007; D. V. Sivukhin 1957). These losses can only be reduced by putting the values of permittivity and permeability as low as possible as per requirement. In second type, non-resonant type metamaterials such as photonic band gap and electromagnetic band gap structures, have very major role to achieve wide bandwidth with high potential applicability in broadband communication networks. PBG and EBG structures have smaller size, which can be varied in sub-wavelengths as well as they have property to resist high tolerance value for the structural changes and deformations. PBG and EBG structures have magical properties compared to resonant metamaterials which has periodicity $\sim\lambda/4$, whereas in resonant type metamaterials have periodicity $\sim\lambda/10$.

Types of Metamaterials

For new and smart wireless communication, PBG structure is the most useful structures for wide bandwidth and it can be used with design of antenna in mm-wave technology. Plane waves with photons when enters into the band gap (stop band) and coming after getting the dispersion diagram for measuring the cut off frequency with phase constant. Metamaterials is also used for artificial magnetic conductors (AMC) (R. A. Silin et al. 2001) with unique application areas. Negative refractive index metamaterials are mostly used for incident waves which is having propagation faster than light propagation and also for angle insensitive devices. According to national research on metamaterials, there are lots of works going on over a decades, in which the most of the works are useful to microwave components and antennas using metamaterials for frequency less than 10 GHz which can be used for high frequency systems. Although metamaterials are not a recently developed research area but from the research and development point of view, works are still left and going on as well. Again one more definition about Metamaterial that it is a material that is artificially engineered/developed to acquire unconventional electromagnetic, thermal, acoustic, and mechanical properties that are not achieved in other naturally occurring materials because of its extraordinary unusual and unique properties such as negative permeability and negative permittivity. When permeability and permittivity both are simultaneously negative, then the electric field, magnetic field, and the propagation vector will be form left-handed medium materials in a double-negative region and the wave propagation are moving critically in reverse signifying material has negative refractive index (L. H. Wen et al. 2018). Different studies and researches are going on by the scientists and researchers all over the world because of maximum benefit can be achieved using this for humanity. The electromagnetic metamaterial properties are the best defined by famous scientist maxwells by giving maxwells equations (D. Yanget al. 2017). Use of metamaterials is more preferable in the design of the microstrip antenna because to minimize the size and increase other parameters such as bandwidth, gain, and return loss as well as efficiency (C. Wang et al. 2018). This will be used and to apply adaptive concepts to innovations in microstrip antenna designs from small laboratories to practical applications of engineering with a good contract (C. Wang et al. 2018). Since it is negative permeability material, the split-ring resonators are used as the unit cell structure for design of metamaterials. The metamaterial having with various applications such as smart antenna, medical devices, optical devices, smart sensors, smart IoT devices, smart solar power, radomes, optical lenses, invisible submarines, radar are being used etc. The metamaterial structures have specific roles in the applications given to enhance the performance in terms of power or energy harvesting, bandwidth, and gain enhancement, reducing the size of the devices for optimum performance and other benefits. The design for microstrip patch antenna in the various field of electronics, especially wireless communication, information technology, and electrical have enormously enhance its characteristics such as bandwidth, return loss, etc. This chapter gives the different designs of metamaterial based antennas with previous works and very new research. Also some future aspects have been discussed at the end of the chapter. 5G and beyond 5G antennas are discussed using metamaterials with several applications.

(b) Equations for metamaterials

All given equations have been used to solved metamaterials properties and also it is used to decide the type of metamaterials as given below,

$\varepsilon_0 \frac{dE}{dt} = \nabla XH - J + f$ (1)

$\mu_0 \frac{dH}{dt} = -\nabla XK - K$ (2)

$\frac{dJ}{dt} + \Gamma_e J = \varepsilon_0 \omega_{pe}^2 E$ (3)

$\frac{dK}{dt} + \Gamma_m K = \mu_0 \omega_{pm}^2 H$ (4)

Basically from equation (1) to (4) are being used to identify the negative index metamaterials which is described by drude model.

METAMATERIAL BASED ANTENNAS

Antennas are one of the important components now a days that makes wireless communications possible in todays era. The work of antennas is to interface the radio system with the external environment (such as LNA, oscillators, receiver etc.). Wireless communication systems require antennas at the both sides like transmitter and receiver to operate properly without causing disturbance. The concept of metamaterials has been widely applied in the design of microwave, millimetre-wave and terahertz devices and antennas now a days to make communications and other important task wisely. With the rapid development of flexible portable devices such as mobile phones, laptops, note pads, watches, wearable devices, etc., antennas with different changing functions, based on variable structures, are in demand for next generation of wireless communication systems such as 6G and beyond. The application of metamaterials related components over the past decade has achievegreat successes in the fields of both science and engineering. Variable metamaterials have been designed from radio frequencies up to optical frequencies (terahertz and infrared etc.), and different functions have been realized, e.g., negative refractive index (NRI), anisotropy and bianisotropy (Lapine et al. 2007). As an interdisciplinary topic that is being used everywhere, metamaterials can be classified into different categories based on different criteria as given different sections. From an operating frequency point of view, they can be classified as microwave metamaterials, terahertz metamaterials, and photonic metamaterials for antenna, polarizers, absorber etc. From a spatial arrangement point of view, there are 1D metamaterials, 2D metamaterials, and 3D metamaterials present. From a material point of view, there are metallic and dielectric metamaterials that is being utilized. One of the most important applications of metamaterials is antenna design that is being discussed here. Due to the unusual properties of metamaterials, we can achieve antennas with novel characteristics which cannot be realized with traditional materials like normal material with positive refractive index. Usually, antennas consist of a combination of conductors, dielectrics and other conventional materials that have certain geometry during design. That design follows either traditional analytical methods and rules that depends on experience. In between, artificial intelligence and powerful optimization techniques used by full-wave simulations (Simulation software) are also utilized to obtain the highest possible performance by fine tuning of the structure's parameters (M. O. Akinsolu et al. 2020). It is fact that the antenna design is primarily focused on determining the optimized geometrical shape of conventional materials and then it is bound by the material characteristics for every material. In an effort to overcome all limitation in normal or conventional material, metamaterials with their unique properties have gained focus in antenna design. Irrespective of the fact that metamaterials are microscopically composed of conventional materials such as conductors and dielectrics, their macroscopic characteristics are completely different due to their smart shapes like SRR etc. The realisation of negative constitutive parameters in the microwave regime is applicable with metamaterials, while their subwavelength size (very small dimension) is another useful feature. As a result, the construction of metamaterials into antennas can offer advanced flexibility and enable novel design strategies that is followed. Therefore, the investigation of the benefits of metamaterial inspired antennas and their capability of having an active role in modern wireless communications is of great significance in advance communication.

Theoretical background of metamaterials:- According to discussion given above, the metamaterial are engineered and manmade materials that can be able to change the electromagnetic waves in many different way compared to conventional as well as natural materials occurring in nature. As discussed, structures have negative permittivity and permeability which are most common examples of metamaterials (A. Alu et al. 2007). There are three major metamaterial categories which can be identified as per finding: a) SNG (Single-Negative) metamaterials having either negative permittivity or negative permeability, b) DNG (Double-Negative) metamaterials have negative permittivity and permeability both and c) ZIM (zero-index materials) have either zero permittivity or zero permeability any one of them. In other like, EBG prohibits the propagation of EM waves and also AMC (Artificial Magnetic Conductors) with zero magnetic field that are also usually known as metamaterials. In the year 1968 as given, Veselago, he was the first researcher who theoretically studied the electrodynamics of DNG media and presented its interesting properties like reversal of the Doppler Effect, negative index of refraction as well as left-handed propagation. On the other hand, in conventional materials which support only the typical forward-wave propagation. The EM wave in SNG media is evanescent since the propagation constant is real ($\gamma \in R$) for that. DNG media supports backward wave propagation, having with the phase velocity and group velocity are opposite to each other and also the wave travels in anti-parallel to the power flux. Some important metamaterial structures are discussed below in the next section.

The presentation and construction with the effect of negative permittivity in the microwave regime is only possible with a periodic arrangement of thin metallic wires of diameter α and periodicity p given in Figure 1 and it was initially proposed by Sir John Pendry and his research group.

Figure 1. Arrangement of thin conductive wires for the representation of negative permittivity with periodic behaviour

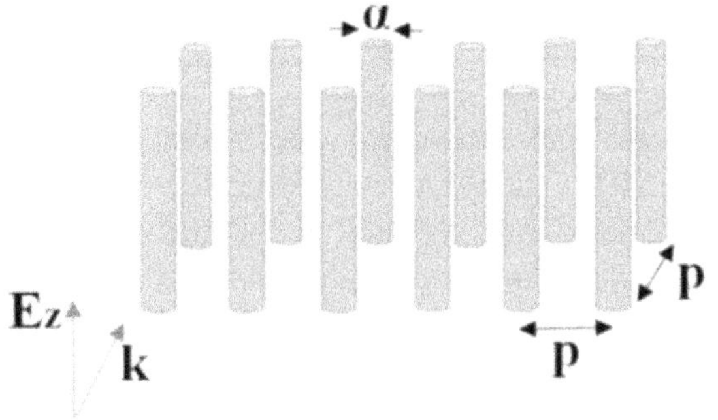

(Milius et al., 2021)

According to Figure 2, the real part of the effective permittivity is negative below the plasma frequency and its medium can be characterized as ENG.

Figure 2. Periodically arranged material with real part of the effective permittivity of infinitely long thin wires

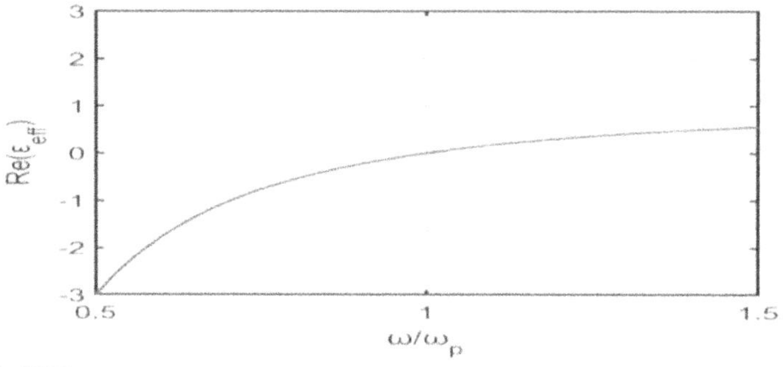

(Milius et al., 2021)

Now, it is important about this material, that there cannot find any materials in nature with negative permeability. Also, it was Pendry who proposed a Split Ring Resonator (SRR) as shown in Figure 3, which represents a negative permeability behaviour around its resonance frequency. An external time-varying magnetic field perpendicular to the SRR's surface, through which currents are induced on both the conductive inner and outer rings and then charges are accumulated across the gaps between them when it is excited with source.

Figure 3. Square split ring resonator geometry of a unit cell. Blue parts represents metallization

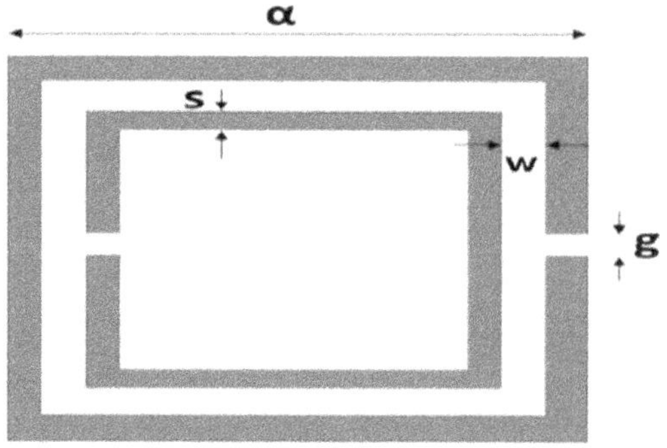

(Milius et al., 2021)

According to Figure 4, the SRR shows a narrow-band behavior as its real permeability is negative. It is important to mention that a single SRR, in other words thin wire cell is similar to an atom in conventional materials. Therefore, in order to generate an effective macroscopic ENG or MNG behavior, lots of number of SRRs structures or thin wire can be arranged periodically.

Figure 4. SRRs or thin wire effective permeability which is real

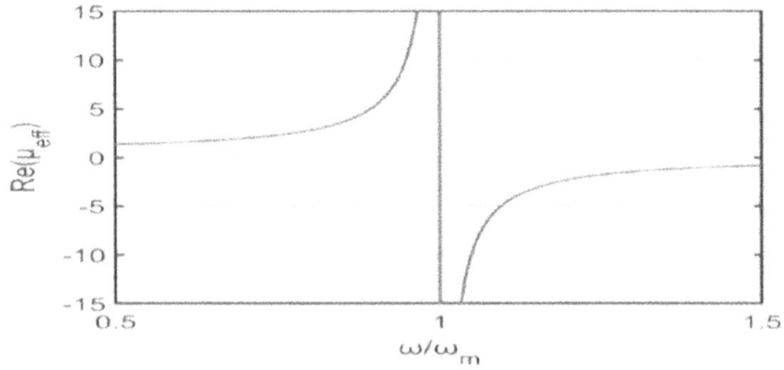

(Milius et al., 2021)

LHM metamaterials can be recognized by combinations of ENG and MNG layers of materials. The experimental representation of this unique material was first performed by Smith et al. (D. R. Smith et al. 2000) in 2000 and he has used a composite structure consisting of wires and SRRs, which is mentioned in Figure 5.

The unit cell of a natural, right-handed transmission line made of of a series inductor L_R and a shunt capacitor C_R, when its losses due to dielectrics and conductors (equivalent to shunt admittance and series resistance) is neglected in the circuit. Using an additional shunt inductance L_L and series capacitance C_L, as given in Figure 6, the dispersion characteristics are found and analyzed. Thiis type of transmission lines is known as composite right/left handed (CRLH) TL (C. Caloz et al. 2006). The enhancement of the system has used vast challenges to wireless systems, especially in the field of antennas. The major role of design and development of antenna can be considered cautiously and comprehensively as the antenna is a major element in the communication system especially wireless and also all type of application which is associated with radio engineering and technology. Today the telecommunication system comes with various operating frequency ranges and several standards for the wireless on a single stage having frequency from low to terahertz range. Hence, multiband antennas are majorly used in mobile technologies enhancing from 2G to 5G and beyond, coming with IEEE 802.11 standards. On the other way, enhancement in LTE uses lower frequency operating ranges upto 4G.

Figure 5. Representation pf DNG medium consisting of unit cells that are synthesized by wires (ENG) placed in front of SRRs (MNG)—both blue and grey parts show metallization

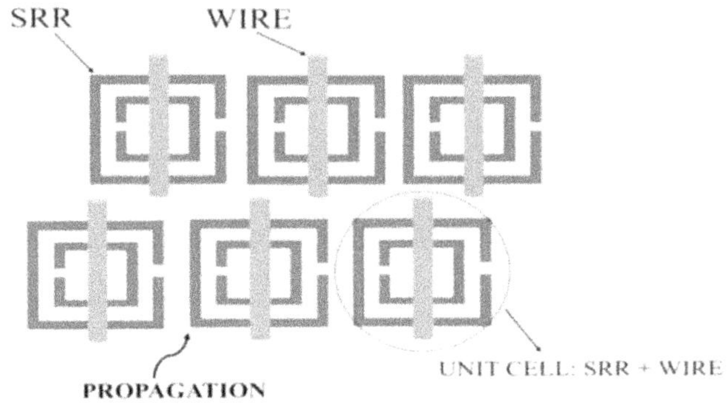

(Milius et al. 2021)

Figure 6. Representation of Unit cell of a composite right/left-handed transmission line consisting of series inductor and shunt capacitor (RH), and also series capacitor and shunt inductor (LH)

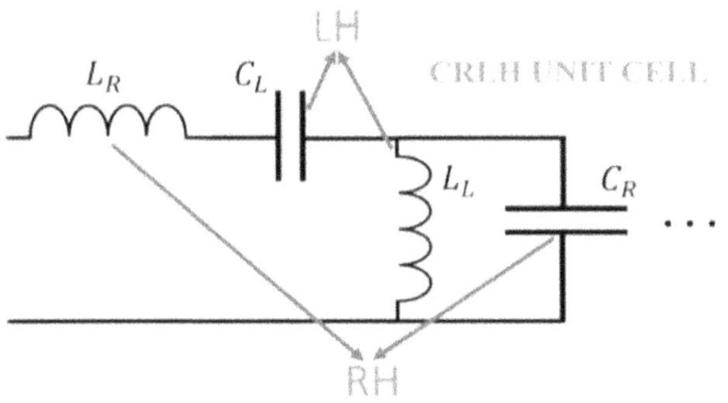

(Milius et al. (2021)

Therefore, it generates a new antenna design with operation over a wide bandwidth as well as frequency and it is also integrated to different types of handheld devices as per given assigned dimensions. Design of these type of antennas basically uses different strategies and methods to achieve novel structures, some of which are briefly given here. Usually, microstrip antennas are used just because their TM01 and TM10 orthogonal modes uses port isolation and reduce cross-polarization factor (Z. Tang et al. 2018). To increase the bandwidth, high gain, radiation pattern, dipole antenna are comprised of electric and magnetic poles are used given previously (H. Huang et al. 2018; C. Ding et al. 2018). There are multiple PCB layers used to extend bandwidth for multiband operation by exciting resonant modes on the plates for range of frequency (E. Tatartschuk et al. 2012). Then insert parasitic elements and various other coupling techniques over the antenna to improve antenna parameters like radiation pattern, bandwidth, and isolation and gain (N. Gneiding et al. 2014). Resonant cavity antennas are developed to increase the antenna directivity, front to back ratio, and gain for better bandwidth (C. Saha et al. 2012). Recently, the metamaterial inspired antenna caught substantial attention from every antenna researchers. The start of metamaterial has significantly represented new design methodologies of materials, particularly materials having unusual properties which can further compared to available materials in nature as given in Figure 8 shown below (R. K. Saraswat et al. 2020). One of the significant classes of metamaterial is electromagnetic metamaterials which are engineered sub-wavelength structures. Mostly, metamaterials are defined as engineered artificial dielectric materials comprising of periodic or non-periodic structures of unit cells with its diameter less than the wavelength of light being propagated through it (Z. Troudi et al. 2019). Such an artificially implemented array of unit cells can create an electromagnetic output that can be customized evenhandedly by magnetic and electric wave components. Hence, a periodic array of unit cells can be considered as an efficient material described by its permittivity and permeability parameters. This arrangement agrees to constitute electromagnetic responses at the desired frequency; such response will not exhibit in material presents in nature. For instance, magnetic materials are much bounded at microwave frequencies. Hence, array periodic structures of metallic rings demonstrate such magnetic response at high frequencies. Rotman demonstrated negative effective permittivity by assembling an array of metal wires (Almpanis E et al. 2017). In Table 1, special characteristics of materials have been given. Also materials classification has mentioned in Figure 9. Negative Permeability Material: The negative permeability materials are the materials having μ negative and ε positive and represent magnetic plasma as well. The absence of magnetic response at the optical range is due to the weak coupling of the magnetic field component of light with an atom which is equal to the Bohr magneton. Some of the arrangements which exhibit magnetic response at microwave and higher frequencies with negative permeability are coupled like nanorods, nanoplates, and strips, an array of metallic staple, and split ring resonator (SRR).

Figure 7. Electromagnetic properties classification of engineered material

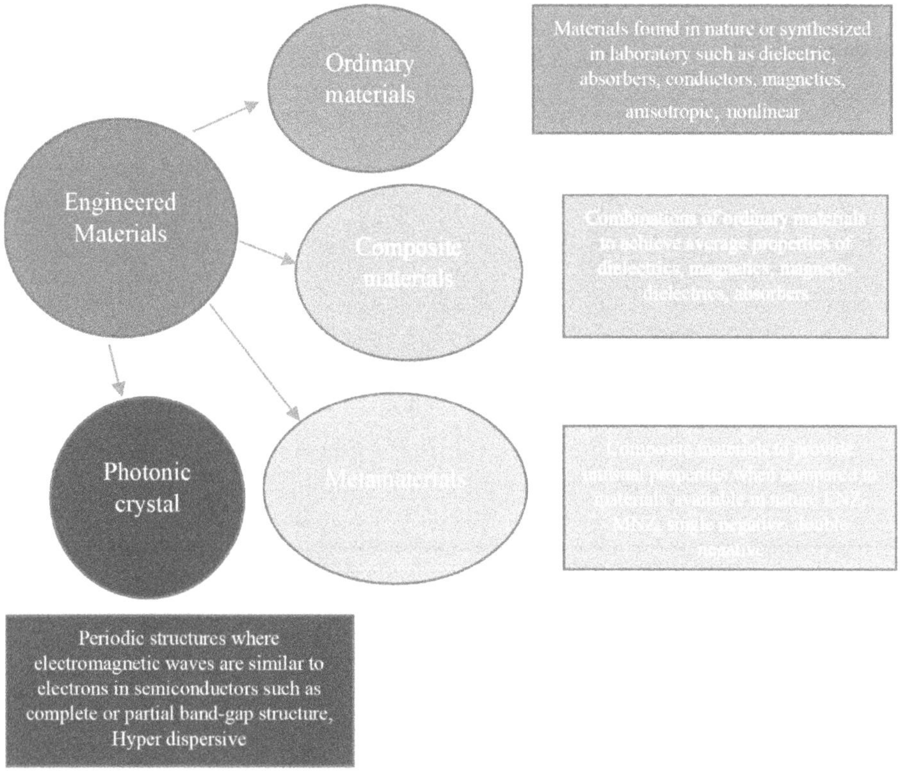

(Caloz et al. (2004)

Table 1. All special material properties in ε-μ domain

Permittivity (ε)	**Permeability (μ)**	**Description**
$\varepsilon = -\varepsilon_0$	$\mu = -\mu_0$	Characterizes anti air in left hand medium yields to the perfect lens
$\varepsilon = 0$	$\mu = 0$	Characterizes anihility yields to the perfect tunneling effects
$\varepsilon > \varepsilon_0$	$\mu = \mu_0$	Characterizes materials available in nature
$\varepsilon = \mu$		Characterizes perfect impedance matching in right hand medium and left hand medium ensuing no reflections

The most general structure used for achieving negative permeability is an SRR (Sihvola, A 2007). Pendry and his team established a structure comprising of two opposite facing concentric split-ring resonators at sub-wavelength dimensions as shown in Figure 9. The gap in each metallic ring circularly prevents the development

Figure 8. Classification of material based on permittivity and permeability

of current inside it. The charges gather at the edge of the gap in the metal ring and exhibit capacitance there. Therefore, the presence of an inductor and capacitor in a metal ri ... ssions for determ ... ven in

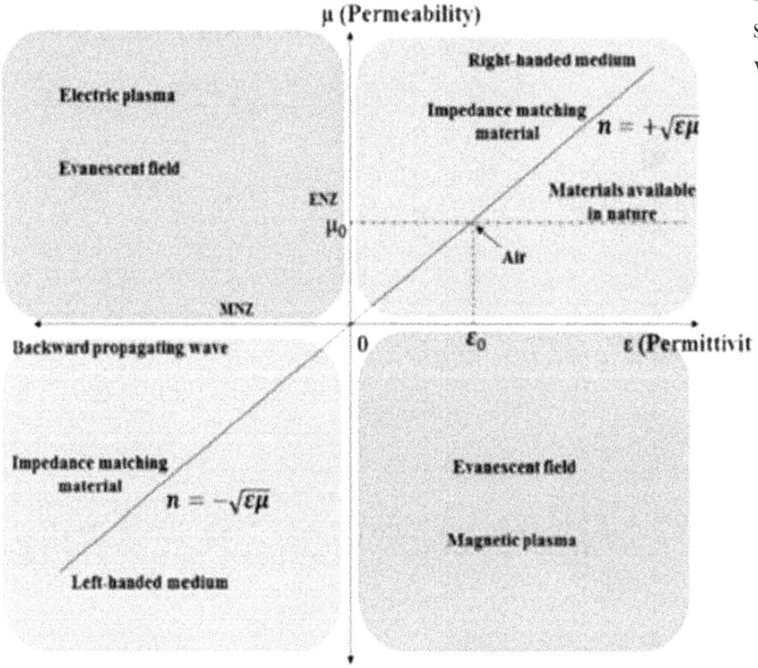

(Caloz et al. 2004)

Figure 9. Single ring split (left side) and double ring split (right side) with Magnetic material resonator structure

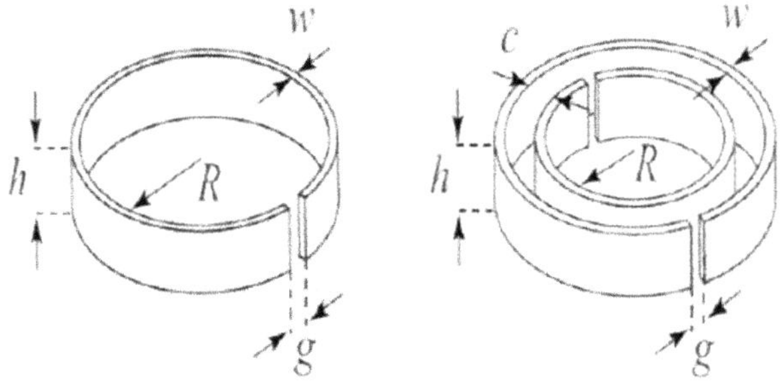

(Caloz et al. 2004)

(Sihvola, A 2007). Several other analytical expressions are available in many literatures for appropriately determining inductance, capacitance, and the resonant frequency of different resonator structure such as SRR, loop gap, and CSRR (Tamayama, et al. 2014). The negative permeability material structures are not just limited to circle/ring structures, but various other geometrical structures can also be used to create negative permeability for example spiral, swiss roll, square, hexagonal (Thippeswamy et al. 2021), (Zhang, S et al. 2019). The transmission and reflection characteristics of the SRR depend on shape and structure. For single SRR with two varying split gaps along the ring gives rise to a change in the resonant frequency of the SRR over a wide range of frequency available. For double split-ring resonators, the influence of the geometrical parameters on SRR is listed in Table 2 by varying one parameter at a time and keeping other parameters constant (Zhang, Y et al. 2008). So far numerous design structures have been available in literature in the field of MTMs technology. The most commonly used are rectangular and circular SRR. The rectangular SRR dominates circular SRR in terms of miniaturization and dense packing.

Negative Permittivity Material: The negative permittivity materials are the materials having ε negative and μ positive and represent electric plasma. To stimulate and design the surface Plasmons, it is necessary to have roughness in the surface, a dielectric coupler to vacuum like hemisphere or prism, and a metallic strip structure. An array of thin metallic wires are organized periodically in a dielectric medium or in vacuum to show the low-frequency stop band characteristics from zero to a certain cut-off frequency. Rotman (W. Rotman 1962) systematically investigated various arrangements of 1D, 2D, and a 3D array of wire network structures to induce resonance in the polarized electric field at lower than the plasma frequency in design.

Table 2. Representation of the effect of geometrical properties on SRR

Geometrical Parameters	Effect on SRR
Increasing the split gap of SRR	As the split gap increases, the value of the capacitance decreases leads to a decrease in total capacitance, and a decrease in capacitance value increase the resonant frequency.
Increasing in spacing between two adjacent rings of SRR	The increase in spacing between the rings reduces the mutual inductance and capacitance, respectively. Due to a decrease in inductance and capacitance resonant frequency increase as analytically expressed.
Increasing the side length of SRR	Increase the side length of SRR increases the effective value of inductance and thereby shifts cutoff frequency towards the lower frequency.
Increasing in metal width	The increase in metal width reduces mutual inductance and mutual capacitor and thereby increase the resonant frequency

Geometrical Parameters	Effect on SRR
Multiple split gaps in SRR	Multiple gaps in SRR intercept magnetic resonance because of the induced current by an electric field and shift resonant frequency towards the higher frequency

(Sihvola (2007))

Such an artificial composite wire structure creates a negative electric permittivity with manageable strength.

Now, one of the most important applications of metamaterials is antenna design. Due to the unusual properties of metamaterials, antennas with novel characteristics can be achieved which cannot be realized with traditional materials. Effect has been discussed below of metamaterials over antenna design.

Electrically small antennas: Now a day's electrically small antennas (ESA) are highly desired for wireless applications as they can be easily integrated with devices without compromising the system form factor. The performance of antennas in terms of bandwidth, gain and directivity etc. is extremely critical and is governed by fundamental limitations in size, i.e., Chu's limit. Although modern integrated circuit technology can miniaturize electronic circuits to a very small size however, in traditional designs, the performance of the antenna is related to its physical size always comparable with wavelength. The antenna usually has dimensions in the order of the operating wavelength. Consequently, this sets boundaries for the size of the whole wireless system is useful of this issue makes metamaterials a hot topic for research in the development of electrically small antennas. This is an opportunity for using a zero refractive index metamaterial (ZIM) medium as its operating wavelength is infinite at the given design frequency. Since the wavenumber in ZIM antennas is zero, but in theory, the physical size of such antennas can be made independent of its operating frequency. Multiband antennas:- Mobile communication systems need to support multiple communication standards, such as 5G, Bluetooth, WiFi, NFC etc. To save space it is desirable to have a single antenna that can simultaneously accommodate these standards just like reconfigurable property (Shivani Chandra et al. 2021). Multiband antennas are normally designed with different resonant structures using different mateials. The disadvantage of this technique is the large size of the resulting antenna, which is dictated by the lowest operating frequency. Because of this, metamaterial structures can support negative and positive modes along with a zeroth-order mode as well, this property can be exploited to realize a singular antenna that operates over multiple bands and is miniature to boot the antenna.

Antenna lenses and polarizers: Directivity and gain of an antenna can be improved with the use of dielectric lenses. The only drawback of this technique is the high cost of the 3D lens. Moreover, the location of the lens needs to be carefully selected in relation with the phase center of the antenna in advance. The high cost of the lens can be offset by using a 2D lens made from metamaterial, which can be easily integrated with the planar antenna structure to reduce the profile and size of the antenna system which is the best feature. Polarizers are universal components employed in diverse application including antennas, imaging, display and microscopy etc. A polarizer based on a chiral medium are used to transform a linearly polarized wave into a circularly polarized wave, and to date various polarization convertors have been proposed as given (H. L. Zhu et al. 2013). Non-alignment in polarization will result in signal fading. This issue can be eliminated by utilizing circularly polarized electromagnetic waves as given (J. D. Kraus 2003). Conventional methods to fabricate polarization convertors are to use optical gratings, anisotropic media, the Brewester as well as Faraday effects.

Other types of antennas based on metamaterials: As other examples of radiating structures based on metamaterials include leaky-wave antennas, magneto dielectric microstrip antennas, ultra-wideband (UWB) antennas with notched, reconfigurable antennas etc. All these designs exhibit better performance than the corresponding conventional designs.

Already designed antenna discussion from the published paper using metamaterials for 5G antenna:

(a) Author has presented this paper with use of metamaterial surface as a reflector and because of this whatever the in-phase reflection characteristic and high-impedance nature occurred, is being improvement in the gain of an antenna. Here, radiating element of the given metamaterial based antenna is made up of copper material which is associated by the dielectric as presented in figure given below, i.e., Rogers-4003 with a standard thickness, loss tangent and a relative permittivity of 1.524 mm designed parameters, 0.0027 and 3.55, respectively. The designed antenna with and without metamaterial surface operates at the central frequency of 3.32 GHz and 3.60 GHz. The given antenna yields gain of 2.76 dB which can be further improved to 6.26 dB, using the metamaterial surface (Fazal Muhammad et. al. 2020).

Figure 10. Design of antenna with the geometry: (a) Front view without metamaterial, (b) back view, (c) 4X4 array design using metamaterial

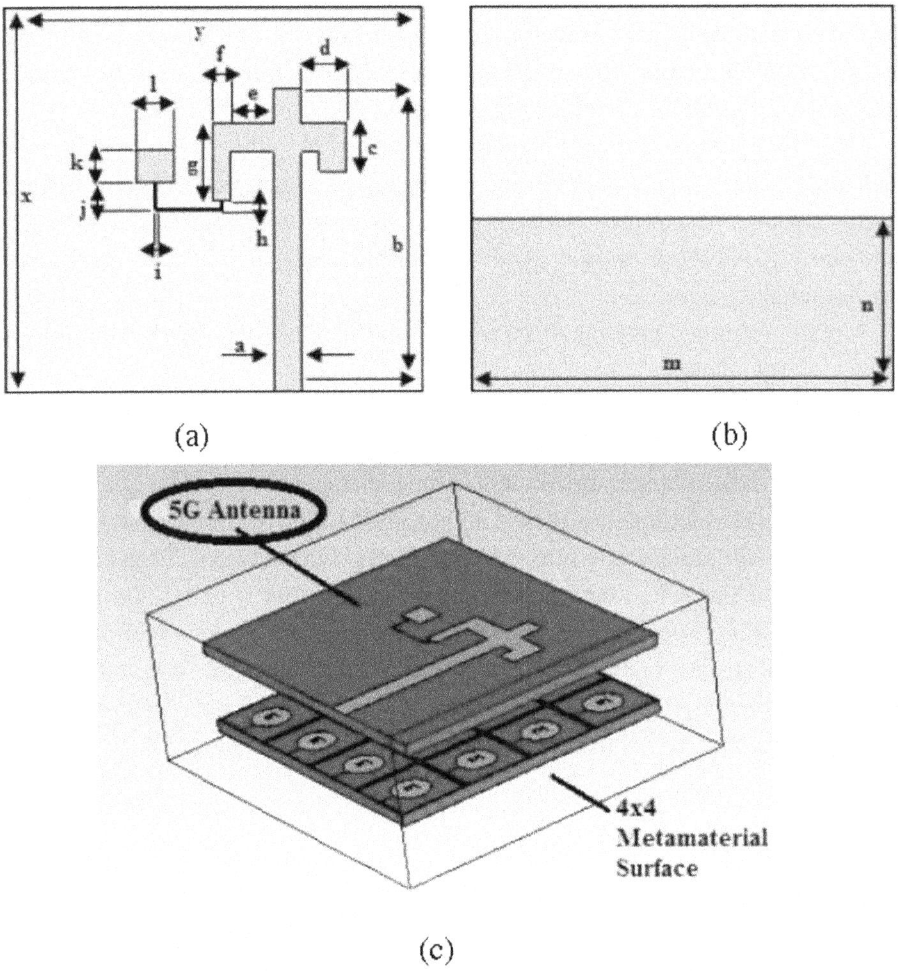

(b) Basically, author has has used low dielectric constant of 2.2, which is of Rogers RT/Duroid substrate, and a dielectric loss tangent of 0.0010 in the given design as presented in figure below. Paper deals with a metamaterial-based multi- This paper gives the value of good return loss of -34.4 dB at 26 GHz, -13.49 dB at 40 GHz,

-13.63 dB at 53.5 GHz, high bandwidth of 5.368 GHz at 26 GHz, 3.76 GHz at 40 GHz, 2.88 GHz at 53.5 GHz, needed voltage standing wave ratio, 1⩽VSWR⩽ 2, high gain of 10 dBi at 26 GHz, 5 dBi at 40 GHz, and antenna radiation efficiency of 99.7% at 26 GHz, and 61% at 40 GHz, 50% at 53.5 GHz. Also the bandwidth, return loss, antenna radiation efficiency and power density indicate an improvement of 5.368 GHz to 5.630 GHz, -34.82 dB to -57.10 dB, 99.7% to 99.8% and 2208 kW/m2 to 2800 kW/m2 respectively after loading and incorporating artificial magnetic split-ring resonator-based metamaterial on the patch (Calico et al. 2021).

Figure 11. (a) 3D view of split-ring resonator, (b) unit cell of proposed split-ring resonator

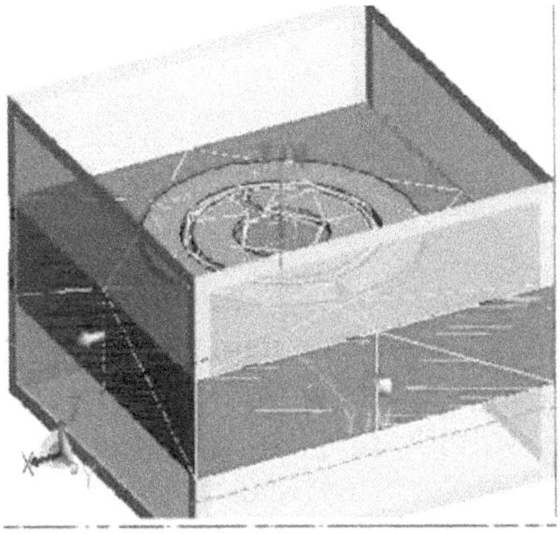

(a)

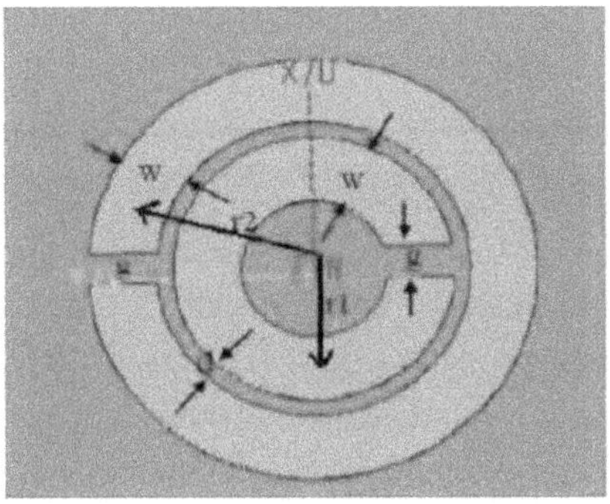

(b)

(c)Author has used a symmetrical monolayer metamaterial superstrate which is kept on a microstrip patch antenna resonant at 3.5 GHz in this paper. metamaterial based CSRR, constituting two rectangular rings is opposite splits on both ends, which are printed on Rogers RT 5880 type substrate with relative permittivity of 2.2, tangential loss of 0.0009, and a thickness of 0.508mm. This resonator is giving negative permittivity as shown in figure given below. Design with 2x3 is giving a gain increase from 4.66 dBi to 5.67 dBi and 4x4 is giving 9.84 dBi (Abderrahim Bellekhiri et al. 2022).

Figure 12. (a) Representation of CSRR, (b) the geometry of the antenna with metamaterial

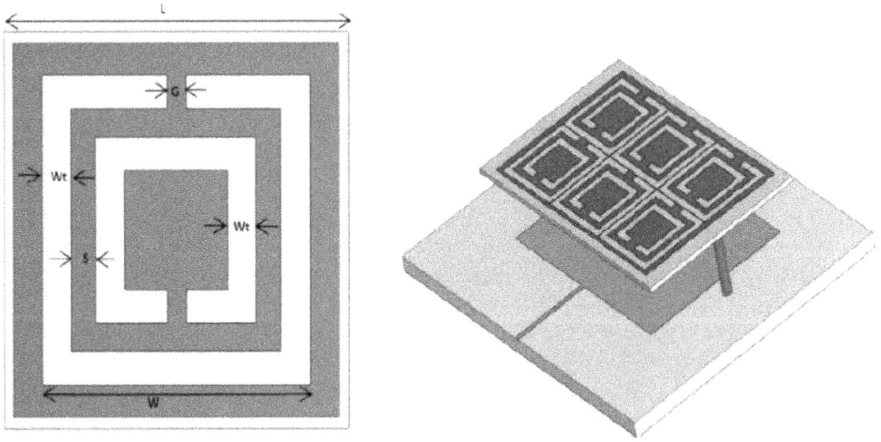

(a) (b)

(d)this paper has presented a high-isolation metamaterial based dual band multiple-input multiple-output antenna for fifth generation millimeter-wave communication networks. The proposed and designed antenna is a pentagon-shaped monopole that provides a dual-band response with a wide operating bandwidth at 5G. The proposed design is capable of covering both 28/28 5G bands and has the merits of broad bandwidth, low profile, high gain (>5 dB), improved isolation (– 38 dB), low envelope correlation coefficient (ECC) (9.99 dB) (Bashar Ali Esmail, 2023).

Figure 13. (a) Representation of the configuration of the proposed metamaterial unit cell, (b) the configuration of the proposed two-port MIMO antenna with metamaterials

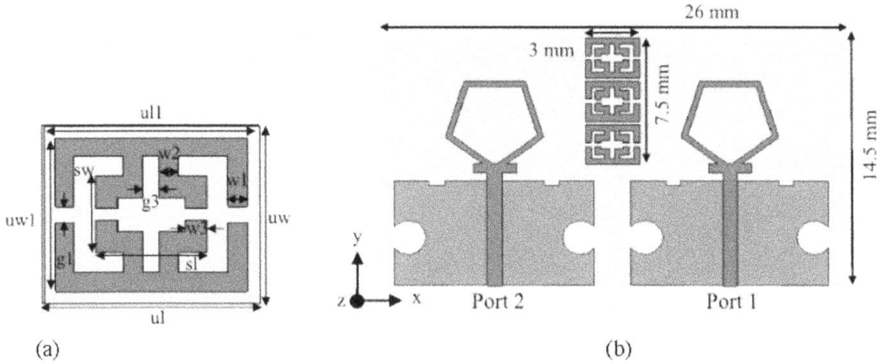

(e) Metamaterial based another antenna has been discussed with two pin diodes and four modes have been calculated with different gain and bandwidth as given in figure 14 with full results and discussion. It resonates at four different frequencies – 5.32 GHz, 5.662 GHz, 6.469 GHz and 6.547 GHz with a maximum gain of 6.44 dBi, 7.75 dBi, 8.54 dBi and 8.39 dBi respectively. Gain values has been mentioned in table 3 (Shivani Chandra and Smrity Dwivedi, 2019), (Shivani Chandra and Smrity Dwivedi, 2021), (Smrity Dwivedi, 2022), (Shivani Chandra and Smrity Dwivedi, 2021), (Smrity Dwivedi, 2022), (Smrity Dwivedi, 2023).

Figure 14. (a) Representation of of CPW feed antenna without split ring resonators, (b), The proposed design structure of the antenna

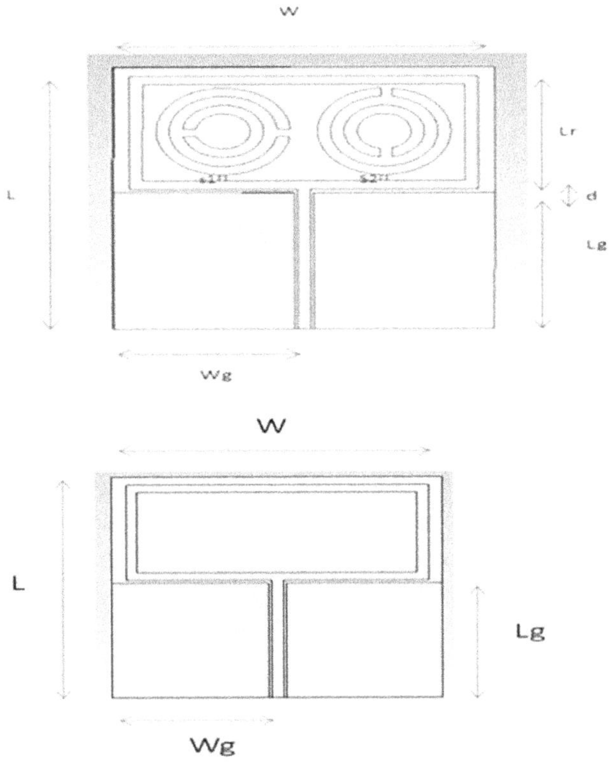

Table 3. Representation of configuration of switches

States	Mode	S1	S2	Resonant Frequency (GHz)	Gain (dBi)
State 1	00	OFF	OFF	6.469	8.54
State 2	01	OFF	ON	6.547	8.39
State 3	10	ON	OFF	5.32	6.44
State 4	11	ON	ON	5.66	7.75

(f) Using CST microwave studio, author has designed and drawn for the structure mentioned. Figure 15 (a) represents the front view and (b) represents the back view of structure where BC-SRR is clearly visible. In this design, the C band waveguide is coupled as transition structure and other side is air box to excite the structure. From Figure 15, it is clear that the maximum gain from the given antenna ~ 10 dBi.

Figure 15. Representation of Schematic view of metamaterial waveguide antenna: (a) Front view, (b) back view, (c) Radiation pattern 3D plot, (b) radiation pattern in 2D (polar) plot

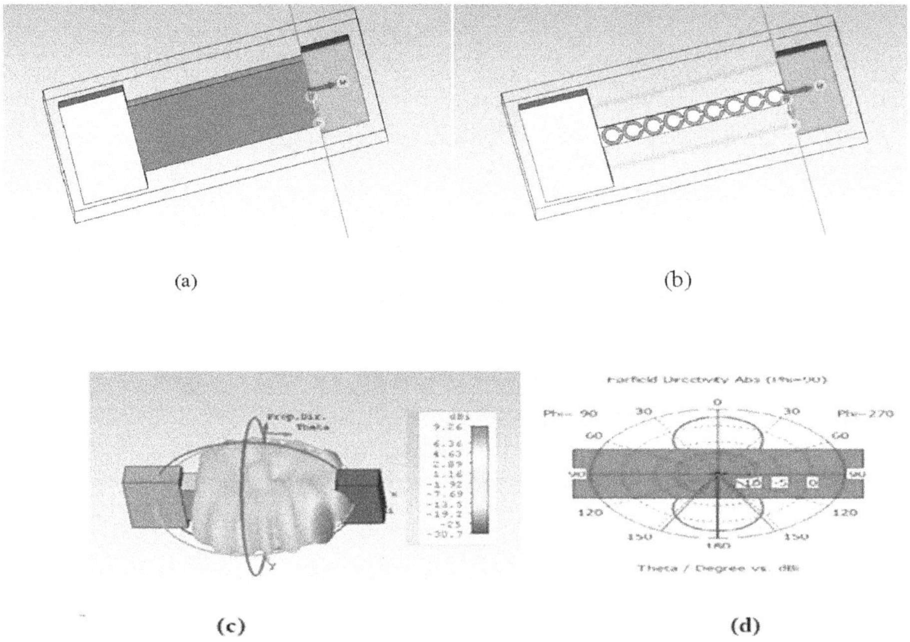

(g) This is author own paper in which multilayer antenna is designed with different materials and air gaps. All designs are having same parameters and but the patch, ground and substrate are changed according to results. Also it is basically high frequency antenna have terahertz frequency and dimensions are in micrometer. This design has given 10.9 Db gain with 52.9 GHz bandwidth as given in figure. Under this design, frequency is increased by 0.3 THz to 0.6 and accordingly gain is improved 3.02 dB to 10.09 dB (Smrity Dwivedi, 2023).

Figure 16. Antenna design with different layering of materials and mentioned with different colors

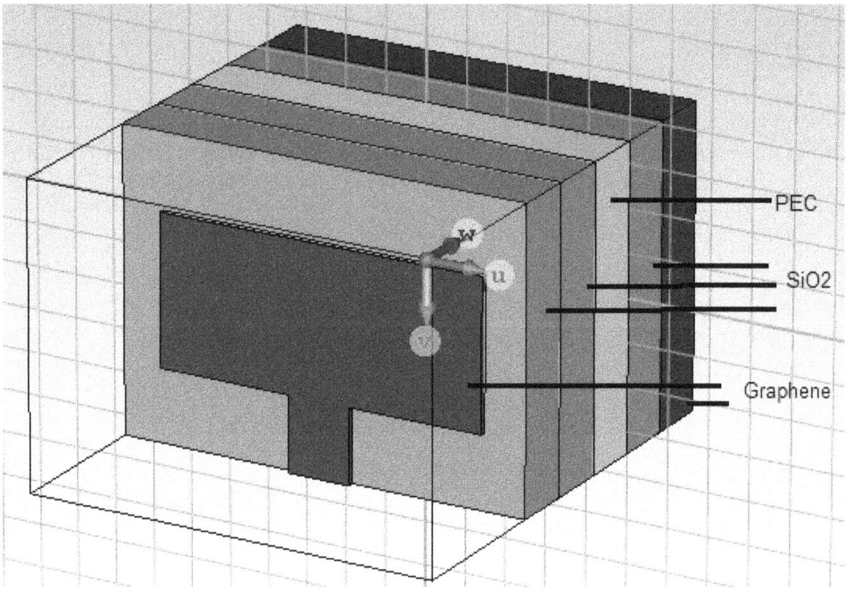

CONCLUSION

In conclusion of metamaterial opened a new dimension in antenna design because of its unique electromagnetic characteristics such as negative permittivity and permeability, negative refractive index, and its manipulation. On the other hand, the evolution of electronic gadgets reserves a compact area for antennas and looking for better performance. Several different approaches are examined for metamaterial based antenna, in specific a)DNG, b)MNG, c)ENG, d)AMC, e)EBG. Some results are discussed which is already published and also author has discussed own design based on metamaterials. Few more papers have been discussed to recognize the field with other options too.

REFERENCES

Akinsolu, M. O., Mistry, K. K., Liu, B., Lazaridis, P. I., & Excell, P. (2020). Machine learning-assisted antenna design optimization: A review and the state- of-the-art. *Proc. 14th Eur. Conf. Antennas Propag. (EuCAP),* (pp. 1–5). IEEE. 10.23919/EuCAP48036.2020.9135936

Ali Esmail, B., & Koziel, S. (2023). High isolation metamaterial-based dual-band MIMO antenna for 5G millimeter-wave applications. *AEÜ. International Journal of Electronics and Communications*, 158, 154470. 10.1016/j.aeue.2022.154470

Almpanis, E., Pantazopoulos, P. A., Papanikolaou, N., Yannopapas, V., & Stefanou, N. (2017). A birefringent etalon enhances the Faraday rotation of thin magneto-optical films. *Journal of Optics*, 19(7), 075102. 10.1088/2040-8986/aa7420

Alu, A., Engheta, N., Erentok, A., & Ziolkowski, R. W. (2007, February). Single-negative, double-negative, and low-index metamaterials and their electromagnetic applications. *IEEE Antennas & Propagation Magazine*, 49(1), 23–36. 10.1109/MAP.2007.370979

Balanis, C. A. (2012). *Antenna theory: Analysis and design.* John Wiley & Sons., 10.1109/LMWC.2004.828009

Bellekhiri, A., Chahboun, N., Laaziz, Y., & El, A. (2022). A New Design of 5G Planar Antenna based on metamaterials with a high gain using array antenna. *ITM Web of Conferences, 48.*

Caloz, C., & Itoh, T. (2004). Array factor approach of leaky-wave antennas and application to 1-D 2-D composite right left-handed (CRLH) structures. *IEEE Microwave and Wireless Components Letters*, 14(6), 274–276. 10.1109/LMWC.2004.828009

Caloz, C., & Itoh, T. (2006). *Electromagnetic Metamaterials Transmission Line Theory and Microwave Applications.* Wiley.

Chandra & Dwivedi. (2020). Simulation analysis of High Gain and Low Loss Antenna with Metallic Ring and spoof SPP Transmission Line. In *7th International Conference on Signal Processing & Integrated Networks.* IEEE Explore.

Chandra & Dwivedi. (2021). Graphene based Radiation Pattern reconfigurable antenna. *International Conference for convergence in Technology.* IEEE.

Colaco & Lohan. (2021). Metamaterial Based Multiband Microstrip Patch Antenna for 5G Wireless Technology-enabled IoT Devices and its applications. *J. Phys.: Conf. Ser.*

Ding, C., Sun, H.-H., Ziolkowski, R. W., & Jay Guo, Y. (2018, December). A dual layered loop array antenna for base stations with enhanced crosspolarization discrimination. *IEEE Transactions on Antennas and Propagation*, 66(12), 6975–6985. 10.1109/TAP.2018.2869216

Dwivedi, S. (2022). *ReconfigurableArray antenna*. Intech Open., 10.5772/intechopen.106343

Dwivedi, S. (2023). *Metamaterials-Based Antenna for 5G and Beyond*. IGI Global. 10.4018/978-1-6684-8287-2.ch001

Dwivedi, S. (2023). Proposed Design for Beyond 5G Antenna for Upgraded Applications with review. *2023 14th International Conference on Computing Communication and Networking Technologies (ICCCNT)*. IEEE. 10.1109/ICCCNT56998.2023.10307737

Dwivedi. (2022). *Design and Analysis of Metamaterial Waveguide Antenna for Broadband Applications*. Springer. 10.1007/978-981-19-5224-1_42

Dwivedi. (2023). *Antenna Array for reconfiguration*. IGI Global. 10.4018/978-1-6684-5955-3

Ejaz, J. W., Anpalagan, A., Imran, M. A., Jo, M., Naeem, M., Bin Qaisar, S., & Wang, W. (2016). *2016 Internet of Things (IoT) in 5G Wireless Communications*. IEEE Access.

Gneiding, N., Zhuromskyy, O., Shamonina, E., & Peschel, U. (2014, October). Circuit model optimization of a nano split ring resonator dimer antenna operating in infrared spectral range. *Journal of Applied Physics*, 116(16), 164311. 10.1063/1.4900479

GSMA. (2019). *5G the internet of things and wearable devices: Radiofrequency exposure*. GSMA Public Policy.

Holloway, D., Dienstfrey, A., Kuester, E. F., O'Hara, J. F., Azad, A. K., & Taylor, A. J. (2009). A Discussion on the Interpretation and Characterization of Metafilms/Metasurfaces: The Two Dimensional Equivalent of Metamaterials. *Metamaterials (Amsterdam)*, 3(2), 100–112. 10.1016/j.metmat.2009.08.001

Holloway, K., Kuester, E. F., Baker-Jarvis, J., & Kabos, P. (2003). A Double Negative (DNG) Composite Medium Composed of Magneto-Dielectric Spherical Particles Embedded in a Matrix. *IEEE Transactions on Antennas and Propagation*, 51(10), 2596–2603. 10.1109/TAP.2003.817563

Huang, H., Li, X., & Liu, Y. (2018, June). A novel vector synthetic dipole antenna and its common aperture array. *IEEE Transactions on Antennas and Propagation*, 66(6), 3183–3188. 10.1109/TAP.2018.2819894

Kraus, J. D., & Marhefka, R. J. (2003). *Antennas for all applications* (3rd ed.). McGraw-Hill.

Lapine, M., & Tretyakov, S. (2007). Contemporary notes on metamaterials. *IET Microwaves, Antennas & Propagation*, 1(1), 3–11. 10.1049/iet-map:20050307

Milius, C., Andersen, R. B., Lazaridis, P. I., Zaharis, Z. D., Muhammad, B., & Jes, T. B. (2021). Metamaterial-Inspired Antennas: A Review of the State of the Art and Future Design Challenges. *IEEE Access : Practical Innovations, Open Solutions*, 9, 89846–89865. 10.1109/ACCESS.2021.3091479

Helena. (2020). *IEEE Access : Practical Innovations, Open Solutions*, 8, 177064–177083.

Rotman, W. (1962, January). Plasma simulation by artificial dielectrics and parallelplate media. *IRE Transactions on Antennas and Propagation*, 10(1), 82–95. 10.1109/TAP.1962.1137809

Saha, C., & Siddiqui, J. Y. (2012, July). Theoretical model for estimation of resonance frequency of rotational circular split-ring resonators. *Electromagnetics*, 32(6), 345–355. 10.1080/02726343.2012.701540

Saraswat, R. K., & Kumar, M. (2020, October). A quad band metamaterial miniaturized antenna for wireless applications with gain enhancement. *Wireless Personal Communications*, 114(4), 3595–3612. 10.1007/s11277-020-07548-z

Sehrai, D. A., Muhammad, F., Kiani, S. H., & Kim, S. (2020). Gain-Enhanced Metamaterial Based Antenna for 5GCommunication Standards. *Computers, Materials & Continua*, 64(3), 1587–1599. 10.32604/cmc.2020.011057

Shamonina, E., & Solymar, L. (2007). Metamaterials: How the Subject Started. *Metamaterials (Amsterdam)*, 1(1), 12–18. 10.1016/j.metmat.2007.02.001

Sihvola, A. (2007). Metamaterials in electromagnetics. *Metamaterials (Amsterdam)*, 1(1), 2–11. 10.1016/j.metmat.2007.02.003

Silin, R. A., & Chepurnykh, I. P. (2001). On Media with Negative Dispersion. *Commun. Technol. Electron.*, 46, 1121–1125.

Sivukhin, D. V. (1957). The Energy of Electromagnetic Fields in Dispersive Media. *Opt. Spektrosk.*, 3, 308–312.

Smith, D. R., Padilla, W. J., Vier, D. C., Nemat-Nasser, S. C., & Schultz, S. (2000). Composite Medium with Simultaneously Negative Permeability and Permittivity. *Physical Review Letters*, 84(18), 4184–4186. 10.1103/PhysRevLett.84.418410990641

Tamayama, Y., Yasui, K., Nakanishi, T., & Kitano, M. (2014). A linear-to-circular polarization converter with half transmission and half reflection using a single-layered metamaterial. *Applied Physics Letters*, 105(2), 021110. 10.1063/1.4890623

Tang, Z., Liu, J., Cai, Y.-M., Wang, J., & Yin, Y. (2018, April). A wideband differentially fed dual-polarized stacked patch antenna with tuned slot excitations. *IEEE Transactions on Antennas and Propagation*, 66(4), 2055–2060. 10.1109/TAP.2018.2800764

Tatartschuk, E., Gneiding, N., Hesmer, F., Radkovskaya, A., & Shamonina, E. (2012, May). Mapping inter-element coupling in metamaterials: Scaling down to infrared. *Journal of Applied Physics*, 111(9), 094904. 10.1063/1.4711092

Thippeswamy, M. C., Kuchibhatla, S. A. R., & Rajagopal, P. (2021). Concentric shell gradient index metamaterials for focusing ultrasound in bulk media. *Ultrasonics*, 114, 106424. 10.1016/j.ultras.2021.10642433819870

Troudi, Z., Machac, J., & Osman, L. (2019, October). Miniaturised planar band-pass filter based on interdigital arm SRR. *IET Microwaves, Antennas & Propagation*, 13(12), 2081–2086. 10.1049/iet-map.2018.5708

Veselago, V. G. (1968). The Electrodynamics of Substances with Simultaneously Negative Values of ε and μ. *Physics Uspekhi*, 10, 509–514. 10.1070/PU1968v010n04ABEH003699

Wang, C., Chen, Y., & Yang, S. (2018, November). Dual-band dual-polarized antenna array with flat-top and sharp cutoff radiation patterns for 2G/3G/LTE cellular bands. *IEEE Transactions on Antennas and Propagation*, 66(11), 5907–5917. 10.1109/TAP.2018.2866596

Wen, L.-H., Gao, S., Mao, C.-X., Luo, Q., Hu, W., Yin, Y., & Yang, X. (2018). A wideband dual-polarized antenna using shorted dipoles. *IEEE Access : Practical Innovations, Open Solutions*, 6, 39725–39733. 10.1109/ACCESS.2018.2855425

Yang, D., Liu, S., & Zhao, Z. (2017, May). A broadband dual-polarized printed dipole antenna with low cross-polarization and high isolation for base station applications. *Microwave and Optical Technology Letters*, 59(5), 1107–1111. 10.1002/mop.30467

Zhang, S., Chen, X., & Pedersen, G. F. (2019). Mutual coupling suppression with decoupling ground for massive MIMOantenna arrays. *IEEE Transactions on Vehicular Technology*, 68(8), 7273–7282. 10.1109/TVT.2019.2923338

Zhang, Y., Hong, W., Yu, C., Kuai, Z. Q., Don, Y. D., & Zhou, J. Y. (2008). Planar ultrawideband antennas with multiple notched bands based on etched slots on the patch and or split ring resonators on the feed line. *IEEE Trans. Antennas Propag.*, *56*(9), 3063–3068. doi:.928815 doi:10.1109/tap.2008

Zhu, H. L., Cheung, S. W., Chung, K. L., & Yuk, T. I. (2013). Linear-to-circular polarization conversion using metasurface. *IEEE Transactions on Antennas and Propagation*, 61(9), 4615–4623. 10.1109/TAP.2013.2267712

Chapter 3
Metamaterials and Metasurfaces for Sensor and Biosensor Applications

Maryam Ghodrati
http://orcid.org/0000-0002-7956-4911
Lorestan University, Iran

Ali Mir
Lorestan university, Iran

Jingda Wen
http://orcid.org/0000-0001-5827-4360
Tsinghua University, China

ABSTRACT

Metamaterials are efficiently homogenizable arrangements of artificial structural components engineered to achieve beneficial and exotic electromagnetic (EM) properties not found in natural materials. Metasurfaces are the two-dimensional analogue of metamaterials consisting of single-layer or multi-layer stacks of planar structures. Both metamaterials and metasurfaces have great potential to be used in a wide range of applications, e.g., antennas, absorbers, slow light devices, photo-catalysis, optical modulation devices, and ultra-sensitive sensors and biosensors. This chapter highlights the recent advances in sensing and biosensing technology based on metamaterials and metasurfaces as well as its main applications, focusing on the effect of material selection on device performance. This chapter offers a detailed and comprehensive analysis of optical sensors and biosensors based on

DOI: 10.4018/979-8-3693-2599-5.ch003

metamaterials and metasurfaces for medical diagnoses, biomolecule detection like viruses, and bacteria, and disease diagnosis such as cancer.

INTRODUCTION

Recently, advancements in the field of nanotechnology have opened up new possibilities for manipulating light at the subwavelength level (Lee et al., 2017; Zhang et al., 2018). When an electromagnetic (EM) wave interacts with a composite material, it generates electric and magnetic dipole moments within the material's inclusions (Lie et al., 2016; Tretyakov, 2013; Pendry et al., 1999). These dipole moments are closely linked to the effective permittivity and permeability of the composite material. By controlling the size, density, shape, and orientation of these inclusions, researchers can create materials with unique electromagnetic properties. These specially engineered materials are commonly referred to as metamaterials (MMs) (Tretyakov, 2017; Tan et al., 2022).

The origin of MMs can be traced back to various examples such as the pyramid brick wall, Parthenon columns, and medieval ruby glass (refer to Figure 1). Additional historical instances of MMs include the study of polarization plane rotation through artificial twisted structures in 1898 and the development of artificial dielectric structures for microwave antenna lenses in 1945 (Leonhardt, 2007). The emergence of modern metamaterials gained attention when Pendry et al. (1999) proposed that arrays of conducting wire could exhibit a negative effective permittivity at a relatively low frequency (<200 THz), while split ring resonators (SRRs) could enable a strong magnetic resonance leading to a negative effective permeability. The idea of MMs originated in 1968 from Veselago's theoretical prediction based on Maxwell's equations of the existence of materials with negative permittivity and permeability (Veselago, 1968).

The concept of "metamaterial" originates from the combination of the Greek words "meta" and "material", with "meta" denoting something that goes beyond the ordinary, reorganized, altered, or innovative. It is a specially engineered material designed to possess unique properties and capabilities that are not found in natural materials. The term "metamaterial" was introduced by Walser in 1999 (Walser, 1999). In 2000, the first instance of a metamaterial exhibiting a negative refraction index was presented, where the wave vector and energy flux, known as backward waves, behaved oppositely in a medium with a negative index. These experiments confirmed the predictions made by Veselago in 1968 regarding negative index materials, including negative refraction, the reversed Doppler effect, and reversed Cherenkov radiation (Duan et al., 2017).

Metamaterials are not obtainable from a uniform and continuous medium, which is why they are typically composed materials. In most cases, MMs are created using individual resonant micro- and nanometer-scale components that imitate the electromagnetic behavior of atoms and molecules found in natural substances. This allows them to interact with light and other types of energy in precise and manageable manners (Duan et al., 2017).

MMs are artificial materials that are utilized to control and manipulate waves, sound, and other physical phenomena in unusual ways other than regular materials by utilizing the structure's natural properties (Hu et al., 2019; Kurosawa et al., 2017; Arbabi et al., 2018). They are periodic or aperiodic arrays of sub-wavelength artificial three-dimensional (3D) material units with extraordinary EM properties. MMs have exotic electrical and magnetic properties that are not typically found in nature, for instance, backward propagation, the reverse Doppler Effect, nearly perfect absorption, negative refractive index, diffraction-limit breaking imaging, and etc (Ma et al., 2020; Xu et al., 2021; Wei et al., 2013).

Figure 1. Examples of some "historical" original metamaterials

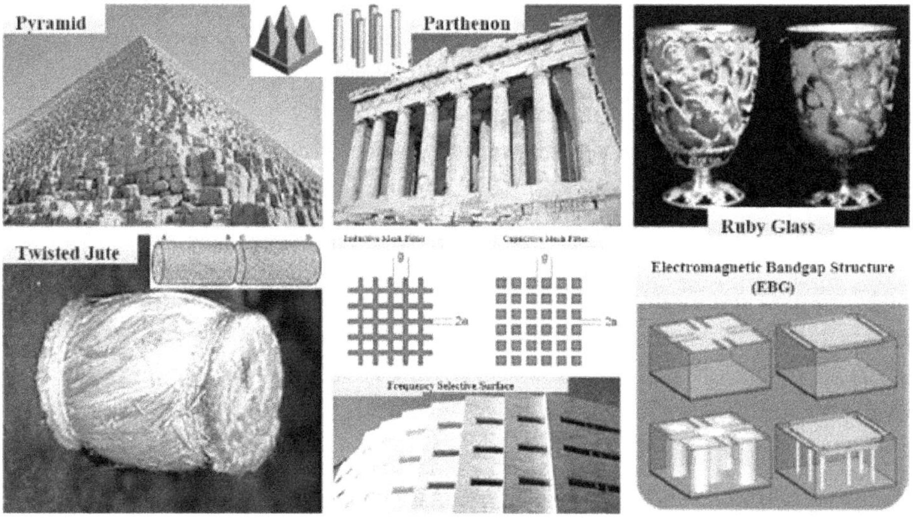

(Leonhardt, 2007)

MMs can also utilize EM wave beams in surprising ways. With the progress of micro/nano processing technology and the precise optics of sub-wavelength structure characteristics, MMs are widely used in slow light devices, photocatalysis, optical modulation devices, and highly sensitive sensors and biosensors (Rifat et al., 2018; Faraji-Dana et al., 2019).

Metamaterials have the potential to revolutionize sensing and biosensing technologies. They exhibit strong electromagnetic field confinement and sharp resonance features, enhancing the interaction between terahertz waves and matter. This, in turn, improves the performance of bio-detection applications (Li et al., 2020; Kumari et al., 2021; He et al., 2021). However, the widespread adoption of MMs is hindered by several drawbacks. These include the complexity of their three-dimensional systems, large device sizes, and the high loss of metal structures, particularly in the high-frequency bands such as terahertz, infrared, and optical. These challenges make it difficult to process MMs effectively (Li et al., 2020; Kumari et al., 2021).

With the emergence of metasurfaces (MSs) in 2011, they have steadily grabbed the attention of researchers. MSs are planar structures with low loss, essentially an extension of the two-dimensional (2D) form of MMs (Yuan et al., 2018; Naderi and Ghodrati, 2017). MSs address the challenges associated with processing difficulties and high loss in MMs, offering convenience for integrating and miniaturizing nano-optical devices. As 2D materials, MSs can be seamlessly integrated into various devices, making them a key feature for nanophotonic circuits (Naderi and Ghodrati, 2018; Yao et al., 2022).

Hence, this characteristic will enable their involvement in photonics for "lab on chip" applications. MSs have demonstrated potential uses in optics, opening up new possibilities for controlling light in a variety of fields, including optical imaging, sensing, displays, and communication (Islam et al., 2022; Naderi and Ghodrati, 2018). Various sensing and biosensing platforms utilizing optical, electrical, and mechanical principles have displayed promising applications across a wide range, from research in laboratories to clinical diagnostics, drug development, and the fight against emerging infectious diseases (Naderi et al., 2020; Wang et al., 2021).

Optical detection platforms have garnered significant attention in recent years among the various genres of sensors and biosensors. These sensors rely on light–matter interactions to identify small entities through absorption, emission, fluorescence, and other principles (Wang et al., 2021). In today's world, optical sensing technology is widely used in applications such as temperature, humidity, distance, refractive index (RI) measurements, material composition recognition, detection of small particles like biomolecules, gases, ions, and their manipulation (Kretschmann et al., 1965; Ghodrati et al., 2020; Ghodrati et al., 2019; Otta, 1968).

Optical sensors with micro/nanostructures that have dimensions comparable to working wavelengths have gained popularity in various demonstrations due to their real-time capabilities, high integration, high sensitivity, and cost-effectiveness. These sensors utilize miniaturized optical elements, integrated photonic circuits, micro/nano-cavities, and artificially engineered optical materials (Desouza et al., 2017;

Ghodrati et al., 2021). One of the advantages of optical sensors and biosensors is their ability to remotely diagnose biomolecular binding signals without the need for physical connections between the excitation source and the detection media. Unlike mechanical and electrical sensors, optical sensors are compatible with physiological solutions and remain unaffected by changes in the ionic strengths of the solutions (Ghodrati et al., 2021; Jiang et al., 2019).

Sensing and biosensing play a crucial role in biomedical engineering, highlighting the urgent need for a fast and dependable method to detect microorganisms like bacteria and viruses. Current techniques are laborious and time-consuming, with the growth of these microorganisms often taking days or even weeks. Moreover, existing detection methods for diseases such as cancer are costly and time-intensive (Jiang et al., 2019; Ghodrati et al., 2020). MSs-based biosensors offer a promising solution for swift, label-free detection of various microorganisms and cancer cells. The unique characteristics of MMs and MSs have the potential to revolutionize numerous research areas, particularly in sensing and biosensing. Sensors utilizing these specially engineered materials excel in high sensitivity, selective detection, and precise measurement of biomarkers, making them invaluable for accurate and early disease diagnosis (Ghodrati et al., 2022; Qiu et al., 2020).

Optical sensors offer several advantages over traditional technologies due to their speed, low-cost, high sensitivity, selectivity, and label-free analysis capabilities. In the literature, a wide array of electromagnetic sensors can be found, such as interferometers (Qi et al., 2002), waveguides (Dell'Olio et al., 2007), surface plasmon resonance (SPR)-based structures (Larsson et al., 2007), cavities/resonators (Yalcin et al., 2006), and all-dielectric MMs (La Spada et al., 2016). While these sensors exhibit many benefits, they also come with limitations like electrically large structures, high losses, dispersive behavior, polarization dependence, and challenges in manufacturing (La Spada et al., 2014; Yotter et al., 2004).

Due to their unique properties resulting from their geometry, material composition, and arrangements, MSs offer exceptional characteristics that are not commonly found in traditional materials or existing technology. These properties have been extensively utilized in previous research to enhance various devices, including antennas (Maci et al., 2011), absorbers (Pozar et al., 1997), guiding structures (Chen et al., 2006), polarizers and modulators (Jiang et al., 2013), lenses for imaging systems (Yu et al., 2011), and cloaking devices (Sun et al., 2012). Despite significant advancements, these technologies still face inherent limitations, such as limited control over their response, highly confined near-field, narrow bandwidth, and manufacturing constraints specific to certain geometries, sources, and polarization dependencies.

In recent years, plasmonic MMs and MSs have emerged as promising options for innovative RI sensors and biosensors. They excel in their ability to confine the electromagnetic field at the nanoscale and enhance the interactions between light and matter, resulting in heightened sensitivity to changes in ambient RI (Qiu et al., 2021; Ghodrati et al., 2023). Additionally, sensors based on MMs and MSs offer several advantages over traditional SPR-based biosensors. Firstly, since RI changes are detected through macroscopic optical responses, such as reflection or transmission of focused input beams, MM and MS-based RI sensors outperform SPR-based ones in terms of fabrication tolerance and stability of the output signal (Chen et al., 2012; Khanikev et al., 2013). Secondly, the periodic arrangements of meta-atoms allow for lower radiative damping and higher quality factor, showcasing intriguing physical phenomena like plasmonically induced transparency and Fano resonances (Chen et al., 2012; Khanikev et al., 2013). Lastly, the incorporation of MM or MS technology can enhance the capabilities of a single nanophotonic RI sensor. Intricate combinations of various meta-atoms within a unit cell or supercell can produce multiple resonances and broad-spectrum slow light effects that are challenging to achieve with SPR-based sensors. These remarkable characteristics position MSs as strong contenders for sensing applications (Ghodrati et al., 2022; Al-Naib, 2022). Through meticulous engineering and optimization of nanostructure designs, highly efficient plasmonic sensors with exceptional sensitivity, high figure-of-merit, and low detection limits can be realized (Sun et al., 2021; Ghodrati et al., 2022).

Recently, 2D materials such as graphene, MXenes, and black phosphorus have attracted considerable attention as potential materials due to their unique electrical and optical properties (Ghodrati et al., 2024; Fallahi et al., 2012). 2D materials are considered as promising fundamental units in state-of-the-art nanophotonic devices due to their outstanding electrical and optical characteristics. Initially, due to their varied band structures, 2D materials enable wide-ranging spectral responses from microwave to visible light. Subsequently, these responses can be adjusted through electrical, chemical, and optical methods, providing a basis for actively controlling light. Moreover, the surfaces of 2D materials are inherently passivated, facilitating their integration with resonant MMs or the creation of heterostructures without encountering 'lattice mismatch'. Lastly, despite being atomically thin, surface plasmons and excitons in 2D materials can amplify light-matter interactions with remarkable efficiency (Ghodrati et al., 2024; Fallahi et al., 2012).

Sensors based on MMs and MSs utilizing 2D materials have shown remarkable sensitivity and detection accuracy. Additionally, they offer excellent dynamic tunability through electrostatic gating and lower intrinsic losses in resonance. These characteristics make them a versatile option compared to traditional sensing techniques relying on common and noble metals (Jiang et al., 2014; Hlali et al., 2021).

In this chapter, a brief introduction to the concept of MMs and MSs as well as their properties and characteristics will be presented. Moreover, it highlights the recent advances in sensing and biosensing technology based on MMs and MSs as well as its main applications, focusing on the effect of material selection on device performance. This chapter offers a detailed and comprehensive analysis of optical sensors and biosensors based on MMs and MSs for medical diagnoses, biomolecule detection like viruses, and bacteria, and disease diagnosis such as cancer. Also, it identifies potential areas for future research in this field.

METAMATERIALS AND METASURFACES

The origin of metamaterial can be traced back to Viktor Veselago's influential paper. After three decades, Smith et al. developed an artificial material that exhibited negative permittivity and negative permeability within the same frequency range (Pendry et al., 1999; An et al., 2009; Parvin et al., 2021).

The negative index materials were created by utilizing two established components: a wire medium capable of producing negative dielectric permittivity and a split ring resonator (SRR) medium capable of producing negative permeability effectively. Experimental evidence demonstrated that electromagnetic waves are unable to travel through a medium where one of the material constants is negative, but propagation resumes when both are negative (Pendry et al., 1999; Ali et al., 2019; Zhang et al., 2021).

The most attractive feature of MMs lies in their ability to create artificial materials with EM properties that do not exist in nature. Moreover, these properties are highly sought after in various practical applications (Zhang et al., 2021; Ma et al., 2022). MMs have the potential to be used in microwave and optical filters, medical devices, remote aerospace applications, sensors, smart solar power management, radomes, EM absorbing materials, high-frequency battlefield communication, and lenses. Similar principles can also be applied to manipulate the behavior of acoustic waves. MMs are typically composed of periodic structures and only meet the effective medium condition within a limited frequency range (Pendry et al., 1999; Chen et al., 2020). A classification of MMs can be found in Figure 2

Figure 2. Classification of metamaterials

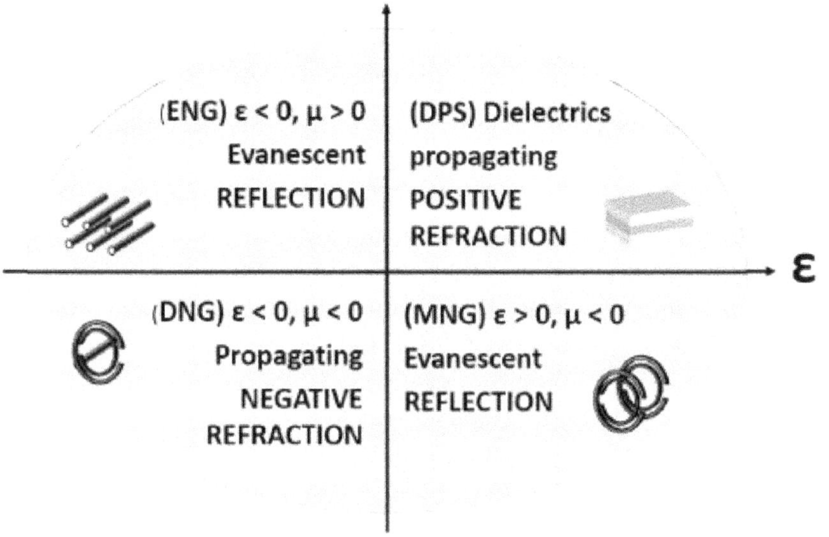

(Pendry et al., 1999)

Double positive mediums (DPS) are found in nature, like naturally occurring dielectrics where both permittivity and magnetic permeability are positive, resulting in forward wave propagation. Epsilon negative media (ENG), on the other hand, have a negative ε_r but a positive μ_r. This characteristic is often seen in plasmas, with noble metals like gold or silver exhibiting ENG properties in the infrared and visible spectrums. MMs behaving like ENG can be synthesized by using a wire medium (Pendry et al., 1999; Xu et al., 2021).

Mu-negative media (MNG) exhibit a positive ε_r and negative μ_r. Gyrotropic or gyromagnetic materials, which are influenced by a quasistatic magnetic field, demonstrate this property. Materials mimicking MNG can be created by arranging SRR or spiral resonators (SR) periodically. In negative-index metamaterials (NIM), Double Negative (DNG) mediums have both negative permittivity and negative permeability, resulting in a negative index of refraction. MMs behaving like DNG medium have been produced by combining thin wires and split ring resonators (Pendry et al., 1999; Asgari et al., 2020).

MMs exhibit unique electrical and magnetic characteristics that are rarely observed in nature, such as reverse propagation, the inverse Doppler Effect, near-total absorption, negative refractive index, surpassing the diffraction limit in imaging, and more. These exceptional properties are contingent upon the material's geometry, allowing for customization as needed. Typically, precise repetitive structures are engineered onto the material, with each structure being smaller in size than the wavelength it aims to manipulate (Hu et al., 2019; Kurosawa et al., 2017).

MSs are typically constructed using an assortment of artificial meta-atoms (MAs), usually consisting of metal structures that are carefully crafted in terms of size, shape, and alignment to produce specific effects. In contrast to 3D bulk MMs, MSs have an extremely thin profile relative to the operational wavelength, resulting in a reduced physical footprint and lower insertion loss. The characterization methods for bulk MMs differ from those used for thin MSs, with MMs often relying on effective medium theory while MSs utilize the generalized sheet transition condition due to their ultra-thin nature (Ma et al., 2020; Arbabi et al., 2018).

The planar configuration of MSs offers a notable advantage in its ease of implementation at both the terahertz and visible light spectra through the standard photolithography process. The behavior of light or EM waves when they pass through two homogeneous and isotropic media is governed by Snell's law (Ma et al., 2020; Xu et al., 2021). In 2011, a breakthrough was made with the introduction of the generalized Snell's laws of reflection and refraction on a metasurface, achieved by incorporating an abrupt phase shift. This phase discontinuity enabled the manipulation of light wavefronts. Since then, MSs have experienced rapid development, leading to the creation of intriguing devices capable of manipulating microwaves, terahertz waves, and visible light (Rifat et al., 2018; Xu et al., 2021). Among the potential applications of MSs, one can mention holography, polarization control, lenses, sensing, cloaking, beam steering, perfect absorbers and hyperspectral imaging (see Figure 3).

Figure 3. Some of the potential applications of metasurfaces, (a) Holography (Hu et al., 2019); (b) Polarization control (Kurosawa et al., 2017); (c) Lenses (Arbabi et al., 2018); (d) Sensors (Ma et al., 2020); (e) cloaking (Xu et al., 2021); (f) Beam steering (Wei et al); (g) Absorbers (Rifat et al., 2018); and (h) Hyperspectral imaging (Faraji-Dana et al., 2019)

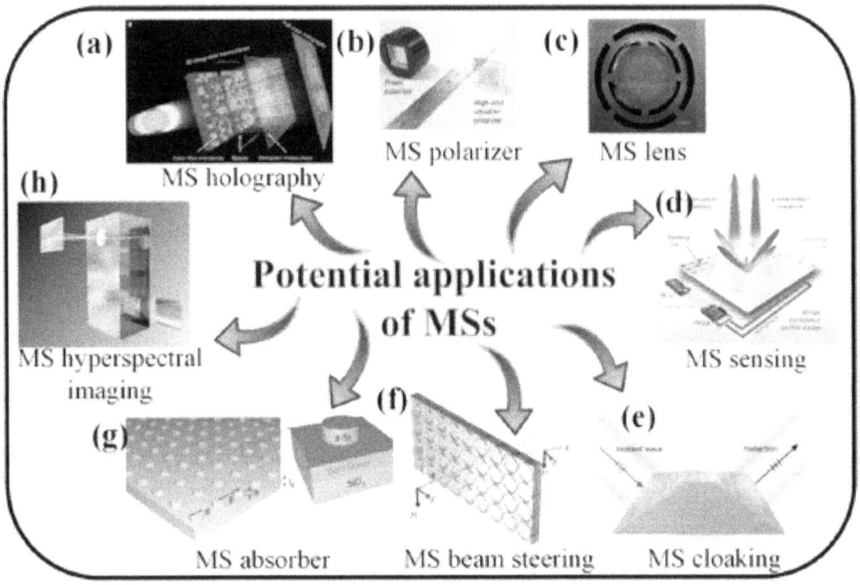

An intriguing use of MSs involves transforming spatially propagating waves (PW) into surface waves (SW) by utilizing a gradient phase profile to efficiently couple PW into SW. Surface plasmon polaritons (SPPs), a type of surface waves confined to a surface, can be easily generated by applying a suitable gradient phase profile to MSs. Geometric MSs have the ability to control light polarization by adjusting the orientation angle of MAs to encode a geometric phase (Kurosawa et al., 2017).

Holography is an advanced method employed to capture and reconstruct wavefronts through the storage and retrieval of phase information. In recent times, researchers have made significant progress in the field of holographic imaging by utilizing highly efficient, broadband, and 3D holography with MSs. Moreover, they have successfully demonstrated various techniques for multiplexing information into a single metasurface hologram, enabling the storage of multiple images at the same frequency. Furthermore, the utilization of a metasurface composed of an array of C-shaped SRRs has been proposed to effectively control both the phase and amplitude profiles of transmitted waves. This remarkable advancement sets the stage for the realization of high-quality holographic imaging (Hu et al., 2019).

MSs possess the capability to create planar devices such as flat lenses. These lenses are equipped with a specially designed phase profile that can convert planar wavefronts into spherical ones and converge incoming waves at a distance away from the lenses. It is important to note that the optical wavefronts in transmission or reflection types will only remain spherical if the incident plane wave is normally incident on the flat lenses. This characteristic leads to high numerical-aperture focusing (Arbabi et al. in 2018). Polarization is a fundamental characteristic of EM waves. Traditional methods for controlling polarization tend to require a significant amount of materials and have limited performance. However, MSs have demonstrated their effectiveness in transforming the polarization of EM waves. Additionally, they offer the advantages of being ultra-thin, having a wide bandwidth, and exhibiting high efficiency (Kurosawa et al. in 2017).

Refractive index (RI) sensing and biosensing are valuable applications of MMs and MSs. This technique identifies alterations in the RI of analyte layers caused by biomolecular interactions. EM RI sensors are especially beneficial for precise and label-free biochemical assays, rendering them appropriate for a range of chemical and biological sensing purposes. Through the manipulation of distinct MAs and their arrangements, it becomes feasible to alter the resonant electromagnetic spectrum significantly, which is influenced by the nearby surroundings (Ma et al., 2020; Yuan et al., 2018; Wei et al., 2013).

The resonant characteristic may induce changes in the output spectrum, aiding in the identification of the RI of adjacent biomolecular analytes. To ensure precise outcomes, particular wavelengths and sensitivity thresholds must be integrated into the mass setups. Furthermore, RI sensors utilized in MM- and MS-based sensing systems present benefits over conventional SPP-based biosensors. When contrasted with SPP-based sensors, MM- and MS-based RI sensing tools exhibit superior performance owing to their manufacturing precision and signal reliability (Ghodrati et al., 2022; Qiu et al., 2020).

These sensors are capable of detecting changes in RI by analyzing the optical responses, such as reflection or transmission, of focused input beams. MMs enable the detection of biomolecules in the THz-GHz frequency ranges, which would otherwise be challenging due to the small scattering cross-section of microorganisms like fungi, bacteria, and viruses at these wavelengths (Ma et al., 2020). The use of THz-GHz detection provides advantages such as non-destructive, non-contact, and label-free sensing. Furthermore, the unique properties of MMs, including their ability to regulate amplitude, phase, polarization, and impedance, make terahertz metamaterial sensing highly promising for various applications in biomedical and other fields (Ma et al., 2020; Yuan et al., 2018).

PLASMONIC METAMATERIALS AND METASURFACES

SPR phenomenon is the result of EM waves resonantly coupling with charge density oscillations at the dielectric-metal interface. Due to the optical momentum mismatch between SPR mode and light in free space, optical excitations in SPR are typically achieved using the attenuated total reflection (ATR) method introduced by Kretschmann (Kretschmann et al., 1965) and Otto (Otta, 1968).

Additionally, compact coupling structures are commonly utilized in SPR-integrated circuits, along with optical gratings and waveguides. In the case of surface plasmon waves (SPWs), only the TM-polarized electric field is present, and these waves decay exponentially at the dielectric-metal interface. The SPW is characterized by the propagation constant as (Kretschmann et al., 1965; Otta, 1968):

$$k_{sp} = \frac{\omega}{c}\sqrt{\frac{\varepsilon_d \varepsilon_m}{\varepsilon_d + \varepsilon_m}} \tag{1}$$

The velocity of light in a vacuum, denoted as c, the angular frequency of the incident light, represented by ω, and ε_m and ε_d are the dielectric constants of metal and dielectric medium, respectively. It is important to note that the real part of ε_m must be negative, and its absolute value should be smaller than ε_d in order to ensure that the metals can SPW. The equation (1) indicates that SPW is highly sensitive to any changes in the properties of the metal and the dielectric media. To initiate the surface plasmon oscillation, it is necessary to excite the electrons in the metal, which can be achieved by imposing light on the surface. The propagation constant of a light wave propagating with a frequency ω in free space is given by (Kretschmann et al., 1965; Otta, 1968):

$$k_d = \frac{\omega}{c}\sqrt{\varepsilon_d} \tag{2}$$

According to the equations provided, the permittivities of the metal and dielectric have opposite signs, indicating that the wave propagation constant of SPW should always be greater than that of the wave propagating in the dielectric (Kretschmann et al., 1965; Otta, 1968). As a result, direct light cannot excite surface plasmons; it requires light with additional momentum or energy that shares the same polarization state as the SPW. Additionally, the propagation constant must be in line with the SPW. Surface plasmon resonance structures are highly sought-after platforms for optical sensors due to their ability to confine and enhance the electromagnetic field near the surface (Kretschmann et al., 1965; Otta, 1968).

In recent times, the utilization of metamaterial-based plasmonic sensors has been employed to enhance the sensitivity of plasmonic sensors. The use of metamaterial-based sensors offers several advantages, including the ability to utilize various geometric structures and different sensing principles that were previously not feasible with conventional plasmonic sensors (Lu et al., 2015; Wang et al., 2022; Meng et al., 2012; Mun et al., 2019). MSs can be classified into two categories: plasmonic (metallic) and dielectric metasurfaces. In plasmonic MSs, the collective oscillations of electrons in metal result in a resonance phenomenon known as localized surface plasmon resonance (LSPR). Propagating surface plasmon polariton (SPP) and LSPR are commonly employed plasmonic RI sensing techniques. Plasmonic MSs offer advantages such as the ability to directly sense analytes at the metal surface, where the field is strongly confined (Aissa et al., 2021; Wang et al., 2022; Barron, 2004; Yoo et al., 2019).

The strong confinement of the field significantly boosts the interaction between light and the analyte, leading to a significant change in the spectral response. MSs are highly promising for sensing purposes due to their remarkable characteristics. By carefully designing and optimizing nanostructures, plasmonic sensors can achieve exceptional sensitivity, high figure-of-merit, and a low detection limit. Researchers have extensively investigated sensors utilizing MMs and MSs that rely on SPPs generated on common and noble metal surfaces (such as Au, Ag, Cu, and Al). Nevertheless, these sensors are hindered by substantial intrinsic losses and lack of tunability (Wang et al., 2021; Kowerdziej et al., 2019; Jiang et al., 2017; Leitis et al., 2019).

In recent times, there has been a growing interest in 2D materials such as graphene, MXenes, transition metal dichalcogenides (TMDs), and black phosphorus (Bp) due to their distinct electrical and optical properties. These materials have the ability to support well-confined SPPs in the terahertz frequency range, particularly at high doping levels (Solntsev et al., 2021; Shrekenhamer et al., 2013; Vasic et al., 2013; Ye et al., 2013).

Notably, sensors utilizing 2D materials in the form of MMs and MSs have shown remarkable sensitivity and detection accuracy. They also offer dynamic tunability through electrostatic gating and exhibit lower intrinsic losses in resonance compared to conventional sensing methods that rely on common and noble metals (Wang et al., 2022; Qiu et al., 2023). As a result, SPR-based sensors are highly desirable for various applications, particularly in the field of medical diagnostics. These sensors can be employed for detecting proteins, human blood groups, DNA, glucose, viruses, analyzing living cells, as well as chemical and gas sensing, among others (refer to Figure 4).

Figure 4. The typical sensing structure of plasmonic metasurface

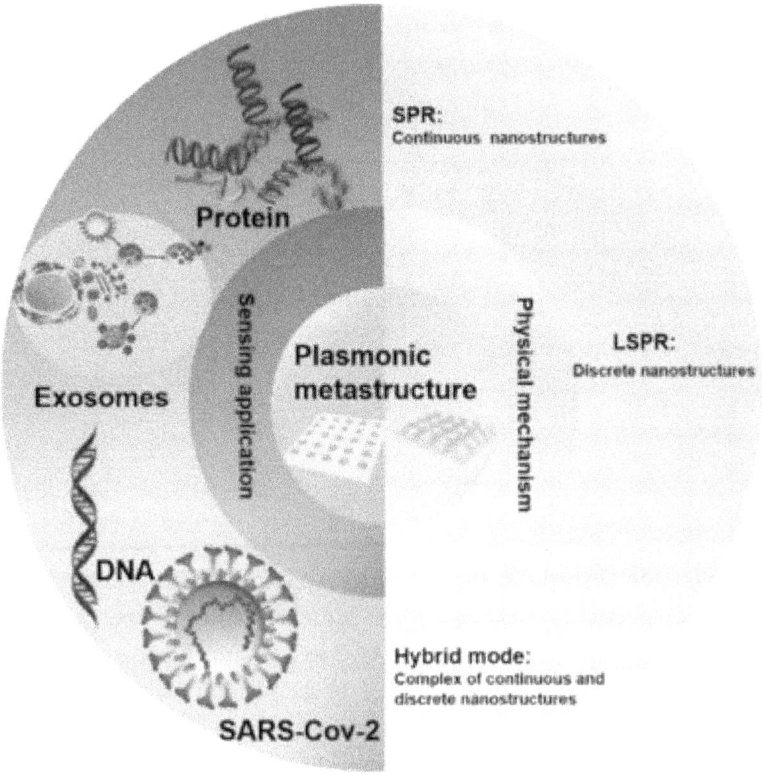

(Wang et al., 2022)

METAMATERIAL AS SENSOR OR BIOSENSOR

Lee et al. introduced an innovative all-metal-grating structure for an ultra-narrow band infrared MM absorber (Lee at al., 2017). The all metal-grating structure is depicted in Figure 5. This absorber achieves an exceptional absorption efficiency of more than 98% with an incredibly narrow bandwidth of 0.66 nm under normal incidence. The remarkable absorption performance is attributed to SPR. Additionally, the SPR-induced strong surface electric field enhancement makes it highly suitable for biosensing applications. When utilized as a plasmonic RI sensor, this ultra-narrow band absorber exhibits a wavelength sensitivity of 2400 nm/RIU and an exceptionally high figure of merit (FOM) of 3640, surpassing the performance of most similar plasmonic sensors reported to date (Lee at al., 2017).

Figure 5. Schematic of the proposed metamaterial structure of a unit cell

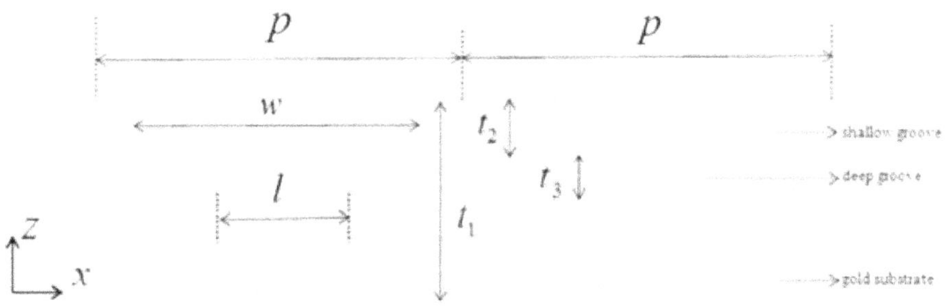

(Lee at al., 2017)

In Figure 6a, the absorption and reflection spectra are represented by the green and red lines, respectively. The green line indicates that the peak absorption can surpass 98% at the resonance wavelength of 2403.15 nm. The distribution patterns of the electric field E_1 and magnetic field H_1 at the resonant wavelength of 2403.15 nm are illustrated in Figure 6b and c, respectively. Figure 6d and e display the snapshots of the electric field distribution E_0 and the magnetic field distribution H_0 at the nonresonant wavelength of 2423.31 nm for comparison. The proposed structure exhibits enhanced field intensity on its surface, attributed to the excitation of SPR. The electric field intensity at the resonant frequency is 40 times greater than that of the incident waves, a crucial aspect in biosensing applications (Lee et al., 2017).

Figure 6. (a) Absorption and reflection spectra of the proposed sensor. Distributions of (b) the electric field E1, and (c) the magnetic field H1 at the resonant wavelength. Distributions of (d) the electric field E0, and (e) the magnetic field H0 at the nonresonant wavelength

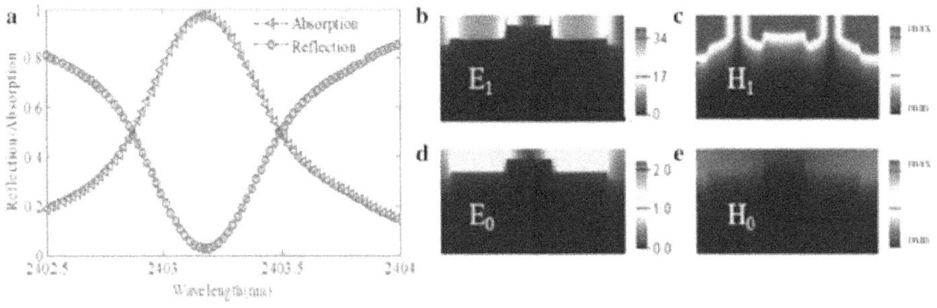

(Lee at al., 2017)

Yao and colleagues have conducted an experimental study showcasing a plasmonic anapole MM sensor for environmental RI in the optical spectrum. The schematic diagram of the plasmonic anapole MM is illustrated in Figure 7a, comprising a planar array of vertical SRRs suspended in a spin on glass layer and covered with a dumbbell-perforated gold film. The apertures in the gold film are aligned with the vertical SRRs (Yao et al., 2022). Upon illumination by a normal x-polarized plane wave, circulating currents in the circular apertures and the vertical SRR are induced, leading to the generation of a circulating magnetic field and the excitation of x-directed toroidal dipole moments, as depicted in Figure 7b. The researchers demonstrated that the sensor displays a high sensitivity to ambient RI at 330 nm/RIU. This innovative structure holds promise for applications of anapole MMs in biosensing and spectroscopy (Yao et al., 2022).

Figure 7. (a) Schematic diagram of the proposed anapole metamaterial array for refractive index sensing; (b) Schematic depiction for the excited anapole mode, which originates from the destructive interference between electric dipole (blue arrow) and toroidal dipole (red arrow with pink torus and green circulating magnetic field) moments

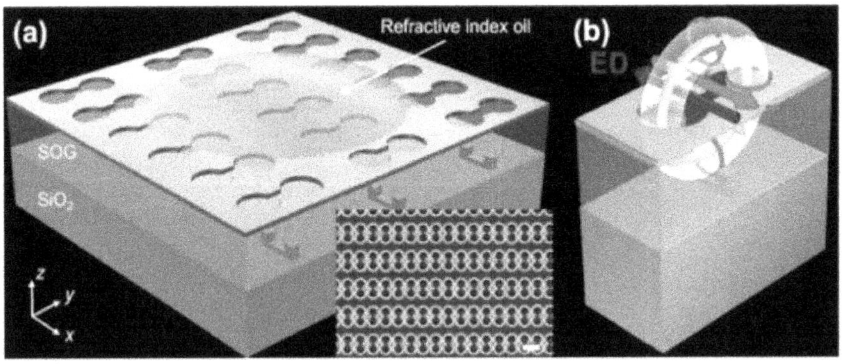

(Yao et al., 2022)

The plasmonic anapole MM's sensing performance was investigated by depositing oil with a RI ranging from 1.30 to 1.39 in increments of 0.01 on the sample. The transmission and reflection spectra were measured in Figure 8a with varying ambient refractive indices. As the ambient RI changed from 1.30 to 1.39, the resonant wavelength redshifted from 1435.5 to 1468.5 nm. The simulated data in Figure 8b closely matched the measured results, showing a resonant wavelength redshift from 1473.5 to 1513.4 nm while maintaining the same line shape. Both sets of results demonstrated a linear relationship between the resonant wavelength and refractive index, consistent with the theoretical model. The slight variation between the two sets of results was attributed to the sample's roughness. Figure 8c displayed sensitivities of approximately 330 nm/RIU and 445 nm/RIU for the measurement and simulation results, respectively (Yao et al., 2022).

Figure 8. Refractive index sensing application of the plasmonic anapole metamaterial. (a) Measured and (b) simulated transmission and reflection spectra with variable ambient refractive index from 1.30 to 1.39 with a step of 0.01. Dark (light) blue and red correspond to transmission and reflection at refractive index n = 1.30 (1.39). (c) The resonant wavelengths of the anapole mode from experimental and simulation results as functions of the ambient refractive index

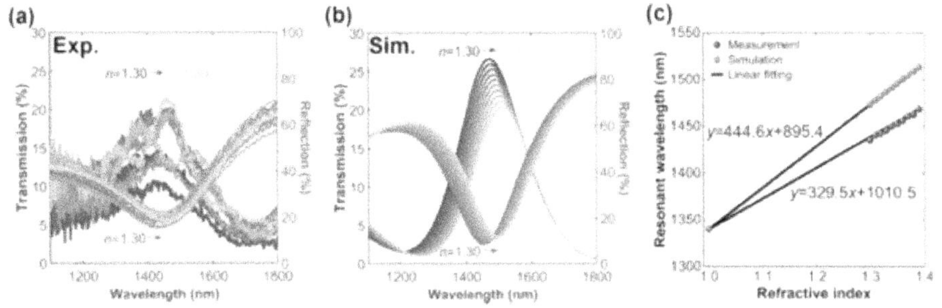

(Yao et al., 2022)

Islam et al. introduced a new MM sensor with a unique shape to detect different oils, fluids, and chemicals using microwave frequency (Islam et al., 2022). The performance of this sensor structure has been thoroughly examined through theoretical and experimental studies. Figure 9 illustrates the design of the MM-based sensor. The proposed sensor utilizes a flame retardant substrate material and a copper resonator. In the z-axis, an EM wave is incident normally, while in the x- and y-axes, a perfectly electrified boundary condition is applied, as depicted in Figure 9. The suggested MM sensor exhibits a high-quality factor (QF) and sensitivity in both frequency shifting and amplitude changing. The sensor's QF is 135, sensitivity is 0.56, and FOM is 76, indicating its efficient performance. Due to its high sensitivity, good QF, and excellent performance, the recommended sensor can be effectively employed in chemical, oil, and microfluidic industries for the detection of various liquid samples (Islam et al., 2022).

Figure 9. Design of Metamaterial-based sensor: PEC-PEC open add spaces boundary conditions

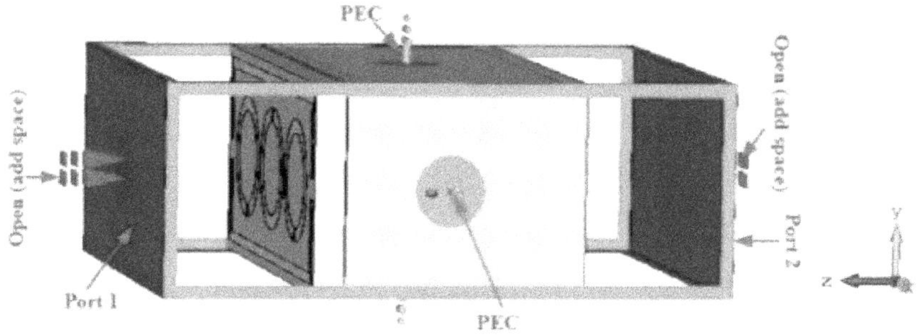

(Islam et al., 2022)

Wang et al. introduced an all-metal MM biosensor operating in the terahertz range. This biosensor utilizes stainless steel materials that are produced using laser-drilling technology. The simulation findings indicate that the maximum RI sensitivity of this MM sensor is 294.95 GHz/RIU, with a FOM of 4.03. To evaluate the effectiveness of this biosensor, bovine serum albumin was selected as the detection substance. The experimental results demonstrate a detection sensitivity of 72.81 GHz/(ng/mm^2) and a limit of detection (LOD) of 0.035 mg/mL (Wang et al., 2021).

The proposed all-metal MM THz biosensor is depicted in Figure 10a. It features a hollow dumbbell pattern on a 50 μm thick stainless-steel plate, arranged periodically along the x and y directions. The unit structure has period sizes of 500 μm (P_x) and 300 μm (P_y). The hollow dumbbell has a length (L) of 294 μm and a gap (H) of 60 μm. The circles at both ends of the hollow dumbbell have a radius (R) of 60 μm. Notably, this biosensor has an all-metal structural design without a traditional dielectric substrate. The THz wave is incident perpendicular to the surface of the MM biosensor (Wang et al., 2021).

Figure 10. (a) Three-dimensional array diagram and cell structure diagram of the biosensor. The structural parameters are Px = 500 μm, Py = 300 μm, L = 294 μm, H = 60 μm, R = 60 μm. (b) Simulated transmission of the biosensor

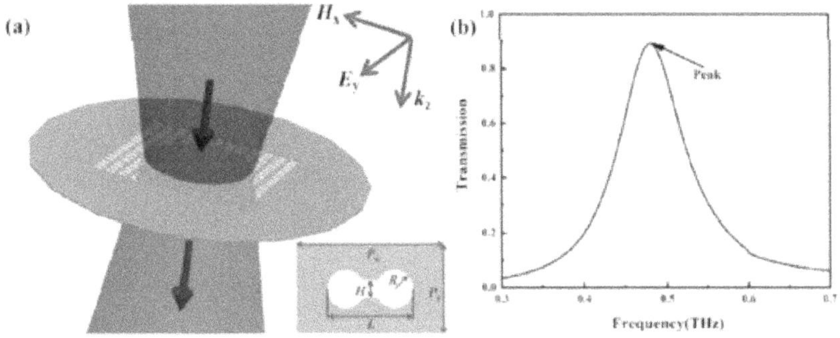

(Wang et al., 2021)

In order to assess the sensing capabilities of the biosensor, a thin layer of analyte measuring 120 μm was introduced to the MM biosensor, as depicted in Figure 11a. Subsequently, simulations were conducted to observe the variation in transmission spectra of the MM biosensor as the RI of the analyte changed, as illustrated in Figure 11b. It was observed that with an increase in the RI of the analyte, the resonance peak frequency underwent a red-shift. The corresponding frequency shift for each RI was recorded. The fitting result displayed in Figure 11c demonstrated a sensitivity of 294.95 GHz/RIU towards the RI. Furthermore, the performance of the biosensor was quantified using FOM, yielding a value of 4.03 (Wang et al., 2021).

Figure 11. (a) Cross-sectional view and top view of the metamaterial biosensor model diagram with a 120-µm thin layer of analyte. (b) Influence of the changes in the analyte's RI on the biosensor's transmission spectra. (c) Corresponding linear fit of the peak's frequency shift with the corresponding RI

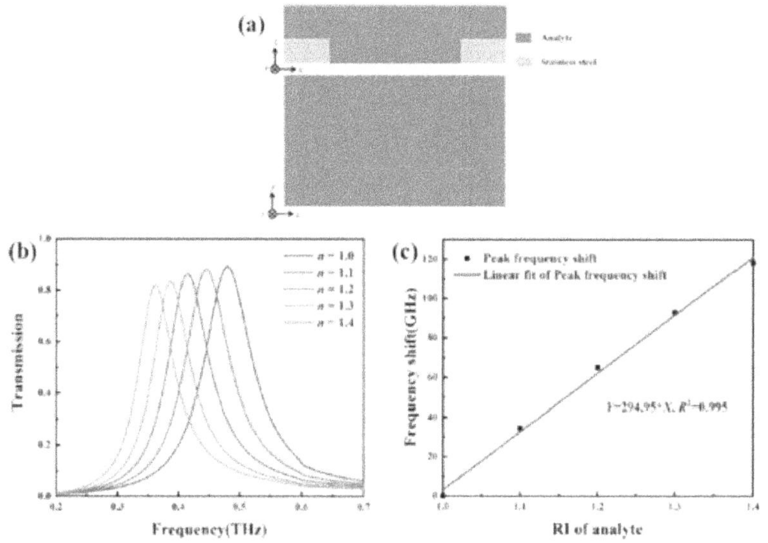

(Wang et al., 2021)

Park et al. developed a MM surface by utilizing a gold rectangular structure on quartz. The structure's diagram is presented in Figure 12a. The MM possesses specific capacitance and inductance parameters. The presence of virus particles in the capacitor gap alters the resonance frequency, which can be explained by a basic LC circuit. The SRR structure proposed was manufactured using e-beam lithography, followed by the deposition of thin films of gold (97 nm) and Cr (3 nm) through e-beam technology (Park et al., 2017). The linewidth of the structure was 4 µm, as depicted in Figure 12b. To measure the sensor's sensitivity, solutions containing PRD1 and MS2 viruses with a density of 10^9/mL were employed. Approximately 10 µL of the virus solution was applied to a surface area of 10 mm^2. It was observed that the sensitivity to the PRD1 virus increased from 6 GHz μm^2/particle to 80 GHz μm^2/particle as the gap width decreased from 3 µm to 200 nm (Park et al., 2017).

Figure 12. (a) Schematic of THz nano-gap Au metamaterial sensing of viruses. (b) SEM image of the viruses deposited on the structure with a gap width (w) of 200 nm. (c) Schematic depiction of the measurement of the dielectric constant of virus layers using the THz metamaterial structure

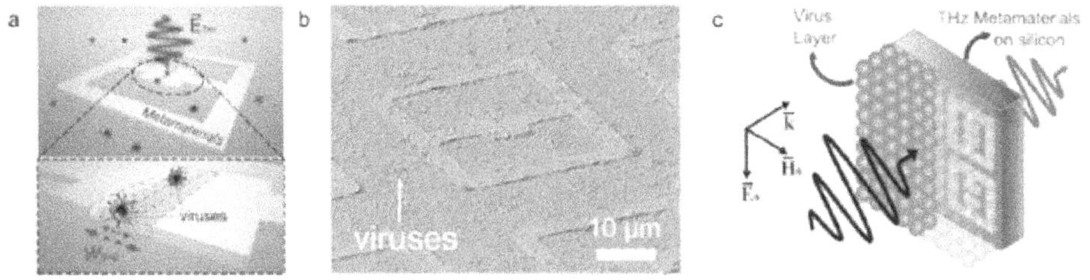

(Park et al., 2017)

Geng et al. successfully combined a microfluidic chip with an SRR-based MM sensor to effectively detect Alpha-fetoprotein (AFP) and glutamine transferase isozymes II (GGT-II) for the early diagnosis of liver cancer. They provided a visual representation of the proposed sensor integrated with a microfluidic chip in Figure 13a, b, along with its corresponding electrical circuit model that incorporates inductance and capacitance. The fabrication process of the SRR structure involved several steps, including RCA standard cleaning, lithography, deposition, and lift-off. Additionally, a 200 nm-thick gold metallic layer was deposited using radio magnetron sputtering (Geng et al., 2017).

Subsequently, a microfluidic channel made of polydimethylsiloxane (PDMS) was utilized on the surface to regulate the required sample volume for surface functionalization. The outcomes at each stage of fabrication can be seen in Figure 13c–g. Prior to the THz testing, the PDMS channel was removed due to its low transmission capability in THz (less than 50%) and the high absorption of THz energy by water. The proposed structure was examined with three different gap widths in the SRRs (2 µm, 4 µm, and 6 µm). As the RI of the surrounding medium increased, a blueshift in the resonance frequency was observed. Additionally, an asymmetry was introduced to enhance the Fano resonance of the SRR sensor, which had a significant impact on the QF. When the SRR had two gaps, total resonance shifts of 19 GHz and 14.2 GHz were observed for GGT-II antigen at a concentration of 5 µm/mL and AFP at a concentration of 0.02524 µg/mL, respectively. The sensitivity exhibited promise in the detection of cancer biomarkers (Geng et al., 2017).

Figure 13. (a) The sketch of THz SRR biosensor integrated with a microfluidics chip. (b) RLC equivalent circuit of the SRR. (c–g) Fabrication steps of the microfluidics integrated SRRs. (c) SRRs on a 4-inch silicon wafer. (d) Four units of the SRR structure with one or two gaps. (e) The SU-8 mould for microfluidics chip fabrication. (f) PDMS microchannel. (g) The final biosensor chip integrated with microfluidics

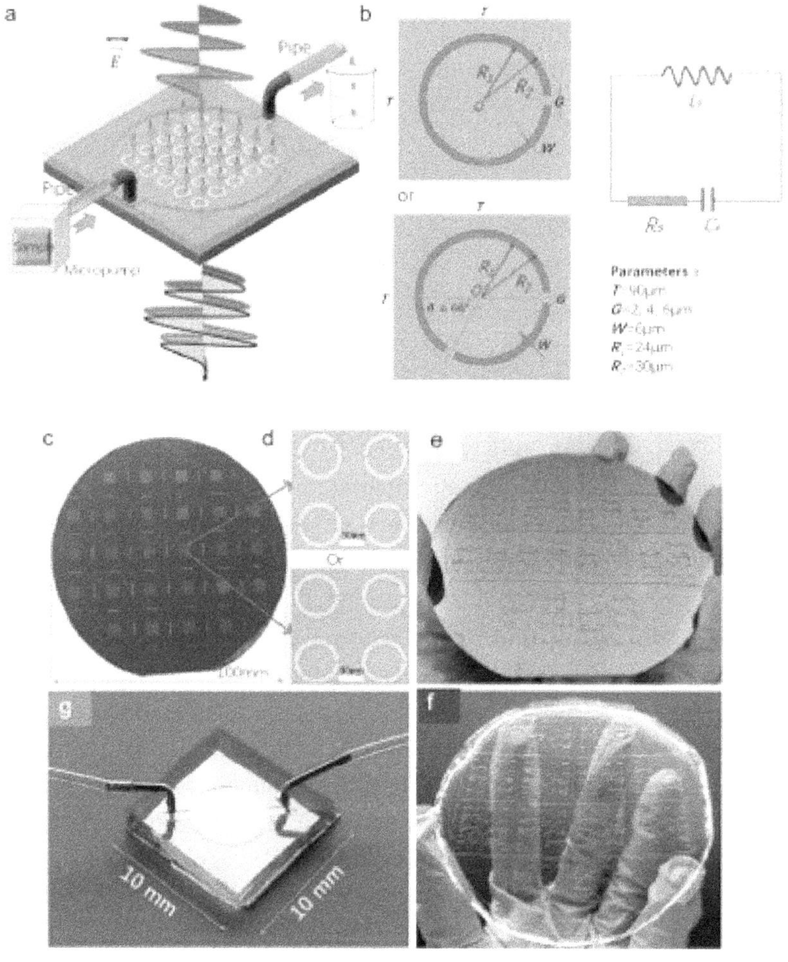

(Geng et al., 2017)

METASURFACES AS SENSOR OR BIOSENSOR

Zhang and colleagues introduced a plasmonic RI sensor that utilizes a metal-insulator-metal (MIM) waveguide connected to a concentric double rings resonator (CCRR) and investigated numerically (Zhang et al., 2018). In Figure 14, the MIM waveguide is depicted along with the concentric double rings resonator. To minimize losses, the silicon substrate is coated with silver, which was selected for its relatively low loss properties. The waveguides are etched onto the silver surface.

Figure 14.(a) The three-dimensional picture of the plasmonic sensor. (b) The two-dimensional picture of the plasmonic sensor

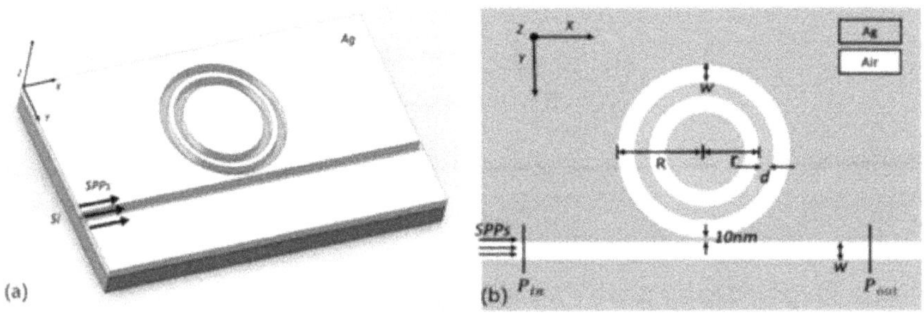

(Zhang et al., 2018)

Figure 15 illustrates the transmission spectra of the CDRR with varying dielectric filling. The RI of the dielectric ranges from 1 to 1.1 in increments of 0.025, resulting in a red shift in the spectra. The sensitivity, denoted as S, represents the change in resonant wavelength per unit change in dielectric. Achieving a high FOM is essential for optical sensors. By analyzing the line fitting in the inset of Figure 15, they determine sensitivities of 708 nm/RIU for mode 1 and 1060 nm/RIU for mode 2. Through the utilization of the innovative supermodes of the CDRR, a sensitivity of 1060 nm/RIU with a remarkable FOM of 203.8 is achieved in the near-infrared spectrum (Zhang et al., 2018).

Figure 15. The transmission spectra with different filling dielectric in the CDRR. The inset shows the relationship between refractive index and resonant wavelength.

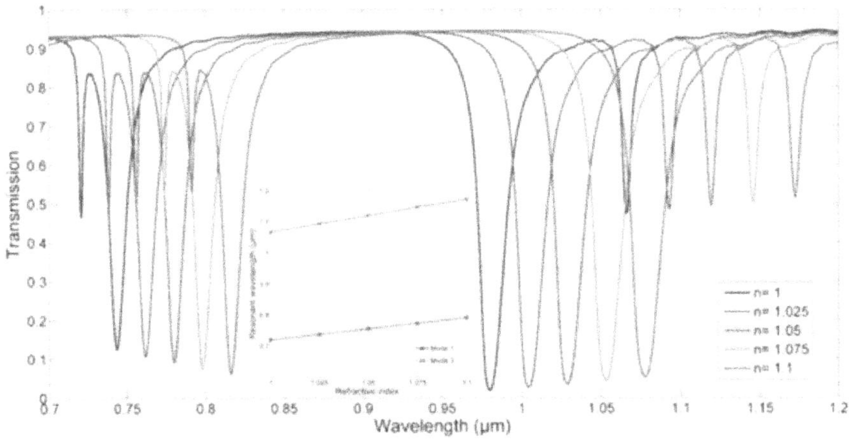

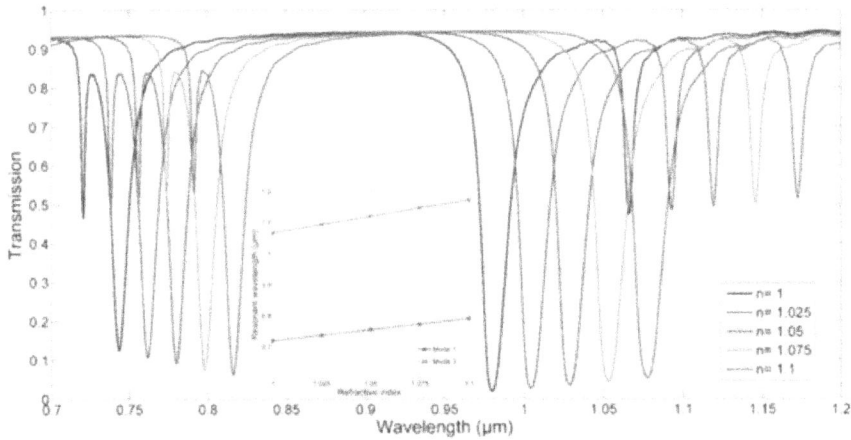

(Zhang et al., 2018)

Lie et al. proposed a plasmonic structure with a network-type MS, which consists of a double-layer metal-dielectric network-type MS on two-layer dielectric films (Lie et al., 2016). Figure 16 illustrates the schematic graph of this plasmonic structure with network-type MS. Inside the double-layer metal-dielectric network structure, there is a rectangular nanohole array. This structure is placed on the two-layer dielectric films. The double-layer metal-dielectric network structure is composed of photoresist spacers coated with Au spacers that contain nanoholes. This configuration facilitates the coupling of LSPRs, photonic modes, and optical cavity mode, leading to the emergence of new optical phenomena. The double-layer dielectric films consist of a polymethyl methacrylate (PMMA) film positioned on a thick silica (SiO2) substrate (Lie et al., 2016).

Figure 16. Schematic graph of the plasmonic structure with network-type metasurface

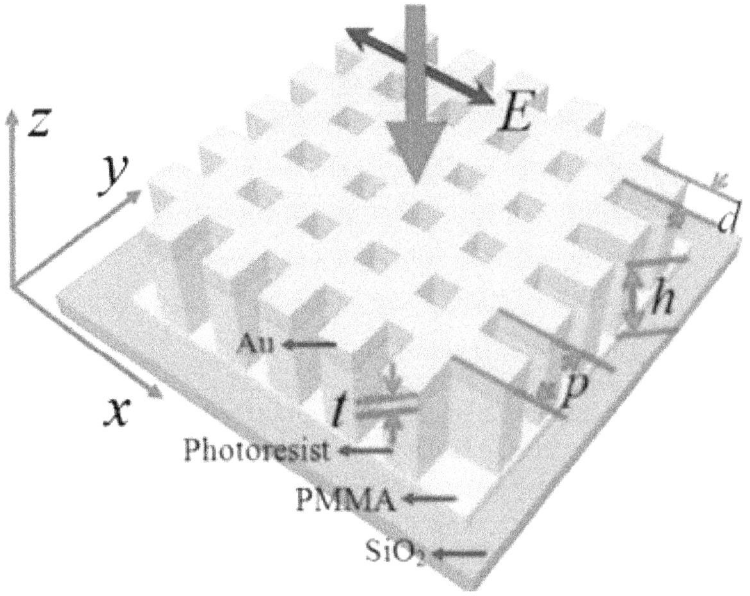

(Lie et al., 2016)

Figure 17a illustrates clear spectral red-shifts as the RI of the filling materials increases from 1.00 to 1.05 with an increment of Δn=0.01. The spectral positions of these reflection dips in relation to the air RI are depicted in Figure 17b. According to the findings, the bulk RI sensitivities of these dips are 145.71, 411.43, 228.57, 160.00, 408.57, and 288.57 nm/RIU, respectively. In Figure 17c, they evaluated the sensing characteristics by substituting the air with a liquid having a similar RI change. Noticeable red-shifts of resonant dips are also observed as the RI increases. The maximum sensitivity reaches up to 596.43 nm/RIU, suggesting that the proposed structure is suitable for liquid sensing as well. Furthermore, the FOM of the proposed structure was determined to be as high as 68.57 (Lie et al., 2016).

Figure 17. Reflection spectra of plasmonic structure filled by air with different RI. Positions of reflection dips versus the RI of b air and c liquid

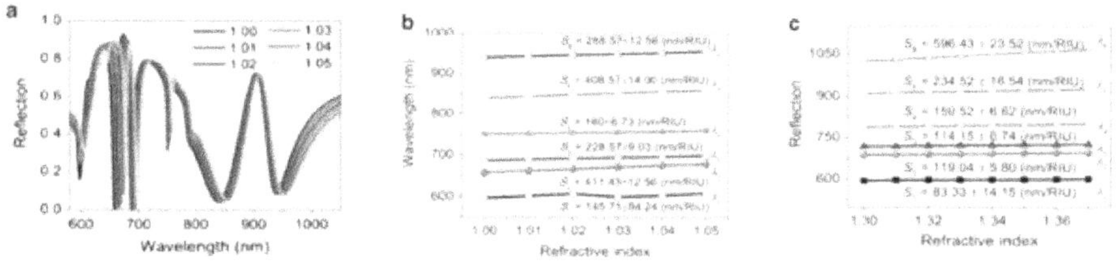

(Lie et al., 2016)

Tan et al. introduced an innovative design for the graphene MS, comprising of an individual graphene ring and an H shaped graphene structure (Tan et al., 2022). Figure 18 provides a visual representation of the schematic structure of the graphene MS. The study focused on investigating the sensing parameters of the graphene MS for cancerous and normal cells, specifically examining cell quantity and position on the MS. The simulated results showed that the theoretical sensitivity, FOM, and quantity of the graphene MS for breast cells were found to be 1.21 THz/RIU, 2.75 RIU^{-1}, and 2.43, respectively (Tan et al., 2022).

Figure 18. Schematic diagram of the proposed graphene-based metasurface on SiO2 substrate. (a) Periodic structure where the incident THz waves with y-polarization is along the z-axis. (b) A unit cell and (c) its top view with geometrical parameters.

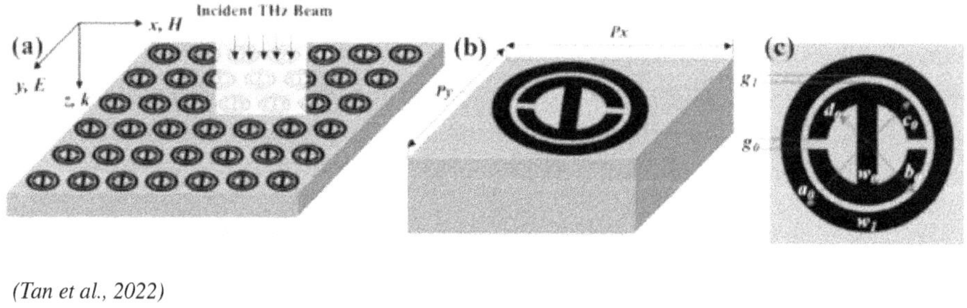

(Tan et al., 2022)

To examine the sensing capabilities of the proposed graphene MS, two types of breast cells, specifically breast normal cell (MCF10A) and breast cancer cell (MCF7), have been selected as the target analytes. Figure 19b provides an illustration of the proposed graphene MS, highlighting four potential sites where MCF10A and MCF7 can be positioned. This analysis aims to uncover the impact of different sites on the transmission response. When considering a single MCF7 cell, the transmission spectrum exhibits a noticeable redshift when the cell is situated at either S_1 or S_2. However, the redshift at S_1 is greater than that at S_2 (Tan et al., 2022).

The transparency window shifts towards a higher frequency range and becomes narrower when MCF7 is present at S_3, as depicted in Figure 19c. Additionally, there is a significant decrease in the transmission peak. When MCF7 is positioned at S_4, the first dip around 0.75 THz nearly disappears, and the second dip around 2.5 THz experiences a noticeable blueshift. Figure 19d demonstrates that the transmission spectrum follows a similar trend as the MCF7 case when MCF10A is located at S_1, S_2, and S_4. However, the transmission spectrum differs when MCF10A is at S_3 compared to MCF7 at the same site, but it is almost identical to the spectrum with MCF10A at S_2. Hence, the transmission spectrum can effectively distinguish breast cancer cells (Tan et al., 2022).

Figure 19. (a) The schematic diagram of the breast cancer cell and their sensing process; (b) The possible sites of MCF10A and MCF7 on the graphene metasurface; (c, d) The transmission spectra of single MCF7 and MCF10A on different sites in panel

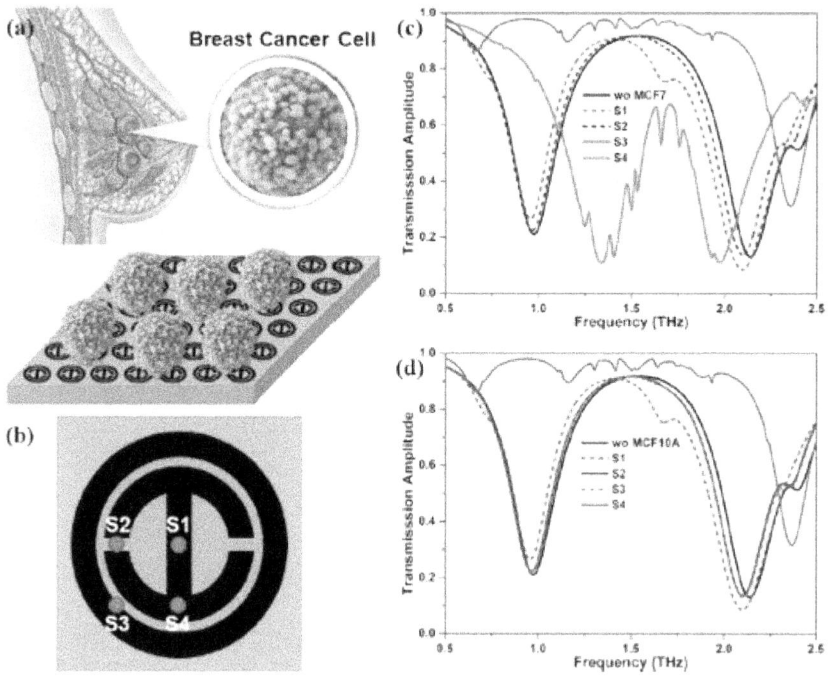

(Tan et al., 2022)

Lie et al. introduced a periodic square array of Au/Co bilayer nanodisks on a thick metal substrate, resulting in the development of a highly efficient magneto-plasmonic sensor (Li et al., 2020). This configuration produces an extremely narrow SPR mode with a full width at half maximum (FWHM) of 7 nm and RI sensitivity reaching 717 nm/RIU. The exceptional resonance properties resemble surface lattice resonance associated with the far-field coupling among nano-objects. The schematic of the all-metallic MS sensor is depicted in Figure 20a, while Figure 20b shows the variations in the transverse magneto-optical Kerr effect (TMOKE) spectrum when the nanosensor is placed in different mediums (Li et al., 2020).

The significant sensitivity of the system results in a notable redshift of the MO signal as the RI increases. The relationship between the resonance position and the RI is depicted in Figure 20c, where the peak positions of each TMOKE spectrum are plotted. Figure 20d showcases the FOM of the magneto-plasmonic sensor in different mediums. In a water environment, the nanosensor exhibits an impressive

FOM close to 7000, which is approximately two orders of magnitude higher than that of gold- or silver-based nanosensors. The proposed nanosystem achieves a substantial TMOKE value of 0.65 in a dielectric medium with an RI of n = 1, thanks to the exceptional SPR effect generated by the hybrid plasmon mode (Li et al., 2020).

Figure 20. (a) Graphical interpretation of a magneto-plasmonic sensor established on a nanodisk array, (b) TMOKE spectrum for various ambient refractive indices, (c) Peak position of TMOKE curves as a function of RI, and (d) FOM as a function of wavelength

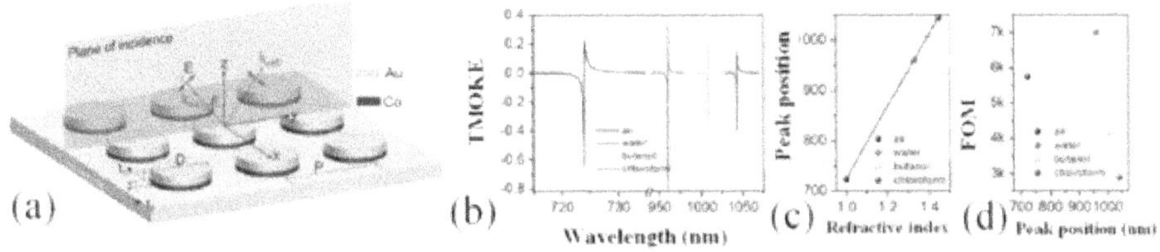

(Li et al., 2020)

Kumari et al. proposed a novel optical MS based on van der Waals's structure. This structure consists of graphene and hBN placed on a silver grating. They investigated this structure for its potential application in biosensing multiple analytes in the mid-infrared region (Kumari et al., 2021). Figure 21 provides a schematic diagram of the device and the absorption curve. The device utilizes the strong and long interaction of hyperbolic phonon polaritons in hBN, while the inclusion of graphene imparts electrical tunability, transforming the device into a multimolecular sensing device. This proposed sensor enhances absorption, enabling surface-enhanced infrared absorption-based detection of the analytes. Through numerical simulations, the biosensor demonstrated high sensitivity and the ability to detect multiple analytes, such as 4,4'-bis(N-carbazolyl)-1,1'-biphenyl (CBP) and nitrobenzene, within a small modal volume (Kumari et al., 2021).

Figure 21. (a) Schematic of the hBN/Graphene van der Waals heterostructure on silver grating; (b) Detection of 50 nm of CBP analyte placed on hBN/silver grating structure, showing its three characteristic fingerprints marked by numbers at (1) 6.65 µm, (2) 6.77 µm and (3) 6.9 µm

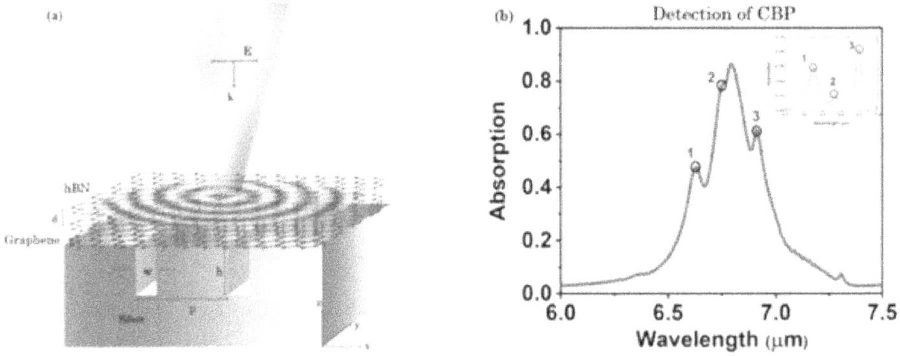

(Kumari et al., 2021)

Yuan et al. have introduced a highly sensitive SPR biosensor that consists of a few layered black phosphorus (BP), monolayer graphene, and an Au thin film. By optimizing the thickness of the gold film and the number of BP layers, they were able to achieve a well-balanced light absorption and energy loss to generate the strongest SPR excitation. The in-plane anisotropy of the BP layer in the device acts as a polarizer, allowing for the detection sensitivity of the biosensor to be optically adjusted by rotating the plasmonic biosensor. The addition of BP-graphene hybrid layers has significantly improved the detection sensitivity of the biosensor, resulting in the highest detection sensitivity of 7.4914×10^4 deg/RIU (Yuan et al., 2018).

Jiang and colleagues conducted research on a graphene-TMDC-graphene hybrid MSs for highly sensitive plasmonic biosensing. They designed optimized graphene-TMDC-graphene/metal sensing substrates for four different types of TMDC materials (WS_2, MoS_2, $MoSe_2$, and WSe_2). Their findings indicated that the graphene-TMDC-graphene hybrid nanostructure could significantly improve SPR sensing performance when operating under ideal conditions (Jiang et al., 2017). In the case of the graphene-WS_2-graphene hybrid MS, monolayer WS_2 was identified as the most effective in transferring electron charge to the metallic sensing surface to enhance the electric field. This graphene-TMDC-graphene hybrid plasmonic MS exhibits promising potential for detecting analytes with low molecular weight or ultra-low concentration levels (Jiang et al., 2017).

A graphene-silica metasurface, which utilizes the tunable plasmon-induced transparency (PIT) effect, has been introduced by He et al. In Figure 22, a schematic illustration of the monolayer graphene-silica MS is presented, along with the transmission spectra of the proposed structure. The metasurface consists of a periodic arrangement, comprising a wide horizontal graphene sheet and two graphene sheets acting as bright and dark modes. This structure exhibits remarkable sensing properties, with a device sensitivity of up to 1.7745 THz/RIU (He et al., 2021).

Figure 22. (a) Schematic illustration of monolayer graphene-silica metasurface; (b) Transmission spectra of our proposed graphene metasurface structure

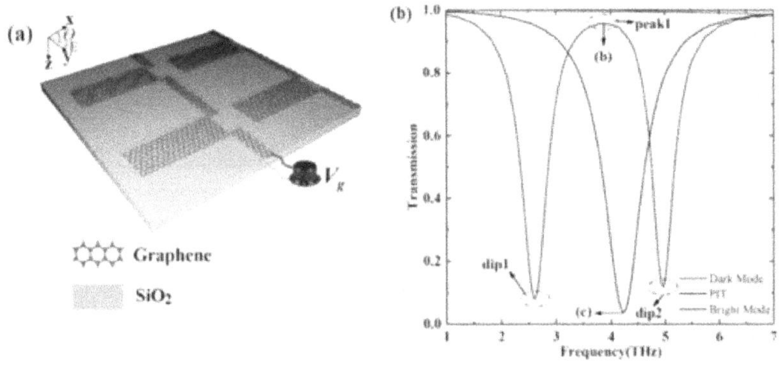

(He et al., 2021)

The SPR-based biosensor is an emerging platform that shows great potential for detecting disease biomarkers. This approach offers several advantages, including non-invasiveness, real-time detection, absence of labels, and rapid detection of cancer markers. In a recent study, Zhu et al. developed a cyclical hard gold nanohole array using nanoimprint and oxygen plasma etching technology. By combining this with soft nano-imprinting lithography, microfluidics, antibody functionalization, and mobile optical spectroscopy, they successfully created a cost-effective plasmonic MSs immune sensing platform. This platform enabled the portable detection of CEA, a cell membrane glycoprotein expressed by normal mucocytes. They observed that the changes in wavelength dip, resulting from the binding of CEA, exhibited a linear decrease or increase within a specific concentration range (Zhu et al., 2020).

The serum of cancer patients was analyzed to determine the actual CEA content, which proved to be highly accurate with minimal error. The sensitivity of the detection reached 490.2 nm/RIU, and the LOD was 5 ng/mL, four times lower than the standard threshold of 20 ng/mL for CEA detection. By employing microfluidic technology and a gold nanohole array in the visible light range, the researchers achieved both stability and high sensitivity in their detection method. This breakthrough offers a new solution for portable medical devices and real-time diagnosis (Zhu et al., 2020). The intricate manufacturing process of the hard substrate poses challenges for the integration of smart devices. In order to streamline this process, Zhu et al. introduced a novel flexible periodic nanopillar design. Through the application of a gold film on the polycarbonate substrate IPS (Polycarbonate), the sensitivity and LOD of the bio-functional IPS were enhanced to 454.4 nm/RIU and 5 ng/mL, respectively. Refer to Figure 23a, b for visual representations (Zhu et al., 2020).

Figure 23. Functionalization of different nanostructures and sensing performance: (a) Biofunctionalization of periodic nanorods; (b) The linear fitting relationship between wavelength dip shift migration and CEA concentration (Zhu et al., 2020); (c) Scanning electron microscope (SEM) image of exosomes captured by functionalized nanohole arrays; (d) Comparison of exosome detection sensitivity between nanohole chip and ELISA (Im et al., 2020)

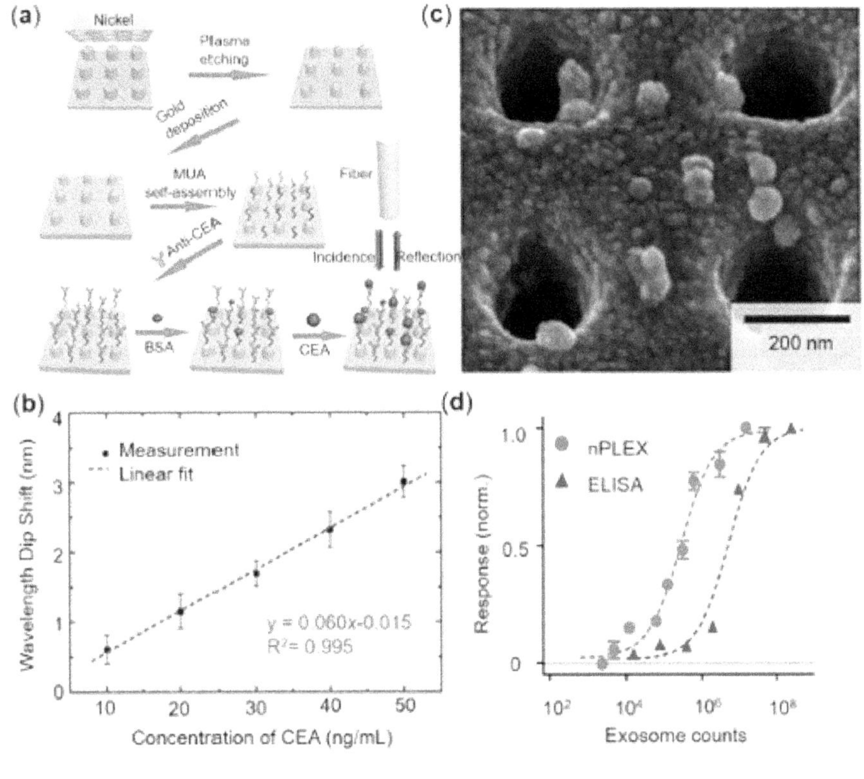

CONCLUSION

In this chapter, we briefly reviewed the applications of MMs and MSs in sensing and biosensing. Both MMs and MSs have great potential to be used in a wide range of applications, especially ultra-sensitive sensors and biosensors. In recent years, plasmonic MMs and MSs have emerged as powerful candidates for novel

RI sensors and biosensors. They possess preeminent capabilities of confining the electromagnetic field at the nanoscale and enhancing the interactions between light and matter, providing high sensitivity to the ambient RI variations. In conclusion, with precise engineering and optimizing the nanostructure designs, performant plasmonic sensors with high sensitivity, high FOM, and low detection limit are achieved. Also, sensors and biosensors need to be developed as a lab-on-a-chip system to make them ubiquitous. Many of the researchers are already working on a lab-on-a-chip configuration of sensors and biosensors. Therefore, MMs and MSs based-on sensors will enable faster and more accurate detection of different analytes which will greatly help medical diagnosis in the future. So, research on MMs and MSs in sensing and biosensing requires more theoretical and practical study and has the potential for further development in the coming years.

REFERENCES

Aïssa, B., Ali, A., & El Mellouhi, F. (2021). Oxide and Organic-Inorganic Halide Perovskites with Plasmonics for Optoelectronic and Energy Applications: A Contributive Review. *Catalysts*, 11(9), 1057. 10.3390/catal11091057

Al-Naib, I. (2022). Terahertz asymmetric S-shaped complementary metasurface biosensor for glucose concentration. *Biosensors (Basel)*, 12(8), 609. 10.3390/bios1208060936005005

Ali, L., Mohammed, M. U., Khan, M., Yousuf, A. H. B., & Chowdhury, M. H. (2019). High-quality optical ring resonator-based biosensor for cancer detection. *IEEE Sensors Journal*, 20(4), 1867–1875. 10.1109/JSEN.2019.2950664

An, J., Lee, J., Lee, S. H., Park, J., & Kim, B. (2009). Separation of malignant human breast cancer epithelial cells from healthy epithelial cells using an advanced dielectrophoresis-activated cell sorter (DACS). *Analytical and Bioanalytical Chemistry*, 394(3), 801–809. 10.1007/s00216-009-2743-719308360

Arbabi, E., Arbabi, A., Kamali, S., Horie, Y., Faraji-Dana, M., & Faraon, A. (2018). MEMS-tunable dielectric metasurface lens. *Nature Communications*, 9(1), 812. 10.1038/s41467-018-03155-629476147

Asgari, S., Granpayeh, N., & Fabritius, T. (2020). Controllable terahertz cross-shaped three-dimensional graphene intrinsically chiral metastructure and its biosensing application. *Optics Communications*, 474, 12608. 10.1016/j.optcom.2020.126080

Barron, L. D. (2004). *Molecular Light Scattering and Optical Activity* (2nd ed.). Cambridge University Press. 10.1017/CBO9780511535468

Chen, H.-T., Padilla, W. J., Zide, J. M., Gossard, A. C., Taylor, A. J., & Averitt, R. D. (2006). Active terahertz metamaterial devices. *Nature*, 444(7119), 597–600. 10.1038/nature0534317136089

Chen, T., Li, S., & Sun, H. (2012). Metamaterials application in sensing. *Sensors (Basel)*, 12(3), 2742–2765. 10.3390/s12030274222736975

Chen, T., Zhang, D., Huang, F., Li, Z., & Hu, F. (2020). Design of a terahertz metamaterial sensor based on split ring resonator nested square ring resonator. *Materials Research Express*, 7(9), 095802. 10.1088/2053-1591/abb496

De Souza, F. A. L., Amorim, R. G., Scopel, W. L., & Scheicher, R. H. (2017). Electrical detection of nucleotides via nanopores in a hybrid graphene/h-BN sheet. *Nanoscale*, 9(6), 2207–2212. 10.1039/C6NR07154F28120993

Dell'Olio, F., & Passaro, V. M. (2007). Optical sensing by optimized silicon slot waveguides. *Optics Express*, 15(8), 4977–4993. 10.1364/OE.15.00497719532747

Duan, Z., Tang, X., Wang, Z., Zhang, Y., Chen, X., Chen, M., & Gong, Y. (2017). Observation of the reversed Cherenkov radiation. *Nature Communications*, 8(1), 14901. 10.1038/ncomms1490128332487

Fallahi, A., & Perruisseau-Carrier, J. (2012). Design of tunable biperiodic graphene metasurfaces. *hysical Review B*, *86* (19), p. 195408.

Faraji-Dana, M., Arbabi, E., Kwon, H., Kamali, S., Arbabi, A., Bartholomew, J., & Faraon, A. (2019). Hyperspectral imager with folded metasurface optics. *ACS Photonics*, 6(8), 2161–2167. 10.1021/acsphotonics.9b00744

Gansel, J. K., Thiel, M., Rill, M. S., Decker, M., Bade, K., Saile, V., Freymann, G., Linden, S., & Wegener, M. (2009). Gold helix photonic metamateri-al as broadband circular polarizer. *Science*, 325(5947), 1513–1515. 10.1126/science.117703119696310

Geng, Z., Zhang, X., Fan, Z., Lv, X., & Chen, H. (2017). A route to terahertz metamaterial biosensor integrated with microfluidics for liver cancer biomarker testing in early stage. *Scientific Reports*, 7(1), 16378. 10.1038/s41598-017-16762-y29180650

Ghodrati, M., Farmani, A., & Mir, A. (2019). Nanoscale Sensor-Based Tunneling Carbon Nanotube Transistor for Toxic Gases Detection: A First-Principle Study. *IEEE Sensors Journal*, 19(17), 7373–7377. 10.1109/JSEN.2019.2916850

Ghodrati, M., & Mir, A. (2022). Improving the Performance of a Doping-Less Carbon Nanotube FET with Dual Junction Source and Drain Regions: Numerical Studies. *Journal of Circuits, Systems, and Computers*, 31(10), 2250182. 10.1142/S0218126622501821

Ghodrati, M., Mir, A., & Farmani, A. (2020). Carbon nanotube field effect transistors-based gas sensors. In *Nanosensors for Smart Cities* (pp. 171–183). Elsevier. 10.1016/B978-0-12-819870-4.00036-0

Ghodrati, M., Mir, A., & Farmani, A. (2021). Non-destructive label-free biomaterials detection using tunneling carbon nanotube-based biosensor. *IEEE Sensors Journal*, 21(7), 8847–8854. 10.1109/JSEN.2021.3054120

Ghodrati, M., Mir, A., & Farmani, A. (2022). Proposing of SPR biosensor based on 2D $Ti_3C_2T_x$ MXene for uric acid detection ımmobilized by uricase enzyme. *Journal of Computational Electronics*, 22, 560–569. 10.1007/s10825-022-01959-w

Ghodrati, M., Mir, A., & Farmani, A. (2022). Sensitivity-Enhanced Surface Plasmon Resonance Sensor with Bimetal/ Tungsten Disulfide (WS_2)/MXene ($Ti_3C_2T_x$) Hybrid Structure. *Plasmonics*, 17(5), 1973–1984. 10.1007/s11468-022-01685-w

Ghodrati, M., Mir, A., & Farmani, A. (2023). Numerical analysis of a surface plasmon resonance based biosensor using molybdenum disulfide, molybdenum trioxide, and MXene for the diagnosis of diabetes. *Diamond and Related Materials*, 132, 109633. 10.1016/j.diamond.2022.109633

Ghodrati, M., Mir, A., & Farmani, A. (2024). *2D Materials/Heterostructures/Metasurfaces in Plasmonic Sensing and Biosensing*.

Ghodrati, M., Mir, A., & Naderi, A. (2020). New structure of tunneling carbon nanotube FET with electrical junction in part of drain region and step impurity distribution pattern. *AEÜ. International Journal of Electronics and Communications*, 117, 153102. 10.1016/j.aeue.2020.153102

Ghodrati, M., Mir, A., & Naderi, A. (2021). Proposal of a doping-less tunneling carbon nanotube field-effect transistor. *Materials Science and Engineering B*, 265, 115016. 10.1016/j.mseb.2020.115016

He, Z., Li, L., Ma, H., Pu, L., Xu, H., Yi, Z., Cao, X., & Cui, W. (2021). Graphene-based metasurface sensing applications in terahertz band. *Results in Physics*, 21, 103795. 10.1016/j.rinp.2020.103795

Hlali, A., Oueslati, A., & Zairi, H. (2021). Numerical simulation of tunable terahertz graphene-based sensor for breast tumor detection. *IEEE Sensors Journal*, 21(8), 9844–9851. 10.1109/JSEN.2021.3060326

Hu, Y., Luo, X., Chen, Y., Liu, Q., Li, X., Wang, Y., Liu, N., & Duan, H. (2019). 3D-integrated metasurfaces for full-colour holography. *Light, Science & Applications*, 8(1), 86. 10.1038/s41377-019-0198-y31645930

Im, H., Shao, H., Park, Y. I., Peterson, V. M., Castro, C. M., Weissleder, R., & Lee, H. (2014). Label-free detection and molecular profiling of exosomes with a nano-plasmonic sensor. *Nature Biotechnology*, 32(5), 490–495. 10.1038/nbt.288624752081

Islam, M. R., Islam, M. T., M, M. S., Bais, B., Almalki, S. H. A., Alsaif, H., & Islam, M. S. (2022). Metamaterial sensor based on rectangular enclosed adjacent triple circle split ring resonator with good quality factor for microwave sensing application. *Scientific Reports*, 12(1), 6792. 10.1038/s41598-022-10729-435474227

Jiang, H., Choudhury, S., Kudyshev, Z. A., Wang, D., Prokopeva, L. J., Xiao, P., Jiang, Y., & Kildishev, A. V. (2019). Enhancing sensitivity to ambient refractive index with tunable few-layer graphene/hBN nanoribbons. *Photonics Research*, 7(7), 815–822. 10.1364/PRJ.7.000815

Jiang, L., Zeng, S., Quyang, Q., Dinh, X. Q., Coquet, P., Qu, J., He, S., & Yong, K. T. (2017). Graphene-TMDC-Graphene Hybrid Plasmonic Metasurface for Enhanced Biosensing: A Theoretical Analysis. *Physica Status Solidi. A, Applications and Materials Science*, 214(12), 1700563. 10.1002/pssa.201700563

Jiang, L., Zeng, S., Xu, Z., Quyang, Q., Zhang, D. H., Chong, P. H. J., Coquet, P., He, S., & Yong, K. T. (2017). Multifunctional hyperbolic nanogroove metasurface for submolecular detection. *Small*, 13(30), 1700600. 10.1002/smll.20170060028597602

Jiang, X., Liang, B., Zou, X., Yin, L., & Cheng, J. (2014). Broadband field rotator based on acoustic metamaterials. *Applied Physics Letters*, 104(8), 083510. 10.1063/1.4866333

Jiang, Z. H., Yun, S., Lin, L., Bossard, J. A., Werner, D. H., & Mayer, T. S. (2013). Tailoring dispersion for broadband low-loss optical metamaterials using deep-subwavelength inclusions. *Scientific Reports*, 3(1), 1571. 10.1038/srep0157123535875

Khanikev, A. B., Wu, C., & Shevets, G. (2013). Fano-resonant metamaterials and their applications. *Nanophotonics*, 2(4), 247–264. 10.1515/nanoph-2013-0009

Kodera, T., Sounas, D. L., & Caloz, C. (2011). Artificial Faraday rotation using a ring metamaterial structure without static magnetic field. *Applied Physics Letters*, 99(3), 31114. 10.1063/1.3615688

Kowerdziej, R., Wróbel, J., & Kula, P. (2019). Ultrafast electrical switching of nanostructured metadevice with dual-frequency liquid crystal. *Scientific Reports*, 9(1), 20367. 10.1038/s41598-019-55656-z31889047

Kretschmann. E, H. Raether, & Notizen, (1965). Radiative decay of non radiative surface plasmons excited by light. *Z. Naturforsch. A,23*, 2135–2136.

Kumari, R., Yadav, A., Sharma, S., Gupta, T. D., Varshney, S. K., & Lahiri, B. (2021). Tunable Van der Waal's optical metasurfaces (VOMs) for biosensing of multiple analytes. *Optics Express*, 29(16), 25800–25811. 10.1364/OE.43228434614900

Kurosawa, H., Choi, B., Sugimoto, Y., & Iwanaga, M. (2017). High-performance metasurface polarizers with extinction ratios exceeding 12000. *Optics Express*, 25(4), 4446–4455. 10.1364/OE.25.00444628241647

La Spada, L., McManus, T. M., Dyke, A., Haq, S., Zhang, L., Cheng, Q., & Hao, Y. (2016). Surface wave cloak from graded refractive index nanocomposites. *Scientific Reports*, 6(1), 22045–22322. 10.1038/srep2936327416815

La Spada, L., Tarparelli, R., & Vegni, L. (2014). Spectral Green's function for SPR meta-structures. *Mater. Sci,Forum 792*, 110–114. 10.

Larsson, E. M., Alegret, J., Käll, M., & Sutherland, D. S. (2007). Sensing characteristics of nir localized surface plasmon resonances in gold nanorings for application as ultrasensitive biosensors. *Nano Letters*, 7(5), 1256–1263. 10.1021/nl070161217430004

Lee, Y., Kim, S.-J., Park, H., & Lee, B. (2017). Metamaterials and Metasurfaces for Sensor applications. *Sensors (Basel)*, 17(8), 1726. 10.3390/s1708172628749422

Leitis, A., Tittl, A., Liu, M., Lee, B. H., Gu, M. B., Kivshar, Y. S., & Altug, H. (2019). Angle-multiplexed all-dielectric metasurfaces for broadband molecular fingerprint retrieval. *Science Advances*, 5(5), eaaw2871. 10.1126/sciadv.aaw287131123705

Leonhardt, U. (2007). Invisibility cup. *Nature Photonics*, 1(4), 207–208. 10.1038/nphoton.2007.38

Li, C., Yu, P., Huang, Y., Zhou, Q., Wu, J., Li, Z., Tong, X., Wen, Q., Kuo, H.-C., & Wang, Z. M. (2020). Dielectric metasurfaces: From wavefront shaping to quantum platforms. *Progress in Surface Science*, 95(2), 100584. 10.1016/j.progsurf.2020.100584

Li, L., Zong, X., & Liu, Y. (2020). All-metallic metasurfaces towards high-performance magneto-plasmonic sensing devices. *Photonics Research*, 8(11), 1742. 10.1364/PRJ.399926

Li, Q., Bao, W., Nie, Z., Xia, Y., Xue, Y., Wang, Y., Yang, S., & Zhang, X. (2021). A non-unitary metasurface enables continuous control of quantum photon-photon interactions from bosonic to fermionic. *Nature Photonics*, 15(4), 267–271. 10.1038/s41566-021-00762-6

Liu, G., Yu, M., Liu, Z., Pan, P., Liu, X., Huang, S., & Wang, Y. (2016). Multi-Band High Refractive Index Susceptibility of Plasmonic Structures with Network-Type Metasurface. *Plasmonics*, 11(2), 677–682. 10.1007/s11468-015-0101-5

Lu, C., Hu, X., Shi, K., Hu, Q., Zhu, R., Yang, H., & Gong, Q. (2015). An actively ultrafast tunable giant slow-light effect in ultrathin nonlinear metasurfaces. *Light, Science & Applications*, 4(6), e302. 10.1038/lsa.2015.75

Ma, Q., Hong, Q., Gao, X., Jing, H., Liu, C., Bai, G., Cheng, Q., & Cui, T. (2020). Smart sensing metasurface with self-defined functions in dual polarizations. *Nanophotonics*, 9(10), 3271–3278. 10.1515/nanoph-2020-0052

Ma, S., Zhang, P., Mi, X., & Zhao, H. (2022). Highly sensitive terahertz sensor based on graphene metamaterial absorber. *Optics Communications*, 528, 129021. 10.1016/j.optcom.2022.129021

Maci, S., Minatti, G., Casaletti, M., & Bosiljevac, M. (2011). Metasurfing: Addressing waves on impenetrable metasurfaces. *IEEE Antennas and Wireless Propagation Letters*, 10, 1499–1502. 10.1109/LAWP.2012.2183631

Meng, X., Depauw, V., Gomard, G., El Daif, O., Trompoukis, C., Drouard, E., Jamois, C., Fave, A., Dross, F., Gordon, I., & Seassal, C. (2012). Design, fabrication and optical characterization of photonic crystal assisted thin film monocrystalline-silicon solar cells. *Optics Express*, 20(S4), A465–A475. 10.1364/OE.20.00A46522828615

Mun, J., & Rho, J. (2019). Importance of higher-order multipole transitions on chiral nearfield interactions. *Nanophotonics*, 8(5), 941–948. 10.1515/nanoph-2019-0046

Naderi, A., & Ghodrati, M. (2017). Improving band-to-band tunneling in a tunneling carbon nanotube field effect transistor by multi-level development of impurities in the drain region. *Eur. Phys. J. Plus,132*.

Naderi, A., & Ghodrati, M. (2018). An efficient structure for T-CNTFETs with intrinsic-n-doped impurity distribution pattern in drain region. *Turk J. Electr. Eng*, 26(5), 2335–2346. 10.3906/elk-1709-180

Naderi, A., & Ghodrati, M. (2018). Cut Off Frequency Variation by Ambient Heating in Tunneling p-i-n CNTFETs. *ECS Journal of Solid State Science and Technology : JSS*, 7(2), M6–M10. 10.1149/2.0241802jss

Naderi, A., Ghodrati, M., & Baniardalani, S. (2020). The use of a Gaussian doping distribution in the channel region to improve the performance of a tunneling carbon nanotube field-effect transistor. *Journal of Computational Electronics*, 19(1), 283–290. 10.1007/s10825-020-01445-1

Otto, A. (1968). Excitation of nonradiative surface plasma waves in silver by the method of frustrated total reflection. *Zeitschrift für Physik*, 216(4), 398–410. 10.1007/BF01391532

Park, S., Cha, S., Shin, G., & Ahn, Y. (2017). Sensing viruses using terahertz nano-gap metamaterials. *Biomedical Optics Express*, 8(8), 3551–3558. 10.1364/BOE.8.00355128856034

Parvin, T., Ahmed, K., Alatwi, A. M., & Rashed, A. N. Z. (2021). Differential optical absorption spectroscopy-based refractive index sensor for cancer cell detection. *Optical Review*, 28(1), 134–143. 10.1007/s10043-021-00644-w

Pendry, J. B. (2000). Negative Refraction Makes a Perfect Lens. *Physical Review Letters*, 85(18), 3966–3969. 10.1103/PhysRevLett.85.396611041972

Pendry, J. B., Holden, A. J., Robbins, D. J., & Stewart, W. (1999). Magnetism from conductors and enhanced nonlinear phenomena. *IEEE Transactions on Microwave Theory and Techniques*, 47(11), 2075–2084. 10.1109/22.798002

Pozar, D. M., Targonski, S. D., & Syrigos, H. D. (1997). Design of millimeter wave microstrip reflectarrays. *IEEE Transactions on Antennas and Propagation*, 45(2), 287–296. 10.1109/8.560348

Qi, Z.-M., Matsuda, N., Itoh, K., Murabayashi, M., & Lavers, C. (2002). A design for improving the sensitivity of a mach–zehnder interferometer to chemical and biological measurands. *Sensors and Actuators. B, Chemical*, 81(2-3), 254–258. 10.1016/S0925-4005(01)00960-1

Qiu, G., Gai, Z., Saleh, L., Tang, J., Gui, T., Kullak-Ublick, G. A., & Wang, J. (2021). Thermoplasmonic-assisted cyclic cleavage amplification for self-validating plasmonic detection of SARS-CoV-2. *ACS Nano*, 15(4), 7536–7546. 10.1021/acsnano.1c0095733724796

Qiu, G., Gai, Z., Tao, Y., Schmitt, J., Kullak-Ublick, G. A., & Wang, J. (2020). Dual-functional plasmonic photothermal biosensors for highly accurate severe acute respiratory syndrome coronavirus 2 detection. *ACS Nano*, 14(5), 5268–5277. 10.1021/acsnano.0c0243932281785

Rifat, A., Rahmani, M., Xu, L., & Miroshnichenko, A. (2018). Hybrid Metasurface Based Tunable Near-Perfect Absorber and Plasmonic Sensor. *Materials (Basel)*, 11(7), 1091. 10.3390/ma1107109129954060

Shrekenhamer, D., Chen, W.-C., & Padilla, W. J. (2013). Liquid Crystal Tunable Metamaterial Absorber. *Physical Review Letters*, 110(17), 177403. 10.1103/PhysRevLett.110.17740323679774

Solntsev, A. S., Agarwal, G. S., & Kivshar, Y. S. (2021). Metasurfaces for quantum photonics. *Nature Photonics*, 15(5), 327–336. 10.1038/s41566-021-00793-z

Sun, Y., Zhang, L., Shi, H., Cao, S., Yang, S., & Wu, Y. (2021). Near-infrared plasma cavity metasurface with independently tunable double Fano resonances. *Results in Physics*, 25, 104204. 10.1016/j.rinp.2021.104204

Tan, C., Wang, S., Li, S., Liu, X., Wei, J., Zhang, G., & Ye, H. (2022). Cancer Diagnosis Using Terahertz-Graphene-Metasurface-Based Biosensor with Dual-Resonance Response. *Nanomaterials (Basel, Switzerland)*, 12(21), 3889. 10.3390/nano1221388936364665

Tretyakov, S. A. (2017). A personal view on the origins and developments of the metamaterial concept. *Journal of Optics*, 19(1), 013002. 10.1088/2040-8986/19/1/013002

Tretyakov, S. (2003). *Analytical Modeling in Applied Electromagnetics*. Artech House. google-Books-ID: MZ3tpGtadhcC.

Vasic, B., Isic, G. & Gajic, R. (2013). Localized surface plasmon resonances in graphene ribbon arrays for sensing of dielectric environment at infrared frequencies. *Journal of Applied Physics*, 113, 013110-013110-013117.

Veselago, V. G. (1968). The electrodynamics of substances with simultaneously negative values of ε and μ. *Soviet Physics - Uspekhi*, 10, 509–514. 10.1070/PU-1968v010n04ABEH003699

Walser, R. (1999). Metamaterials: What are they? What are they good for? *Acta Obstetricia et Gynecologica Scandinavica*, 97, 388–393.

Wang, G., Zhu, F., Lang, T., Liu, J., Hong, Z., & Qin, J. (2021). All-metal terahertz metamaterial biosensor for protein detection. *Nanoscale Research Letters*, 16(1), 109. 10.1186/s11671-021-03566-334191133

Wang, Z., Chen, J., Khan, S. A., Li, F., Shen, J., Duan, Q., Liu, X., & Zhu, J. (2022). Plasmonic Metasurfaces for Medical Diagnosis Applications: A Review. *Sensors (Basel)*, 22(1), 133. 10.3390/s2201013335009676

Wei, Z., Cao, Y., Su, X., Gong, Z., Long, Y., & Li, H. (2013). Highly efficient beam steering with a transparent metasurface. *Optics Express*, 21(9), 10739–10745. 10.1364/OE.21.01073923669930

Xu, H.-X., Hu, G., Wang, Y., Wang, C., Wang, M., Wang, S., Huang, Y., Genevet, P., Huang, W., & Qiu, C.-W. (2021). Polarization-insensitive 3D conformal-skin metasurface cloak. *Light, Science & Applications*, 10(1), 75. 10.1038/s41377-021-00507-833833215

Xu, T., Xu, X., & Lin, Y. S. (2021). Tunable terahertz free spectra range using electric split-ring metamaterial. *Journal of Microelectromechanical Systems*, 30(2), 309–314. 10.1109/JMEMS.2021.3057354

Yalcin, A., Popat, K. C., Aldridge, J. C., Desai, T. A., Hryniewicz, J., Chbouki, N., Little, B. E., King, O., Van, V., & Chu, S. (2006). Optical sensing of biomolecules using microring resonators. *IEEE Journal of Selected Topics in Quantum Electronics*, 12(1), 148–155. 10.1109/JSTQE.2005.863003

Yao, J., Ou, J. Y., Savinov, V., Chen, M. K., Kuo, H. Y., Zheludev, N. I., & Tsai, D. P. (2022). Plasmonic anapole metamaterial for refractive index sensing. *PhotoniX*, 3(1), 23. 10.1186/s43074-022-00069-x

Ye, Q., Wang, J., Liu, Z., Deng, Z. Ch., Kong, X. T., Xing, F., Chen, X. D., Zhou, W. Y., Zhang, C. P., & Tian, J. G. (2013). Polarization-dependent optical absorption of graphene under total internal reflection. *Applied Physics Letters*, 102(2), 021912. 10.1063/1.4776694

Yoo, S., & Park, Q.-H. (2019). Metamaterials and chiral sensing: A review of fundamentals and applications. *Nanophotonics*, 8(2), 249–261. 10.1515/nanoph-2018-0167

Yotter, R. A., & Wilson, D. M. (2004). Sensor technologies for monitoring metabolic activity in single cells-part ii: Nonoptical methods and applications. *IEEE Sensors Journal*, 4(4), 412–429. 10.1109/JSEN.2004.830954

Yu, N., Genevet, P., Kats, M. A., Aieta, F., Tetienne, J.-P., Capasso, F., & Gaburro, Z. (2011). Light propagation with phase discontinuities: Generalized laws of reflection and refraction. *Science*, 1210713. Sun,S., He, Q., Xiao, S., Xu, Q., Li, X., & Zhou, L. (2012). Gradient-index meta-surfaces as a bridge linking propagating waves and surface waves. *Nature Materials*, 11, 426.

Yuan, Y., Yu, X., Ouyang, Q., Shao, Y., Song, J., Qu, J. & Yong, K.T. (2018). Highly anisotropic black phosphorous-graphene hybrid architecture for ultrassensitive plasmonic biosensing: Theoretical insight. *2D Materials*, 5, 025015.

Zhang, C., Liu, Q., Peng, X., Ouyang, Z., & Shen, S. (2021). Sensitive THz sensing based on Fano resonance in all-polymeric Bloch surface wave structure. *Nanophotonics*, 10(15), 3879–3888. 10.1515/nanoph-2021-0339

Zhang, Z., Yang, J., He, X., Zhang, J., Huang, J., Chen, D., & Han, Y. (2018). Plasmonic Refractive Index Sensor with High Figure of Merit Based on Concentric-Rings Resonator. *Sensors (Basel)*, 18(2), 116. 10.3390/s1801011629300331

Zhong, J., Xu, X., & Lin, Y. S. (2021). Tunable terahertz metamaterial with electromagnetically induced transparency characteristic for sensing application. *Nanomaterials (Basel, Switzerland)*, 11(9), 2175. 10.3390/nano1109217534578491

Zhu, J., Wang, Z., Lin, S., Jiang, S., Liu, X., & Guo, S. (2020). Low-cost flexible plasmonic nanobump metasurfaces for label-free sensing of serum tumor marker. *Biosensors & Bioelectronics*, 150, 111905. 10.1016/j.bios.2019.11190531791874

Chapter 4
Antenna Design for Beyond 5G:
5G and 6G Applications With Different Materials and Techniques Including Metamaterials and Metasurfaces

Smrity Dwivedi
IIT BHU, India

ABSTRACT

In the last few years, economic and social development has been greatly influenced by the advancements in the field of mobile communication and technology. 5G technology has emerged as an important of the future 2020 generation which is already in use. After the development of fifth generation technologies, researchers, scientist, and engineers are looking for wide bandwidth which should be improve wireless systems and devices to provide better services and fast experience. Also, the development of 5G wireless network technology is the response to the crucial factors that lead to this demand because of its ability to provide extremely fast internet speed, high bandwidth, high performance, reduced latency, and high reliability and better gain. Metamaterials (LHM) hold great promises for such devices and for many applications like medical, technologies, communications, defense, etc. Metamaterials have interesting properties, making them more amenable to transmit and receive in small quantities. In this chapter, the authors will review the recent design with advancement in 5G antennas for beyond 5G application and also we will give few design of antenna for 6G applications.

DOI: 10.4018/979-8-3693-2599-5.ch004

INTRODUCTION

As per requirement and demand of wide bandwidth as well as fast network, researchers, scientists and engineers are looking for enhancement and advancement in the new and recent technology such as fifth generation communication systems. As since decades, after coming of fourth generation technologies which was providing to users a fast networking, good bandwidth as well as high data rate, but now, requirement of bandwidth has increased and depending on demand, fifth generation technology has come, in which various parameters have been improved and depending on that designs, they have been developed. Antenna design is one of the major part of fifth generation systems. Researchers have been working to develop the best and suitable design in low cost and high performance. There are different types of designs in which is reconfigurable antenna, antenna with graphene for terahertz application, array antenna to achieve high directivity and gain, MIMO antenna to enhance bandwidth and gain. Also there are several materials are being used to design antenna for fifth generation and beyond fifth generation, in which metamaterial, graphene, gold, SiO2, polymide etc. are the frequency used to achieve high bandwidth, especially these materials can be used for beyond fifth generation and terahertz applications. Few specific properties and design have been given in next sections, using several design properties given above. Their important properties as mentioned are being used to achieve wide bandwidth and high gain using design for beyond fifth generation applications. Due to the high and fast growth of wireless communications as well as the high demand for the integration and mixing of multiple wireless standards into a single and well platform, it is the most important that the operating frequency, radiation pattern, and polarizations of antennas can be reconfigurable and changing as per requirement. Reconfigurable antennas change and modify their operating frequency, impedance bandwidth, polarization, and radiation pattern as per the operating requirements of the usage and all applications. They can easily radiate with multiple patterns at different frequencies and polarizations. Now a days, obtaining the desired functionality for a reconfigurable antenna and mixing it into a complete system to achieve a cost effective and efficient way is a challenging task for antenna designers. As we are moving towards from fifth generation to sixth generation, these analysis is very helpful to design such smart and intelligent sixth generation antennas. Converting an antenna into a reconfigurable one by applying different techniques to change the antenna's internal structure and parameters by applying different techniques has been again challenging task. Multiple factors need to be considered such as getting a better gain, good efficiency, stable radiation pattern, and a perfect impedance match throughout all the antenna's operation in given environment. To achieve a these properties throughout the operation, the reconfigurable antenna designers must focus on the following questions such as which property such as frequency, radiation pattern, or polarization, they must be modified for the antenna and how the radiating elements of the antenna structure are reconfigured to get the required property as it is important. As well as which reconfiguration technique can minimize the negative effects on the antenna performances so that high value parameters can be achieved. A reconfigurable antenna provides the same functionality as that given by multiple single purpose antennas.

As already, researchers are working under fifth generation (5G) technology to achieve wide and good bandwidth, reliability in work, higher channel capacity to get more data and high data rate, so they developed so many theories to apply to wireless communication systems. Also this new technology is helping to establish the connection between machines, objects and all the devices to reduce the latency, enhance the network efficiency as well as to utilize the full spectrum allotted to it and to use high traffic capacity. Fifth generation is also helping to bring vast and drastic changes in the industrial, scientific, and medical sectors with economic growth and developing the new devices at high data rate. For the successful deployment of fifth generation technology, multiple input multiple output is a core and necessary for giving greater network coverage and improved channel capacity as per requirement. MIMO is an important technology that creates simultaneous transmission and reception of huge and maximum data signals over the same radio channel by using more number of transmitting and receiving antennas together. It is being used to enhance and grow the strength of the signal propagating through the given radio channel system. The MIMO technology starts different latest and new wireless standards such as wireless LAN, wifi-MAX, long term evolution, fifth generation, beyond fifth generation and next higher generations supports larger number of users with higher data rates, better system capacity, and less fading effects with the help of various diversity techniques and several antennas for application requirement. The characteristic parameters of a perfect and well-designed antenna give better performance of communication system to great extent of sending signal. Hence, the modelling of a high precision and perfect MIMO antenna with its material selection, design, simulation, analysis, and optimization techniques are catching special attention of the researchers of the antenna community for enhancing performance of wireless communication system by providing improved and enhanced transmission speed, data rate, and channel capacity. In past and recent years, a significant process has been noticed in the design of antenna aspects of planar MIMO antennas due to the several efforts and work done of the antenna designers. Advances and new in recent communication systems require minimal weight, low cost, high performance, and low-profile and high efficiency antenna to meet the demand for next generation wireless communication devices. Due to the saturation velocity, high electrical conductivity and high mobility, graphene patch antennas are preferably used in the beyond fifth generation and Terahertz band frequency region (THz). Apart from that, MIMO antenna is usually required to eliminate and compensate for the losses due to high path and atmospheric attenuation in high frequency band spectrum and to offer higher data rates.

LITERATURE REVIEWS FOR THE DESIGN OF ANTENNA

The part of electromagnetic spectrum in between the microwave and infrared range is defined as Tera Hertz band which varies in the frequency from 0.1 THz – 10 THz (S. Das et al. 2021) in frequency spectrum. The full-fledged utilization of this THz band is still least discovered in application areas because of lack of materials at these terahertz frequencies. But, in recent years, the advanced wireless communication standards demand higher and faster data rates for the users. It has increased the requirements of massive operating frequency range with an very wide bandwidth for the perfect connectivity of next generation wireless communication systems. In this regard, only the massive terahertz frequency band spectrum (0.1–10 THz) fulfils the requirements of multi-functional gadgets for next generation communication systems by connecting machines, objects, and devices with superior reliability with high tera bit data. The quality of terahertz communication systems depends on the successful establishment of high-speed transmission and reception links in communication systems. On the other hand, transmission and reception links are majorly dependent on the characteristics of designed THz antennas. In the terahertz regime, microstrip patch antennas (MPAs) (Ch Krishna et al. 2021), (K. V. Babu et al, 2023) are highly in demand due to its several advantageous features as it has low profile design, light weight-compact dimension, planar structure for easy integration etc.. Hence, the characteristics parameters of MPA with gain, bandwidth, efficiency etc., are very low due to the effect of surface waves. However, the overall performance of the microstrip antenna can be enhanced by limiting the surface wave radiation with the help of integrating the photonic bandgap (PBG) and electromagnetic band gap structures into the planar patch antenna geometry NH Rosla, et al. 2008). In PBG designs, photonic crystals are embedded into the dielectric substrate, which suppress the undesired frequency bands and generates a resistive effect against the surface waves to propagate through the dielectric substrate and improves the characteristics parameters of patch antennas. While comparing to microwave and infrared, terahertz spectrum has several advantages like wider unused frequency bands, less interference, very low power requirement, higher data throughput rate, high gain etc. Irrespective of several advantages, there is a major problem of fading due to reflections and scattering of signals. In most of the cases, the channel gets influenced by problem like fading, which degrades the performance of the terahertz wireless communication systems. As a solution, to improve the performance of terahertz communication system, a more number of radiating elements can be used to overcome the fading issue. MIMO antenna system is capable to minimize the fading and to improve the system capacity for wide bandwidth. MIMO antennas are considered as a key element in terahertz systems to provide higher and extremely high data rates along with better signal reliability and efficiency. Progressively, researchers have reported various design methodologies for single antenna element, array antenna configuration, and MIMO terahertz patch antenna structures, such as photonic crystal THz patch antenna array (Benlakehal ME, 2022), 1 × 2 patch array antenna on periodic and nonperiodic photonic crystal (Benlakehal ME, 2022),, photonic crystal and silicon materials based flexible terahertz antenna (EddineTemmar MN et al. 2021) polyimide substrate based photonic crystal THz antenna (Hocini A et al. 2019), PBG substrate based terahertz antenna with segmented objects guiding technique (Khezzar D et al. 2021), simple rectangular terahertz microstrip patch antennas for the surveillance applications (RH Mahmud et al. 2020), rhombus-shaped single element terahertz antenna (CM Krishna et al. 2021), graphene based terahertz patch antenna with glass substrate R. Goyal et al. 2018)S. Ananda et al. 2014), graphene nano-ribbon based simple rectangular patch antenna for terahertz applications, polyimide based fractal MIMO antenna with coaxial feeding (K. V. Babu et al. 2022), a multi-layer terahertz rectangular patch antenna array (A. Azarbar et al. 2014), trapezoidal shaped microstrip patch antenna by implanting photonic crystals in the substrate layer (A. Singh et al. 2015), THz microstrip antenna with the help of binary particle swarm optimization algorithm based synthesized photonic bandgap substrate (MNE Tenmar, 2019), a tree shaped micro-scaled wide-band MIMO designed for THz systems (K. V. Babu, 2022), U-shaped and elliptical- shaped slots loaded hexagonal terahertz patch antennas (S. M. Shamim et al. 2021), rectangular patch antenna loaded with superstrate M. Younssi et al. 2013), a grapheme based simple rectangular patch antenna (SM Shamim et al. 2021), microstrip antenna with inset-feed loaded with 1D, 2D, 3D photonics crystal substrate (R. K. Kushwaha et al. 2021), dual polarized grapheme based patch antenna (M. Shalini et al. 2019), PBG integrated patch antenna (R. K. Kushwha, 2018) microstrip antenna with frequency selective surface (FSS) and photonic band-gap (PBG) (A. Nejati et al. 2014), proximity coupled array antenna with PBG and SRR (R K Kushwaha et al. 2018), a hexagonal structure of MIMO antenna array (M. Singh et al. 2021), multi-layered THz array antenna structure (Vettikalladi H et al. 2019), microstrip MIMO antenna for terahertz short range wireless communication systems (T Okan, 2021), 2 × 2 patch antenna array with PBG substrate (ME Benlakehel et al. 2022), 1 × 2 patch antenna array with homogeneous and periodic PBG substrates (ME Benlakehel et al. 2022), 2 × 2 MIMO antenna with photonic crystals substrate (ME. Benlakehel et al. 2022), CPW-Fed MIMO antenna using an array of parasitic elements (K. V. Basu et al. 2022), planar antenna backed by PBG substrate with circular shape metallic insertions (I. Ahmad et al. 2021), DGS and PBG based simple microstrip antenna (S. Ullah et al. 2019), simple patch antenna with 5 × 5 type PBG structure (SK Tripathi et al. 2019), graphene based patch antenna with PBG [37], and many more designs which is giving wide bandwidth and gain. THz antennas are used for high-speed wireless communication devices at present. THz MIMO antennas have been used to improve the isolation and give very high gain which is being utilized with its high frequency to meet the high speed and data demands of today's fast-growing world and users. THz MIMO antennas can be designed using metamaterials to improve their parameters. Metamaterials are artificial materials used with different microwave and optical structures to improve their parameters because of their negative refraction. The metamaterial MIMO antennas are getting the researcher's attention because of their parameter-improving capacity and it improves gain, bandwidth, isolation, etc. The metamaterial-loaded antennas are designed way back in 2013 by Patel and Kosta (S. K. Patel et al. 2012)but these antennas are based on a single-element patch antenna. In this research, a split ring resonator is used which is the basic component of metamaterial design. The broadband and high gain antennas are achieved using metamaterials by incorporating them in the patch and in the ground plane. The ground plane is defected to improve gain and bandwidth in metamaterial antennas (R. S. Keerthi, et al. 2021). The metamaterial antenna is designed and an optimization algorithm is used to optimize the parameters and achieved a good response (EI Kenawy et al. 2022). Metamaterial antenna with its negative behavior helped in achieving enhanced bandwidth for ISM and RFID applications. The antenna is designed with low-cost FR4 material to be applicable in wide applications because of its low cost and easy affordability (Najumunnisa, M. et al. 2022). A single-band THz antenna has been designed and simulated using a finite element method-based simulator using metamaterial (A. T. Devpriya et al. 2019). The designed antenna is showing one band with low bandwidth but can be improved further in the future with other techniques. One such antenna which can be considered for the broadband, multiband, and high gain antenna is the MIMO antenna. MIMO antenna can be designed with metamaterials for beyond fifth generation and THz communication to meet the high speed and high data demand of today's world.

Metamaterial MIMO antenna is designed for beyond fifth generation and THz communication application using graphene material. Tunability is achieved by changing the graphene chemical potential of the design (P. Das et al. 2022). Miniaturization of the THz MIMO antenna design is also possible through the inclusion of metamaterials in it. The Split ring resonator (SRR)-C shape is used in this design to achieve miniaturization and broadband results (K. K. Naik et al. 2021). The metamaterial-based THz MIMO antennas are designed using metamaterial components like split ring resonators, thin wires, etc. The inclusion of this metamaterial is in terms of resonating patches. The ground layer is also defected in some research to improve the antenna parameters (R. Kumari et al. 2022). The split ring resonators are also employed in substrate materials to improve performance. The fabrication of such structures is a little difficult compared to the normal patch-based split ring resonators as split ring resonators need to be embedded into the substrate to make them effective (Koutsoupidou, M. et al. 2014) The metamaterial-based THz antenna is designed for its possible applications in 6G and IoT applications. The metamaterial components have been employed to improve the gain and bandwidth parameters and make it application in the 6G high-speed devices (S. A. Khaleel et al. 2022).

CONCEPT OF METAMATERIAL

There has been a great deal of attention given in the field of metamaterials over a past few years (C. L. Holloway et al. 2009; V. G. Veselago 1968). Metamaterials are fully synthetic materials developed to get unique properties not normally found in nature. It was artificial dielectric in terms of electromagnetics. Appropriately designed metamaterials can affect electromagnetic radiation or sound, which is not found in bulk materials. Basically, metamaterials are divided into two major categories, resonant and non resonant types, depending upon the oscillation of waves as well as periodicity of structures. Resonant types of metamaterials have well specific permittivity and permeability, which further divide into sub categories. Double negative (DNG), negative permittivity (ENG), negative permeability (MNG) and negative refractive index types are widely used resonant type metamterials, whereas anisotropic and hyperbolic are the non resonant type metamaterials, which have certain specified bands for electromagnetic waves. Today's, metamaterials based structures are more commonly used in waveguides, antennas design, filters design and lots of other applications. Left handed metmaterials (LHM), which has negative permittivity and negative permeability is also known as DNG materials, are used for design of coplanar waveguides, filters (low pass, high pass) and microstrip lines Antennas (D. R. Smith et al. 2000; C. L. Holloway et al. 2003). LHM based microstrip antennas have high efficiency, low profile and high bandwidth, which is broadly usable in wide band applications. Simple microstrip antennas have radiation not only from patch, but also from substrate material that supports patch and ground plane, which is basically a surface waves, further creates radiation losses. So, the antenna has low bandwidth, low efficiency, poor directivity as well as higher losses (E. Shamonina et al. 2007; D. V. Sivukhin 1957). These losses can only be reduced by keeping the values of permittivity and permeability as low as possible. On the other hand, non resonant type metamaterials such as photonic band gap (PBG) or electromagnetic band gap structures (EBG), plays an important role to achieve higher bandwidth with greatest potential applicability in broadband communication systems. These

PBG structures have smaller size, which varies in sub wavelengths as well as have a greatest tolerance value to resist the structural deformations. PBG structures have magical properties than resonant metamaterials with periodicity ~λ/4, whereas resonant metamaterials have periodicity ~λ/10. For advance communication, PBG structure is more valid for high bandwidth and can be applicable with antenna design in mm-wave technology. Plane waves with photons enter into the band gap (stop band), coming after finding the dispersion diagram for getting the cut off frequency with phase constant. Metamaterials is also used for artificial magnetic conductors (R. A. Silin et al. 2001). Negative refractive index metamaterials is used for incident waves which have propagation faster than light propagation and for angle insensitive devices. According to national research in metamaterials, lots of works have been going on over a decades, in which the most the works are dedicated to microwave components and antennas using metamaterials for frequency less than 10 GHz.

Table 1. Special material properties in ε-μ domain

Permittivity (ε)	Permeability (μ)	Description
ε = -ε0	μ = -μ0	Characterizes anti air in left hand medium yields to the perfect lens
ε = 0	μ = 0	Characterizes anihility yields to the perfect tunneling effects
ε > ε0	μ = μ0	Characterizes materials available in nature
ε = μ		Characterizes perfect impedance matching in right hand medium and left hand medium ensuing no reflections

Although metamaterials are not a new topic, but for the research and development point of view, works are still left. Metamaterial is a material that is artificially engineered to acquire unconventional electromagnetic, thermal, acoustic, and mechanical properties that are not found in other naturally occurring materials because of its extraordinary unusual and unique kind of properties such as negative permeability and negative permittivity.

When the values of permeability and negative permittivity both are simultaneously negative, then the electric field, magnetic field, and the propagation vector form left-handed medium materials in a double-negative region and wave propagation are moving curiously in reverse signifying material has negative refractive index (L. H. Wen et al. 2018). Different studies and researches are being carried out by scientists and researchers all over the world that will benefit humanity through the use of this metamaterial. The metamaterial electromagnetic properties are the best defined by famous maxwells equations given as (D. Yanget al. 2017). The preferences of using metamaterials in the design of the microstrip antenna are to minimize the size and increase other parameters such as bandwidth, gain, and return loss as well as efficiency (C. Wang et al. 2018). This will enable to apply and adapt concepts to innovations in microstrip antenna designs from laboratories for practical applications of engineering with a good contract (C. Wang et al. 2018). Since the permeability negative material, the unit cell split-ring resonators are the most used structure used as metamaterials. The metamaterial has various applications such as smart antenna, medical devices, optical devices, smart sensors, smart IoT devices, smart solar power, radomes, optical lenses, invisible submarines, radar, etc. The metamaterial structures have significant roles in the applications given above for boosting their performance in terms of power or energy harvesting, bandwidth, and gain enhancement, squeezing the size of the devices for optimum performance and other benefits. The design for microstrip patch antenna in the various field of electronics, communication, information technology, and electrical has enormously increased owing to its high characteristics such as bandwidth, return loss, etc. This chapter aims at providing a multi-frequency band and improvement in the performance of the previously proposed antenna applicable for various 5G enabled IoT smart Devices for interrupted quality services. This chapter, therefore, provides all the necessities required to fulfil the demand for next-generation IoT devices.

ANTENNAS DESIGN FOR 5G AND 6G ERA

Antenna Design for Fifth Generation Technologies

In last few years, as per economic and social development which is greatly influenced by the advancements in the field of mobile communication and technology, 5G technology has emerged as an important of the future 2020 generation which is already in use. 5G technology is an enhance technology with evolutionary and revolutionary services for users. It is the recent generation of technology which is providing us ultra-high data rates, very low latency, more capacity, and good quality of service. It is worth mentioning that 5G technology has opened unleash new opportunities to leapfrog traditional barriers to development. As 5G technology supports IoT also, it gives easy major societal transformation in the fields of education, industry, healthcare, and other social sectors. 5G technology is now unlocked an extensive IoT ecosystem wherein many devices will be connected and by maintaining a trade-off between latency, cost, and speed a network can suffice the communication needs as per requirement. As 5G has come, the major 3GPP standards undergo continual change as it investigates an organized release of new functionality and is responsible for new releases of standards as per the planned schedules. It has stated three different usage scenarios of 5G communications which are as follows, (1) Enhanced Mobile Broadband (eMBB): gives ultra high speed indoor and outdoor connection and supports good and uniform quality of service at the edge of the cell, on highways, in aircraft, and train. Also, it provides high data rates upto 20 Gbps in the indoor area and 2 Gbps in the outdoor area. (2) Massive Machine Type Communications (mMTC): It helps to support IoT to interconnect a very large number of devices in which a single base station can support 10000 or more devices for different applications like smart power grids, smart cities, sensors. (3) Ultra-reliable and Low Latency Communications (uRLLC): It is used for the requirement such as low latency (below 1 ms) and low packet loss (1 in 10000 packets). At some instances it supports remote medical surgery, safety in transportation, and wireless control of manufacturing process. As given in figure 4, 5G technology is used by eight specification requirements namely frequency bands, mobility, data rate, forward error correction, access technology, latency, spectral efficiency, and connection density taking into account the connection reliability (H. C. Huang, 2018), (W. Wang et al. 2017). The use of shorter frequencies (millimeter waves between 30 GHz and 300 GHz) for 5G networks is the reason that why 5G can be faster from previous network. It can operate in both lower bands (e.g., sub 6 GHz) as well as mmWave. 5G is comparatively faster as compared to 4G, giving up to 20 Gbps peak data rates and 100 plus Mbps average data rates. To get low latency, low density parity check codes are used as a forward error correction code in 5G technology. Mobility represents maximum mobile station speed at which a defined quality of service can be achieved and is more than 500 km/h. As per International Telecommunications Union Radio communication Sector, the average spectral efficiency also termed as spectrum efficiency is up to 9 bit/s/Hz. The 5G technology supports beam division multiple access and filters bank multicarrier. The highly directive beams of signals help to achieve space division multiple access which can be termed as BDMA. It can handle many users in 5G systems thereby increasing the system capacity very fast. In FBMC, a bank of filters is being used which provides better spectral efficiency. The low latency feature of 5G technology is driving new ways of using high quality video in real time for user interface. The use of artificial intelligence (AI) along with video analytics is more likely to turn high definition camera streams into actionable information. One of the pivotal parts of the 5G device is an antenna which is required to work at an enhanced gain, bandwidth with lesser radiation losses. So, antenna design for 5G devices becomes very crucial while maintaining the above mentioned parameters for 5G communication as per requirement and use. There are several work on this topic have already been done and still has been continuing to design the best antennas in low cost with high gain and bandwidth. So, in this chapter, we have tried to explore few a5G antennas in a holistic way that were proposed in recent years considering their performance enhancement techniques. The chapter also aims to direct the researcher for further advancement in the 5G antenna design as per their applications and to take it for 6G applications (Sumit Kumar et al. 2020).

Figure 1. Specifications of 5G technology

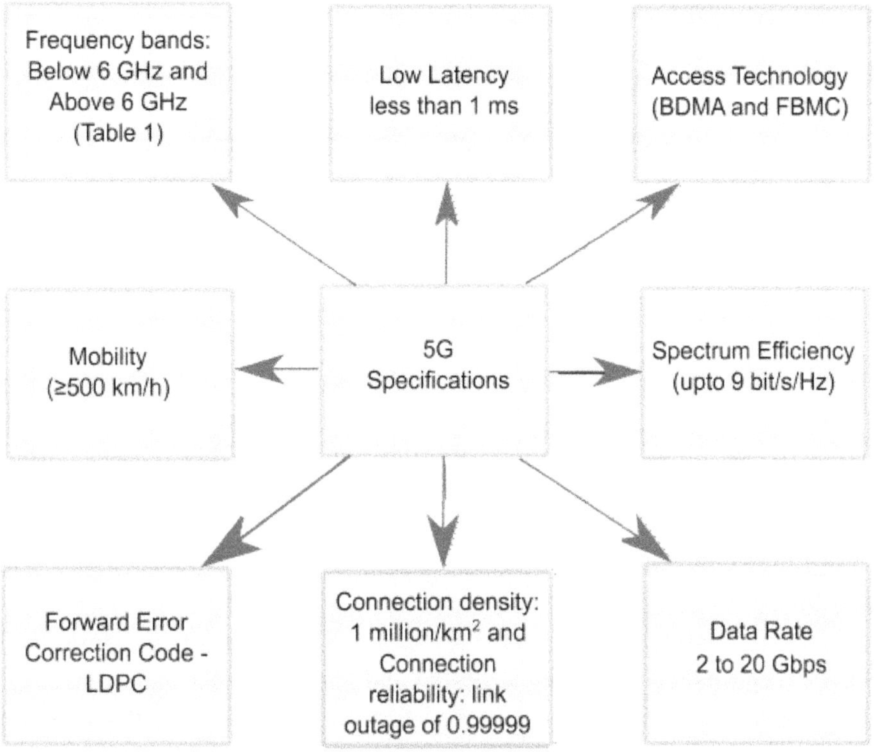

(Sumit Kumar et al. 2020)

Antenna Design for Beyond 5G Technologies

The microstrip patch antenna having with high demand and their compactness as well as ease of fabrication are widely used (Sarthak Singhal, 2019) (Lin Guo, 2014). Many universities has already published many research items related to 6G technologies, its advantages and few designs but that time there was no use and even they were not properly written Samsung research, 2020).Then many research institute in collaboration with industries, are working and has produces few important results in this area (K. B. Letaeif et al. 2019). This new technology for beyond 5G will basically work in terahertz frequency range which is already defined for range 0.1 and 10 THz, and then it is divided into two more categories according to frequency range as given in Fig.1. (B. Duan, 2020), (L. Chi et al. 2020). Here, from 0.1–0.3 THz is known as a sub-TeraHertz region and next part is known as pure terahertz band from later from 0.3–10 THz (M. K. Imara et al. 2020).

Figure 2. Range of frequency in beyond 5G and sixth generations

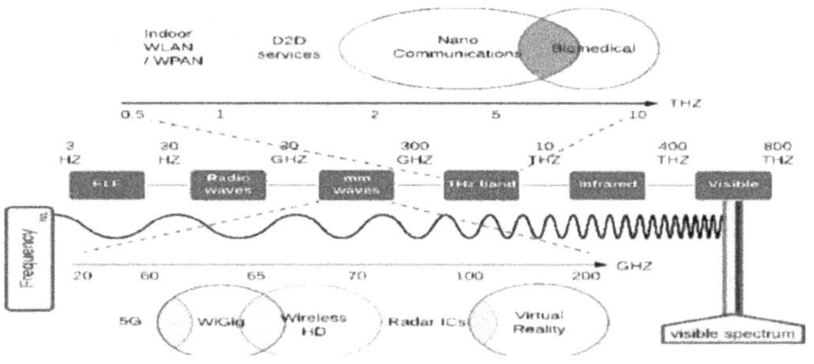

(I. Fitri et al. 2017)

This high frequency band is also known as the terahertz gap just because of lack of materials which could respond to these high frequencies. This frequency band may also be known as Tera waves. As per fig.2 which represents the frequency range in sixth generation wireless technology and system. As previously reported by many researchers that beyond 5G are the smart system giving secure information and it will run using artificial intelligence technology which will connect every smart device with smart intelligence system. In today research area, there are many types antenna which are being used according to design and application for high frequency like reflect array antenna are giving better performance with a new substrate material like polymide at high speed and useful in many applications such as spectroscopy in research and biomedical field. Also, few more design like elliptical pattern antenna which has high gain and good radiation pattern with ease of fabrication and can be used for any distance whether it is short or long. Due to increase demand of changing frequency, polarization and radiation pattern, reconfigurable antennas have been introduced with many smart material like graphene, metamaterials, metasurfaces etc giving MIMO applications with less loss and high gain with enhanced bandwidth at high frequency. Depending on distance covered, fading effect and introduction of solid state devices with these antenna are creating new interest between researchers as dielectric resonator antenna, MIMO antenna with cmos technology are the best suited for getting proper dispersion and less fading effect. Nano technology is also being used to get high bandwidth and less isolation effects (I. Fitri et al. 2017), (M. kokkonen et al. 2020), (Zhuo Wei Miao et al. 2018), (Binbin Cheng et al. 2018).

Hence, to develop future generation beyond 5G, frequency is increased into THz region and for that frequency, fabrication is very complex as it has very small size even in micrometer which is very difficult to make and measure as well. For that purpose, very good materials like graphene, polymide, metasurface, gold etc are required, and also well-equipped instruments are needed to perform complex fabrication and its measurement such as high frequency vector network analyzers. This frequency demands small dimensions as per wavelength and also few more additional components in some cases and costly substrate material and for the radiating elements. In other word, this frequency region will face free space loss and molecular absorption as sky wave and ground wave communication system simultaneously. As per high frequency, few materials like perfectly electric conductors show the high loss and cannot work properly at this frequency. Even some dielectric materials as mentioned above, can work at this high frequency but some improvement and advancement are required so that they can properly work without having ohmic loss and dielectric loss. Even few new technology are evolved such as use of surface Plasmon or SPP which is having less loss and behaving like slow wave structures at required and desired modes as they are also use as wave (Shivani Chandra et al., 2019), (Shivani Chandra et al. 2021), (Smrity Dwivedi, 2022), (Shivani Chandra et al. 2020), (Smrity Dwivedi, 2022), (Smrity Dwivedi, 2023), (L. Dussopt et al. 2003), (P. J. Rainville et al. 1992), (L. R. Tan et al, 2013), (G. H. Huff et al. 2007), (Rakhi Patil et al. 2022), (Sudhir Ingle Uria et al. 2022), (Krishna Dharavathu et al. 2022), (Peng Yang, 2020), (Abd Khadum Ali et al, 2022), (V. P. Eduardo et al. 2022).

Metamaterial Antennas

Antennas are one of the important components now a days that makes wireless communications possible in todays era. The work of antennas is to interface the radio system with the external environment (such as LNA, oscillators, receiver etc.).

Wireless communication systems require antennas at the both sides like transmitter and receiver to operate properly without causing disturbance. The concept of metamaterials has been widely applied in the design of microwave, millimetre-wave and terahertz devices and antennas now a days to make communications and other important task wisely. With the rapid development of flexible portable devices such as mobile phones, laptops, note pads, watches, wearable devices, etc., antennas with different changing functions, based on variable structures, are in demand for next generation of wireless communication systems such as 6G and beyond. The application of metamaterials related components over the past decade has achieved great successes in the fields of both science and engineering. Variable metamaterials have been designed from radio frequencies up to optical frequencies (terahertz and infrared etc.), and different functions have been realized, e.g., negative refractive index (NRI), anisotropy and bianisotropy (Lapine et al. 2007). As an interdisciplinary topic that is being used everywhere, metamaterials can be classified into different categories based on different criteria as given different sections. From an operating frequency point of view, they can be classified as microwave metamaterials, terahertz metamaterials, and photonic metamaterials for antenna, polarizers, absorber etc. From a spatial arrangement point of view, there are 1D metamaterials, 2D metamaterials, and 3D metamaterials present. From a material point of view, there are metallic and dielectric metamaterials that is being utilized. One of the most important applications of metamaterials is antenna design that is being discussed here. Due to the unusual properties of metamaterials, we can achieve antennas with novel characteristics which cannot be realized with traditional materials like normal material with positive refractive index. Usually, antennas consist of a combination of conductors, dielectrics and other conventional materials that have certain geometry during design. That design follows either traditional analytical methods and rules that depends on experience.

In between, artificial intelligence and powerful optimization techniques used by full-wave simulations (Simulation software) are also utilized to obtain the highest possible performance by fine tuning of the structure's parameters (M. O. Akinsolu et al. 2020). It is fact that the antenna design is primarily focused on determining the optimized geometrical shape of conventional materials and then it is bound by the material characteristics for every material. In an effort to overcome all limitation in normal or conventional material, metamaterials with their unique properties have gained focus in antenna design. Irrespective of the fact that metamaterials are microscopically composed of conventional materials such as conductors and dielectrics, their macroscopic characteristics are completely different due to their smart shapes like SRR etc. The realisation of negative constitutive parameters in the microwave regime is applicable with metamaterials, while their subwavelength size (very small dimension) is another useful feature. As a result, the construction of metamaterials into antennas can offer advanced flexibility and enable novel design strategies that is followed. Therefore, the investigation of the benefits of metamaterial inspired antennas and their capability of having an active role in modern wireless communications is of great significance in advance communication.

State of Art (Theoretical background):- As already discussed above about the metamaterial, these are engineered, man made materials that are able of changing the electromagnetic waves in a different way than conventional, natural materials occurring in nature. For instance, structures that exhibit negative permittivity and permeability are most common examples of metamaterials (A. Alu et al. 2007).

Three main metamaterial categories can be identified as per finding: i) SNG (Single-Negative) metamaterials that possess either negative permittivity or negative permeability, ii) DNG (Double-Negative) metamaterials that simultaneously have negative permittivity and permeability both and iii) ZIM (zero-index materials) that have either zero permittivity or zero permeability any one. Also, EBG (Electromagnetic Bandgaps) that prohibits the propagation of EM waves and AMC (Artificial Magnetic Conductors) that have zero magnetic field are also usually regarded as metamaterials. In 1968 as discussed, Veselago was the first one to theoretically study the electrodynamics of DNG media and given their interesting properties such as reversal of the Doppler Effect, negative index of refraction and left-handed propagation. While the conventional materials support only the typical forward-wave propagation. The wave in SNG media is evanescent since the propagation constant is real ($\gamma \in R$). Left-handed or DNG media support backward wave propagation, where the phase velocity and group velocity are opposite to each other and the wave travels anti parallel to the power flux. Some important structures are discussed below using diagram. The demonstration and effects of negative permittivity in the microwave regime is possible with a periodic arrangement of thin metallic wires of diameter α and periodicity p as shown in Figure 3 and was initially proposed by Sir John Pendry and his group.

Figure 3. Periodic arrangement of thin conductive wires for the realisation of negative permittivity

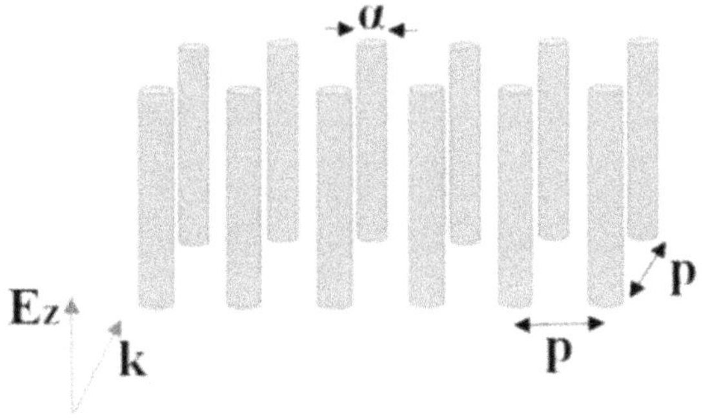

(Milius et al., 2021)

As shown in Figure 4, the real part of the effective permittivity is negative below the plasma frequency and the medium can be characterized as ENG.

Figure 4. Real part of the effective permittivity of periodically arranged infinitely long thin wires

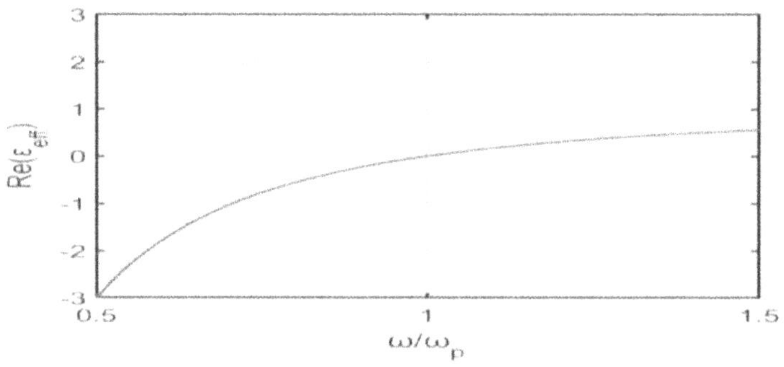

(Milius et al., 2021).

It is the fact that there are no materials in nature with negative permeability. Subsequently, it was Pendry again who proposed a Split Ring Resonator (SRR) as depicted in Figure 4, which shows a negative permeability behaviour around its resonance frequency. Under the presence of an external time-varying magnetic field perpendicular to the SRR's surface, currents are induced on both the conductive inner and outer rings and charges are accumulated across the gaps between them when it is excited.

Figure 5. Geometry of a square split ring resonator unit cell. Blue parts depict metallization

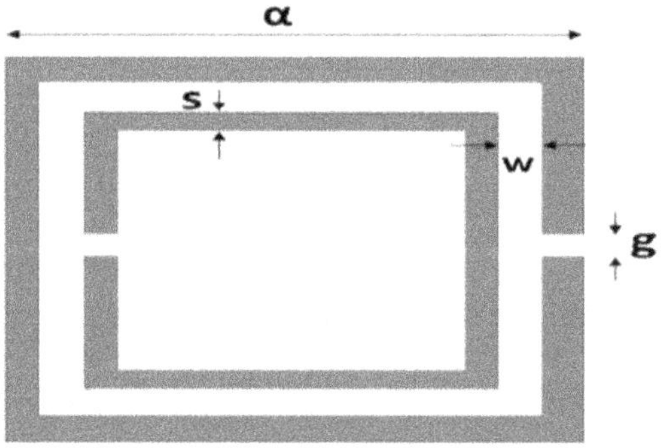

(Milius et al., 2021), (Aneri pandya et al. 2021), (Turkish Upadhyay et al. 2021), (TK Upadhyay et al. 2021).

Table 2. Effect of geometrical properties on SRR

Geometrical Parameters	Effect on SRR
Increasing the split gap of SRR	As the split gap increases, the value of the capacitance decreases leads to a decrease in total capacitance, and a decrease in capacitance value increase the resonant frequency.
Increasing in spacing between two adjacent rings of SRR	The increase in spacing between the rings reduces the mutual inductance and capacitance, respectively. Due to a decrease in inductance and capacitance resonant frequency increase as analytically expressed.
Increasing the side length of SRR	Increase the side length of SRR increases the effective value of inductance and thereby shifts cutoff frequency towards the lower frequency.
Increasing in metal width	The increase in metal width reduces mutual inductance and mutual capacitor and thereby increase the resonant frequency
Multiple split gaps in SRR	Multiple gaps in SRR intercept magnetic resonance because of the induced current by an electric field and shift resonant frequency towards the higher frequency

PRACTICAL DESIGN OF ANTENNA FOR BEYOND 5G SYSTEMS

(**a**) This is author own paper in which I designed and simulated of THz antennas for beyond 5G has been proposed and investigated. This new era of beyond 5G is useful to connect most of the devices and create a proper balance between systems and users as per demand of technology. For that purpose, a large bandwidth is supposed to be required with better gain. Hence, a new frequency region that is at THz band is utilized. Graphene has been taken as base material having capability of operating at high frequency and also to support graphene, polymide and silicon dioxide have been taken as substrate materials along with layering of air and ground as well. Few literatures survey has been discussed with evolution of terahertz Bandwidth is increased from 0.3 THz to 0.6 THz and the gain is increased 3.02 dB to 10.09 dB for the frequency range from 0.1 THz to 1 THz. a technique of gain improvement has been applied with different material layering on the same structure and simulated which is giving the simulated bandwidth greater than previous that is 52.9 GHz and higher simulated gain achieved with 10.9 dB (Smrity Dwivedi, 2023).

Table 3. Comparison table of proposed structures

Proposed structure	Frequency (THz)	Bandwidth (GHz)	Gain (dB)
Basic structure	0.45	30.4	4.21
First structure	0.45	35.6	8.48
Second structure	0.45	37	8.49

Proposed structure	Frequency (THz)	Bandwidth (GHz)	Gain (dB)
Third structure	0.45	32	8.05
Fourth structure	0.45	52.9	10.9

Figure 6. The radiating patch and ground plane are represented by grey color, while substrate is represented by purple color and PEC with white color

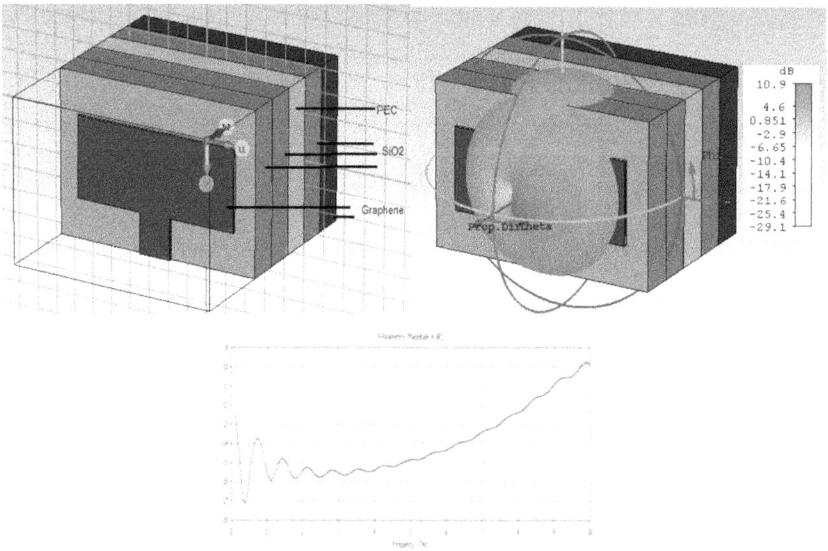

(b) This is author own work in which author has taken inserted H- shaped antenna with partial ground with three major designs and their comparison at terahertz frequency for beyond 5G applications. This new era of beyond 5G is useful to connect most of the devices and create a proper balance between systems and users as per demand of technology. Antenna design is proposed with normal rectangular patch and two more modifications have been done in same patch and it is simulated with the help of CST microwave studio for its important parameters such as bandwidth and gain. For the three structures first design, second design and third design bandwidth is 100 GHz, 250 GHz, 400 GHz with gain 13.5 dBi, 14.4 dBi, 14.8 dBi respectively. A simple PEC material as a radiating element and polyamide as substrate have been taken. Partial ground has been taken with microstrip line feedline (Smrity Dwivedi, 2023), (Utkarsh Gupta et al., 2021),(Smrity Dwivedi, 2023), (Smrity Dwivedi, 2023).

Table 4. Comparison table of proposed structures

Proposed structure	Bandwidth (GHz)	Gain (dBi)
First Design	100	13.5
Second Design	250	14.4
Third Design	400	14.8

Figure 7. Final structure antenna with inserted H- shaped

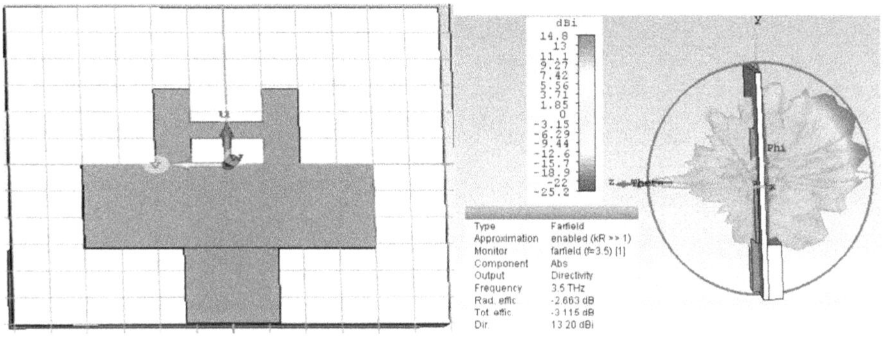

(c) This is author own work in which author has proposed design and simulation of THz antennas for beyond 5G. This new era of beyond 5G is useful to connect most of the devices and create a proper balance between systems and users as per demand of technology. For that purpose, requirement of high bandwidth with maximum gain is one primary constraint for improving performance because of that high frequency regime is used in this paper. Hence, a high frequency has been selected to design the antenna, without using the material like graphene, gold etc. A simple copper and roger have been taken for patch and substrate respectively. Here, array has been adopted which is giving MIMO characteristics which is best suited for high gain extra wide band terahertz frequency. Terahertz MIMO antenna used here, having 2-elements and 3- elements arrays. Antenna structures have been simulated using commercial software. The gain of 3- element arrays is 24.5 dBi and the gain of 2-element arrays is 22.1 dBi. Few literatures survey has been discussed with evolution of terahertz. Bandwidth is increased from 0.2 THz for 2-element arrays and 0.45 THz for 3- element arrays for the frequency range from 2 THz to 6 THz.

Table 5. Designed structures comparison

Proposed Structure	Frequency (THz)	Bandwidth (THz)	Gain (dBi)
2-elements array	2.74	0.2	22.1
3-elements array	5.3502	0.45	24.5

Figure 8. The patch and substrate are mentioned in above figure with front view of 3- elements MIMO antenna

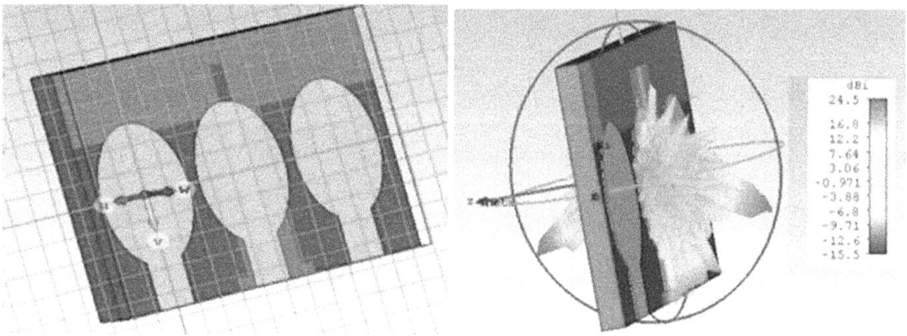

(d) Author has simulated an elliptical leaf pattern six leaves THz antenna in this paper with wide band applications for beyond 5G communication systems. Antenna design is proposed with two elliptical leaves and increased up to six leaves to enhance the performance. Bandwidth and gain of antenna are measured for two leaves, four leaves and six leaves are 300 GHz, 350 GHz, and 500 GHz with 24.9 dBi, 25 dBi and 25.2 dBi gain respectively. Author has got three different gains for three frequencies which is giving three wide bandwidth which is need of beyond 5G applications. Maximum gain 25.2dBi at 500 GHz bandwidth has been obtained for six leaves elements.

Table 6. Proposed structures design comparison

Designed structure	Bandwidth (GHz)	Gain (dBi)
Two leaves pattern	300	24.9
Four leaves pattern	350	25
Six leaves pattern	500	25.2

Figure 9. Design of antenna with six elliptical leaves and radiation pattern

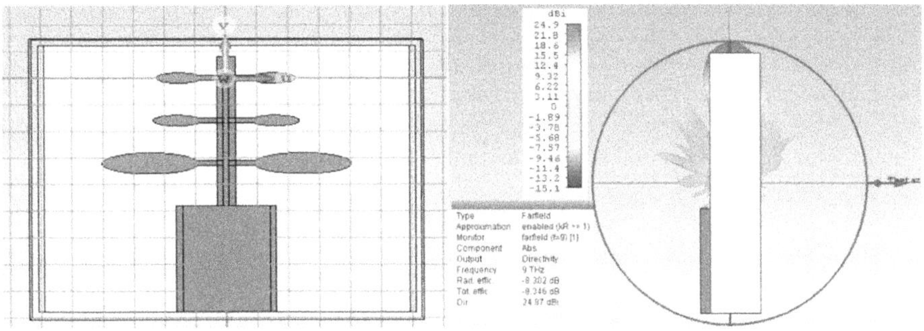

(d) Present paper is especially for the analysis of frequency reconfiguration antenna which removes the problem of achieving multifunctional systems using single antenna at different frequency according to user requirements with enhance performance in low volume. At high frequency from millimetre to terahertz range, data rate and efficient systems are the most important part at less transmission loss to get reliable wireless communication. In order to use an antenna at different frequencies for a particular frequency band, a Frequency Reconfigurable Antenna design is proposed using dual patch of gold material structure. It uses SiO_2 whose dielectric constant 3.6 (60×60 μm^2) as a substrate with a thickness of 1.575 μm. It uses a coaxial feeding with SMA connector and a metal surface at the back of substrate of the same size which act as a reflector. This reflector is used to increase the gain and other characteristics of the antenna by reducing the surface radiation. A varactor diode is used between two patches as a switch for connecting both the patches. Structure resonates at three different frequencies which is 5.6 THz, 6.2 THz, and 14.6 THz with a maximum gain of 6.44 dBi, 7.75 dBi, and 8.54 dBi respectively.

Figure 10. Reconfigurable antenna with switch, S11 parameter, radiation pattern

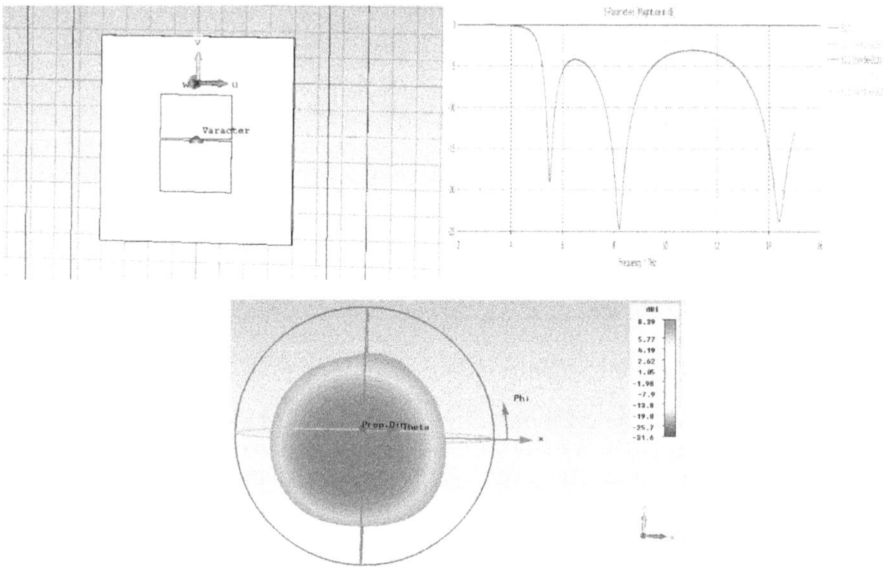

(f) Here, author has used graphene to design the frequency reconfigurable antenna as graphene has unique properties to work with high frequency. This unique tunable property of graphene can be used to design a frequency reconfigurable antenna. When a bias voltage is applied to the designed antenna array as shown in fig. 8, the chemical potential of graphene is varied from 0.4 eV to 0.8 eV. With the increase of chemical potential and change of surface conductivity the dramatical change in antenna properties are observed. Increased chemical potential gives the increase in return loss and shifts of antenna resonance frequency to the higher frequencies. To analyze the property of frequency reconfigurable antenna, the proposed structure is simulated in CST software and the result is shown in figure 16. Chemical potential of all the four patch elements of 1×4 antenna array are changed simultaneously from 0.4eV to 0.8eV. Obtained simulated result is summarized in table 1. Shift in frequency is observed from 0.64 THz to 0.74 THz with change of chemical potential from 0.4eV to 0.8eV. It is clear that the achieved frequency band which is covered by the proposed frequency reconfigurable antenna is 0.64-0.74 THz which gives the bandwidth of 100 GHz which is high enough. Hence, with such reconfigurable antennas we can easily remove the drawback of narrow band antennas.

Figure 11. Orthographic view of 1×4 graphene antenna array

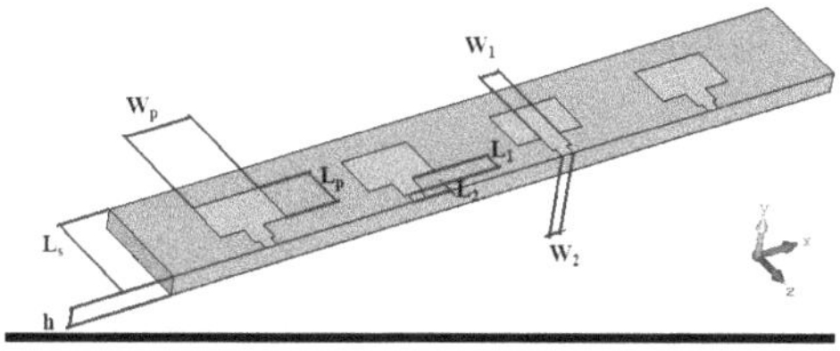

Figure 12. Effect of variation of graphene chemical potential on resonating frequency

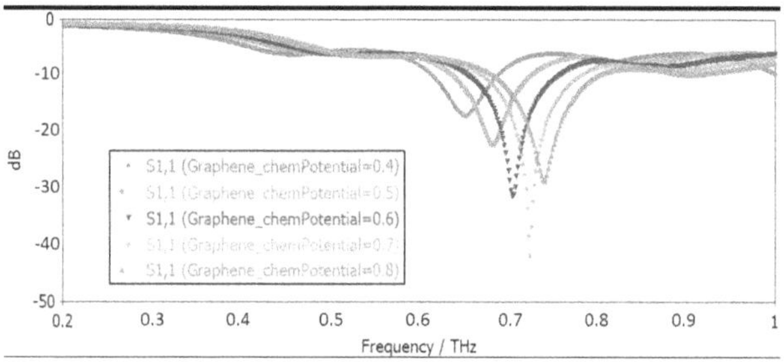

Table 7. Performance parameters of frequency reconfigurable antenna

Graphene Chemical Potential (eV)	Frequency (THz)
0.4	0.64
0.5	0.69
0.6	0.70
0.7	0.71
0.8	0.74

(g) In this paper, author has discussed and used metamaterial surface as a reflector due to its in-phase reflection characteristic and high-impedance nature to improve the gain of an antenna. The radiating element of the proposed metamaterial based antenna is made up of copper material which is backed by the dielectric as shown in figure given below, i.e., Rogers-4003 with a standard thickness, loss tangent and a relative permittivity of 1.524 mm, 0.0027 and 3.55, correspondingly. The proposed antenna with and without metamaterial surface operates at the central frequency of 3.32 GHz and 3.60 GHz, correspondingly. The traditional antenna yields a boresight gain of 2.76 dB which is further improved to 6.26 dB, using the metamaterial surface (Fazal Muhammad et. al. 2020). So, this antenna is well used for 5G applications and to use it for beyond 5G, few modifications will be done.

Figure 13. Geometry of antenna: (a) Front view without metamaterial, (b) back view, (c) 4X4 array design using metamaterial

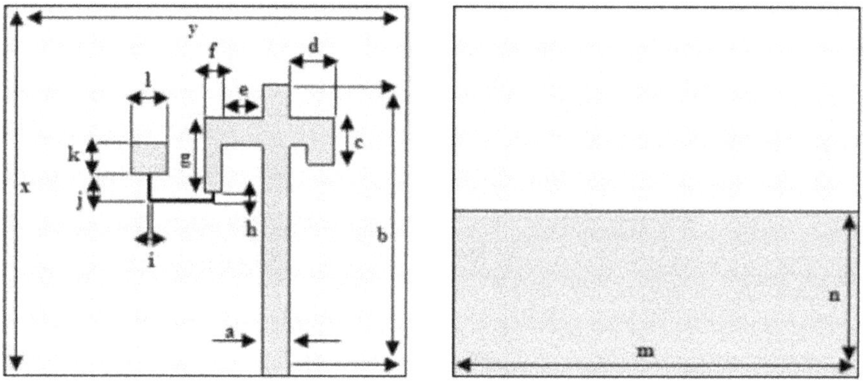

(a) (b)

(i) In this paper, author has used low dielectric constant of 2.2, which is of Rogers RT/Duroid substrate, and a dielectric loss tangent of 0.0010 in the proposed design as given in figure below. Author has developed a metamaterial-based multiband multiband microstrip rectangular shape patch antenna with a wide high-performance bandwidth because of the demand. The design has a resonant frequency of 26 GHz. The simulations have been done using FEKO software to analyzed its performance. The simulation and analysis reveal a good return loss of -34.4 dB at 26 GHz, -13.49 dB at 40 GHz, -13.63 dB at 53.5 GHz, high bandwidth of 5.368 GHz at 26 GHz, 3.76 GHz at 40 GHz, 2.88 GHz at 53.5 GHz, desirable voltage standing wave ratio, $1 \leqslant VSWR \leqslant 2$, high gain of 10 dBi at 26 GHz, 5 dBi at 40 GHz, and high antenna radiation efficiency of 99.7% at 26 GHz, and 61% at 40 GHz, 50% at 53.5 GHz. The bandwidth, return loss, antenna radiation efficiency and power density indicate an improvement of 5.368 GHz to 5.63 GHz, -34.82 dB to -57.10 dB, 99.7% to 99.8% and 2208 kW/m2 to 2800 kW/m2 respectively after loading and incorporating artificial magnetic split-ring resonator-based metamaterial on the patch (Calico et al. 2021).

Figure 14. (a) 3D view of split-ring resonator, (b) unit cell of proposed split-ring resonator

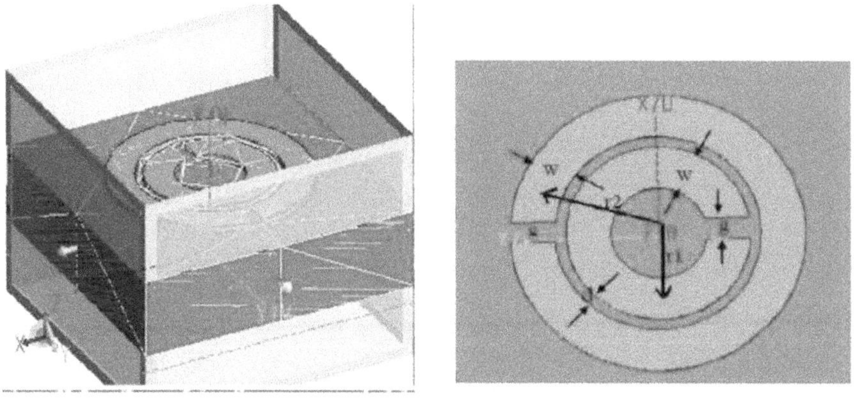

(a) (b)

CONCLUSION

THz bands (0.1-10 THz) provide extreme-wide bandwidth allowing an ultra-high-speed communication for many future 6G applications. Many challenges on the THz band needs to be overcome for the next 6G wireless communication system. In this paper, a detailed review of the most significant THz band antenna challenge is presented. We have highlighted the antenna specifications in 6G technology for ultra-high-speed THz wireless communication applications. The antenna design challenges to meet the 6G antenna specifications are clarified. Rigorous literature survey of beyond 5G antenna has been reviewed and presented completely. There are so many different types of antenna design for beyond 5G with their results have been mentioned to create proper understanding of antenna for high frequency. Each and every work has different objective, so there have different indication of frequency selection, antenna types, material, fabrication, and improvement technique. Here, in section 3, author has presented own work in which high frequency from GHz to THz has been covered. Several designs including reconfigurable antenna, MIMO antenna, different pattern antenna, array antenna with different materials like gold, graphene, metamaterial, SiO2 etc. have been taken. Also, different frequency range from 5GHz to 10 THz has been selected for the design. Maximum gain approximately 24.5 dBi has been found out from the design.

REFERENCES

Ahmad I, Ullah S, Ullah S, Habib U, Ahmad S, Ghaffar A, Alibakhshikenari M, Khan S, Limiti E (2021) Design and analysis of a photonic crystal based planar antenna for THz applications. *Electronics 10*(16), 1941.   ronic s1016 1941.10. 3390/ elect

Ali Kadhum, A. (2022). *Design and Analysis of Novel Reconfigurable Monopole Antenna Using Dip Switch and Covering 5G-Sub-6-GHz and C-Band Applications*. MDPI.

Alu, A., Engheta, N., Erentok, A., & Ziolkowski, R. W. (2007, February). Singlenegative, double-negative, and low-index metamaterials and their electromagnetic applications. *IEEE Antennas & Propagation Magazine*, 49(1), 23–36. 10.1109/MAP.2007.370979

Ananda, S., Sriram Kumara, D., Wub, R. J., & Chavali, M. (2014). Graphene nanoribbon-based terahertz antenna on polyimide substrate. *Optik (Stuttgart)*, 125(19), 5546–5549. 10.1016/j.ijleo.2014.06.085

Azarbar, A., Masouleh, M. S., & Behbahani, A. K. (2014). A new terahertz microstrip rectangular patch array antenna. *Int J Electromagn Appl*, 4, 25–29.

Babu, K. V. (2023). *Design and analysis of fractal-based THz antenna with co-axial feeding technique for wireless applications.Recent advances in graphene nanophotonics*. Springer Nature Switzerland.

Babu, K. V., Das, S., Sree, G. N. J., Madhav, B. T. P., Patel, S. K. K., & Parmar, J. (2022). Design and optimization of micro-sized wideband fractal MIMO antenna based on characteristic analysis of graphene for terahertz applications. *Optical and Quantum Electronics*, 54(5), 281. 10.1007/s11082-022-03671-2

Benlakehal, M. (2022). Design and analysis of novel microstrip patch antenna array based on photonic crystal in THz. *Opt Quantum Electron, 54*(5), 1-16.2

Benlakehal, M. E., Hocini, A., & Khedrouche, D. (2022). Design and analysis of a 1 × 2 microstrip patch antenna array based on photonic crystals with a graphene load in THz. *Journal of Optics*. 10.1007/s12596-022-01006-8

Benlakehal, M. E., Hocini, A., Khedrouche, D., Temmar, M. N., & Denidni, T. A. (2022). Design and analysis of a 1 × 2 microstrip patch antenna array based on periodic and aperiodic photonic crystals in terahertz. *Optical and Quantum Electronics*, 54(10), 672. 10.1007/s11082-022-04076-x

Benlakehal, M. E., Hocini, A., Khedrouche, D., Temmar, M. N., & Denidni, T. A. (2022). Design and analysis of MIMO system for THz communication using terahertz patch antenna array based on photonic crystals with graphene. *Optical and Quantum Electronics*, 54(11), 693. 10.1007/s11082-022-04081-0

Benlakehal, M. E., Hocini, A., Khedrouche, D., Temmar, M. N. E., & Denidni, T. A. (2022). Design and analysis of a 2 × 2 microstrip ratch antenna array based on periodic and non-periodic photonic crystals substrate in THz. *Optical and Quantum Electronics*, 54(3), 190. 10.1007/s11082-022-03563-5

Cheng, B., Cui, Z., Lu, B., Qin, Y., Liu, Q., Chen, P., He, Y., Jiang, J., He, X., Deng, X., Zhang, J., & Zhu, L. (2018). 340-GHz 3-D imaging radar with 4Tx-16Rx MIMO array. *IEEE Transactions on Terahertz Science and Technology*, 8(5), 509–519. 10.1109/TTHZ.2018.2853551

Chi, L., Weng, Z., Qi, Y., & Drewniak, J. L. (2020). A 60 GHz PCB Wideband Antenna-in-Package for 5G/6G Applications. *IEEE Antennas and Wireless Propagation Letters*, 1225(c), 1–1. 10.1109/LAWP.2020.3006873

Das, P., Singh, A. K., & Mandal, K. (2022). Metamaterial loaded highly isolated tunable polarisation diversity MIMO antennas for THz applications. *Optical and Quantum Electronics*, 54(4), 250. 10.1007/s11082-022-03641-8

Das, S., Anveshkumar, N., Dutta, J., & Biswas, A. (Eds.). (2021). *Advances in terahertz technology and its applications*. Springer. 10.1007/978-981-16-5731-3

Devapriya, A. T., & Robinson, S. Investigation on metamaterial antenna for terahertz applications. J. Microw.Optoelectron. Electromagn. Appl. (2019). 10. 1590/ 2179- 10742 019v1 8i315 77

Dussopt, L., & Rebeiz, G. M. (2003, April). Intermodulation distortion and power handling in RF MEMS switches, varactors, and tunable filters. *IEEE Transactions on Antennas and Propagation*, 51(4), 1247–1256.

Dwivedi, S. (2022). *ReconfigurableArray antenna*. Intech Open., 10.5772/intechopen.106343

Dwivedi, S. (2023). *Metamaterials-Based Antenna for 5G and Beyond*. IGI Global. 10.4018/978-1-6684-8287-2.ch001

Eduardo, V. P. (2022). FORMAT: A Reconfigurable Tile-Based Antenna Array System for 5G and 6G Millimeter-Wave Testbeds. *IEEE Systems Journal*, 16(3), 4489–4500. 10.1109/JSYST.2022.3146360

El-Kenawy, E. S. M., Ibrahim, A., Mirjalili, S., Zhang, Y. D., Elnazer, S., & Zaki, R. M. (2022). Optimized ensemblealgorithm for predicting metamaterial antenna parameters. *Computers, Materials & Continua*, 2022, 023884. 10. 32604/ cmc

Emara, M. K., Stuhec-Leonard, S. K., Tomura, T., Hirokawa, J., & Gupta, S. (2020). Laser-Drilled All- Dielectric Huygens' Transmit-Arrays as 120 GHz Band Beamformers. *IEEE Access : Practical Innovations, Open Solutions*, 8, 153815–153825. 10.1109/ACCESS.2020.3018297

Fitri, I., & Akbart, A. A. (2018). A new gridded parasitic patch stacked microstrip antenna for enhanced wide bandwidth in 60 GHz Band. *2017 International Conference on Broadband Communication, Wireless Sensors and Powering, BCWSP 2017*. IEEE. 10.1109/BCWSP.2017.8272571

Goyal, R., & Vishwakarma, D. K. (2018). Design of a graphene-based patch antenna on glass substrate for high-speed terahertz communications. *Microwave and Optical Technology Letters*, 60(7), 1594–1600. 10.1002/mop.31216

Guo, L., Huang, F., & Tang, X. (2014). A novel integrated MEMS helix antenna for terahertz applications. *Optik (Stuttgart)*, 125(1), 101–103. 10.1016/j.ijleo.2013.06.016

Gupta, U., & Dwivedi, S. (2021). Frequency Reconfigurable Antenna Using Metamaterial Split Ring Resonators for Smart Applications", *2021 International Conference on Electrical, Communication, and Computer Engineering (ICECCE)*. IEEE. 10.1109/ICECCE52056.2021.9514067

Hocini, A., Temmar, M. N., Khedrouche, D., & Zamani, M. (2019). Novel approach for the design and analysis of a terahertz microstrip antenna based on photonic crystals. *Photonics and Nanostructures*, 36, 100723. 10.1016/j.photonics.2019.100723

Holloway, D., Dienstfrey, A., Kuester, E. F., O'Hara, J. F., Azad, A. K., & Taylor, A. J. (2009). A Discussion on the Interpretation and Characterization of Metafilms/Metasurfaces: The Two Dimensional Equivalent of Metamaterials. *Metamaterials (Amsterdam)*, 3(2), 100–112. 10.1016/j.metmat.2009.08.001

Holloway, K., Kuester, E. F., Baker-Jarvis, J., & Kabos, P. (2003). A Double Negative (DNG) Composite Medium Composed of Magneto-Dielectric Spherical Particles Embedded in a Matrix. *IEEE Transactions on Antennas and Propagation*, 51(10), 2596–2603. 10.1109/TAP.2003.817563

Hong, W., Jiang, Z. H., Yu, C., Zhou, J., Chen, P., Yu, Z., Zhang, H., Yang, B., Pang, X., Jiang, M., Cheng, Y., Al-Nuaimi, M. K. T., Zhang, Y., Chen, J., & He, S. (2017, December). Multibeam antenna technologies for 5G wireless communications. *IEEE Transactions on Antennas and Propagation*, 65(12), 6231–6249. 10.1109/TAP.2017.2712819

Huff, G. H., Bahukudumbi, P. B., Everett, W. N., Beskok, A., Bevan, M. A., Lagoudas, D., & Ounaies, Z. (2007). Microfluidic reconfiguration of antennas. *Proc. Antenna Appl. Symp.*, (pp. 241-258). IEEE.

Keerthi, R.S., Dhabliya, D., Elangovan, P., Borodin, K., Parmar, J., & Patel, S.K. (2021). Tunable high-gain and multiband microstrip antenna based on liquid/copper split-ring resonator superstrates for C/X band communication. *Phys. B Condens. Matter*. . physb. 2021. 413203.10. 1016/j

Khaleel, S.A., Hamad, E.K.I., Parchin, N.O., & Saleh, M.B. (2022). MTM-inspired graphene-based THz MIMO antenna configurations using characteristic mode analysis for 6G/IoT applications. *Electronics 11*(14), 2152. ronic s1114 2152.10. 3390/ elect

Khezzar, D., Khedrouche, D., & Denidni, T. A. (2021). New design of a broadband PBG-based antenna for THz band applications. *Photonics and Nanostructures*, 46, 100947. 10.1016/j.photonics.2021.100947

Koutsoupidou, M., Karanasiou, I. S., & Uzunoglu, N. (2014). Substrate constructed by an array of split ring resonators for a THz planar antenna. *Journal of Computational Electronics*, 13(3), 593–598. 10.1007/s10825-014-0575-y

Krishna, ChM., Das, S., Lakrit, S., Lavadiya, S., Madhav, B. T. P., & Sorathiya, V. (2021). Design and analysis of a super wideband (0.09–30.14 THz) graphene-based log periodic dipole array antenna for terahertz applications. *Optik (Stuttgart)*, 247, 167991. 10.1016/j.ijleo.2021.167991

Krishna, C. M., Das, S., Nella, A., Lakrit, S., & Madhav, B. T. P. (2021). A micro-sized rhombus shaped THz antenna for high-speed short-range wireless communication applications. *Plasmonics*, 16(6), 2167–2177. 10.1007/s11468-021-01472-z

Krishna, D., Maheshwari, T., Prakash, K. B., Saheb, S. G., & Muhhidin, S. (2022). *Design a Frequency Reconfigurable Antenna for 5G\6G Applications. IJIRSE, 8*(6).

Kumari, R., Tomar, V. K., & Sharma, A. (2022). Miniaturization and performance enhancement of super wide band four element MIMO antenna using DNG metamaterial for THz applications. *Optical and Quantum Electronics*, 54(9), 577. 10.1007/s11082-022-04011-0

Kushwaha, R. K., & Karuppanan, P. (2021). Investigation and design of microstrip patch antenna employed on PCs substrates in THz regime. *Aust J Electr Electron Eng*, 18(2), 118–125. 10.1080/1448837X.2021.1936779

Kushwaha, R. K., Karuppanan, P., & Malviya, L. D. (2018). Design and analysis of novel microstrip patch antenna on photonic crystal in THz. *Physica B, Condensed Matter*, 545, 107–112. 10.1016/j.physb.2018.05.045

Kushwaha, R. K., Karuppanan, P., & Srivastava, Y. (2018). Proximity feed multiband patch antenna array with SRR and PBG for THz applications. *Optik (Stuttgart)*, 175, 78–86. 10.1016/j.ijleo.2018.08.139

Lapine, M., & Tretyakov, S. (2007). Contemporary notes on metamaterials. *IET Microwaves, Antennas & Propagation*, 1(1), 3–11. 10.1049/iet-map:20050307

Letaief, K. B., Chen, W., Shi, Y., Zhang, J., & Zhang, Y. J. A. (2019, August). The Roadmap to 6G: AI Empowered Wireless Networks. *IEEE Communications Magazine*, 57(8), 84–90. 10.1109/MCOM.2019.1900271

Machine learning-assisted antenna design optimization: A review and the stateof-the-art. *Proc. 14th Eur. Conf. Antennas Propag. (EuCAP),* (pp. 1–5). IEEE. .10.23919/EuCAP48036.2020.9135936

Mahmud, R.H. (2020). Terahertz microstrip patch antennas for the surveillance applications. *Kurd J Appl Res (KJAR), 5*, 17–27. ce. 2020.1.2.10. 24017/ scien

Miao, Z.-W., Hao, Z.-C., Wang, Y., Jin, B.-B., Wu, J.-B., & Hong, W. (2018). A 400-GHz High-Gain Quartz-Based Single Layered Folded Reflectarray Antenna for Terahertz Applications. *IEEE Transactions on Terahertz Science and Technology*, 9(1), 78–88. 10.1109/TTHZ.2018.2883215

Milius, C., Andersen, R. B., Lazaridis, P. I., Zaharis, Z. D., Muhammad, B., & Jes, T. B. (2021). Metamaterial-Inspired Antennas: A Review of the State of the Art an Future Design Challenges. *IEEE Access : Practical Innovations, Open Solutions*, 9, 89846–89865. 10.1109/ACCESS.2021.3091479

Najumunnisa, M., Sastry, A. S. C., Madhav, B. T. P., Das, S., Hussain, N., Ali, S. S., & Aslam, M. (2022). A metamaterial inspired AMC backed dual band antenna for ISM and RFID applications. *Sensors (Basel)*, 22(20), 8065. 10.3390/s2220806536298414

Nejati, A., Sadeghzadeh, R. A., & Geran, F. (2014). Effect of photonic crystal and frequency selective surface implementation on gain enhancement in the microstrip patch antenna at terahertz frequency. *Physica B, Condensed Matter*, 449, 113–120. 10.1016/j.physb.2014.05.014

Okan, T. (2021). High efficiency unslotted ultra-wideband microstrip antenna for sub-terahertz short range wireless communication systems. *Optik (Stuttgart)*, 242, 166859. 10.1016/j.ijleo.2021.166859

Pandya, A., Upadhyay, T. K., & Pandya, K. (2021). Design of metamaterial based multilayer antenna for Navigation/Wifi/ Satellite applications. *Progress in Electromagnetics Research M. Pier M*, 99, 103–113. 10.2528/PIERM20100105

Patel, S. K., & Kosta, Y. P. (2012). Meandered multiband metamaterial square microstrip patch antenna design. Waves Rand. *Complex Media*, 2012, 723837. 10. 1080/ 17455 030

Rainville, P. J., & Harackewiez, F. J. (1992, December). Magnetic tuning of a microstrip antenna fabricated on a ferrite film. *IEEE Microw. Guided Wave*, 2(12), 483–485.

Roslan, N. H., Awang, A. H., & Hizan, H. M. (2008). *The effect of photonic crystal parameters on the terahertz photonic crystal cavities microstrip antenna performances*. In: *Proceedings of the 2018 IEEE International RF and Microwave Conference (RFM)*, Penang, Malaysia.

Shalini, M., & Madhan, M. (2019). Design and analysis of a dualpolarized graphene based microstrip patch antenna for terahertz applications. *Optik (Stuttgart)*, 194, 163050. 10.1016/j.ijleo.2019.163050

Shamim, S. M., Das, S., Hossain, M. A., & Madhav, B. T. P. (2021). Investigations on Graphene-Based Ultra-Wideband (UWB) microstrip patch antennas for Terahertz (THz) Applications. *Plasmonics*, 16(5), 1623–1631. 10.1007/s11468-021-01423-8

Shamim, S. M., Uddin, M. S., Hasan, M., & Samad, M. (2021). Design and implementation of miniaturized wideband microstrip patch antenna for high-speed terahertz applications. *Journal of Computational Electronics*, 20(1), 604–610. 10.1007/s10825-020-01587-2

Silin, R. A., & Chepurnykh, I. P. (2001). On Media with Negative Dispersion. *Commun. Technol. Electron.*, 46, 1121–1125.

Singh, A., & Singh, S. (2015). A trapezoidal microstrip patch antenna on photonic crystal substrate for high-speed THz applications. *Photonics and Nanostructures*, 14, 52–62. 10.1016/j.photonics.2015.01.003

Singh, M., Singh, S., & Islam, M. T. (2021). Highly efficient ultra-wideband MIMO patch antenna array for short range THz applications, emerging trends in terahertz engineering and system technologies devices, materials, imaging, data acquisition and processing. Springer. 10.1007/978-981-15-9766-4

Singhal, S. (2019). Elliptical ring terahertz fractal antenna. *Optik (Stuttgart)*, 194, 163129. 10.1016/j.ijleo.2019.163129

Sivukhin, D. V. (1957). The Energy of Electromagnetic Fields in Dispersive Media. *Opt. Spektrosk.*, 3, 308–312.

Smith, D. R., Padilla, W. J., Vier, D. C., Nemat-Nasser, S. C., & Schultz, S. (2000). Composite Medium with Simultaneously Negative Permeability and Permittivity. *Physical Review Letters*, 84(18), 4184–4186. 10.1103/PhysRevLett.84.418410990641

Sumit Kumar, A. S. (2020). Fifth Generation Antennas: A Comprehensive Review of Design and Performance Enhancement Techniques. *IEEE Access, Digital Object Identifier.* IEEE. .10.1109/ACCESS.2020.3020952

Tan, L. R., Wu, R. X., Wang, C. Y., & Poo, Y. (2013, March). Magnetically Tunable Ferrite Loaded SIW Antenna. *IEEE Antennas and Wireless Propagation Letters*, 12, 273–275. 10.1109/LAWP.2013.2248113

Temmar, M. N., Hocini, A., Khedrouche, D., & Denidni, T. A. (2020). Enhanced flexible terahertz microstrip antenna based on modified silicon-air photonic crystal. *Optik (Stuttgart)*, 217, 164897. 10.1016/j.ijleo.2020.164897

Temmar, M. N. E., Hocini, A., Khedrouche, D., & Denidni, T. A. (2021). Analysis and design of MIMO indoor communication system using terahertz patch antenna based on photonic crystal with graphene. *Photonics and Nanostructures*, 43, 100867. 10.1016/j.photonics.2020.100867

Temmar, M. N. E., Hocini, A., Khedrouche, D., & Zamani, M. (2019). Analysis and design of a terahertz microstrip antenna based on a synthesized photonic bandgap substrate using BPSO. *Journal of Computational Electronics*, 18(1), 231–240. 10.1007/s10825-019-01301-x

Tripathi, S. K., & Kumar, A. (2019). High gain miniaturised photonic band gap terahertz antenna for size-limited applications. *Aust J Electr Electron Eng*, 16(2), 74–80. 10.1080/1448837X.2019.1602944

Upadhyaya, T. K., Kosta, S. P., Jyoti, R., & Palandoken, M. (2014, October 07). Negative refractive index material-inspired 90-deg electrically tilted ultra wideband resonator. *Optical Engineering (Redondo Beach, Calif.)*, 53(10), 107104–107104. 10.1117/1.OE.53.10.107104

Urja Sudhir Ingle. (2022). *Reconfigurable Antenna for 5G Applicationsx*. ACCAI.

Vasu Babu, K., Das, S., Varshney, G., Sree, G. N. J., & Madhav, B. T. P. (2022). A micro-scaled graphene-based treeshaped wideband printed MIMO antenna for terahertz applications. *Journal of Computational Electronics*, 21(1), 289–303. 10.1007/s10825-021-01831-3

Vasu Babu, K., Sree, G. N. J., Kumari, S. V., & Das, S. (2022). Design and analysis of a CPW-fed fractal MIMO THz antenna using an array of parasitic elements. In Das, S., Nella, A., & Patel, S. K. (Eds.), *Terahertz devices, circuits and systems*. Springer. 10.1007/978-981-19-4105-4_4

Veselago, V. G. (1968). The Electrodynamics of Substances with Simultaneously Negative Values of ε and μ. *Soviet Physics - Uspekhi*, 10(4), 509–514. 10.1070/PU1968v010n04ABEH003699

Wang, C., Chen, Y., & Yang, S. (2018, November). Dual-band dual-polarized antenna array with flat-top and sharp cutoff radiation patterns for 2G/3G/LTE cellular bands. *IEEE Transactions on Antennas and Propagation*, 66(11), 5907–5917. 10.1109/TAP.2018.2866596

Wen, L.-H., Gao, S., Mao, C.-X., Luo, Q., Hu, W., Yin, Y., & Yang, X. (2018). A wideband dual-polarized antenna using shorted dipoles. *IEEE Access : Practical Innovations, Open Solutions*, 6, 39725–39733. 10.1109/ACCESS.2018.2855425

Yang, D., Liu, S., & Zhao, Z. (2017, May). A broadband dual-polarized printed dipole antenna with low cross-polarization and high isolation for base station applications. *Microwave and Optical Technology Letters*, 59(5), 1107–1111. 10.1002/mop.30467

Yang, P. (2020). *Reconfigurable 3-D Slot Antenna Design for 4G and Sub-6G Smartphones with Metallic Casing*. MDPI. 10.3390/electronics9020216

Younssi, M., Jaoujal, A., Yaccoub, M. D., El Moussaoui, A., & Aknin, N. (2013). Study of a microstrip antenna with and without superstrate for terahertz frequency. *International Journal of Innovation and Applied Studies*, 2(4), 369–371.

Chapter 5
Metamaterial and Metasurface for Wireless Power Transfer and Magneto-Inductive Waveguide

Thanh Son Pham
http://orcid.org/0000-0002-3608-5929
Vietnam Academy of Science and Technology, Vietnam

Xuan Thanh Pham
http://orcid.org/0000-0001-9979-3220
Hanoi University of Industry, Vietnam

Manh Kha Hoang
http://orcid.org/0000-0002-0938-5969
Hanoi University of Industry, Vietnam

ABSTRACT

Metamaterials, specifically in their two-dimensional form known as metasurfaces, have emerged as a focal point of extensive investigation within contemporary scientific discourse. This chapter serves to elucidate the intricacies of the theoretical underpinnings and the practical implementation of a metasurface architecture engineered to function within the frequency domain of megahertz (MHz). The metasurface under consideration exhibits considerable promise in the realm of wireless power transfer (WPT) systems, showcasing its potential to develop this domain. Furthermore, its intrinsic ability to propagate magneto-inductive waves (MIWs)

DOI: 10.4018/979-8-3693-2599-5.ch005

imbues it with a multifaceted utility, enabling its application in diverse scenarios such as near-field information transmission and the development of structures that facilitate the simultaneous conveyance of both energy and information.

INTRODUCTION

Artificially engineered composites known as metamaterials exhibit unique electromagnetic properties that surpass those of conventional materials (Wen et al., 2022a; Zheng et al., 2024). These properties include a negative refractive index, amplification of evanescent waves, and the ability for backward wave propagation (Pham et al., 2019). The diverse range of distinctive characteristics found in metamaterials enables their utilization in various applications such as energy harvesting, perfect absorption, and the enhancement of antenna efficiency (Fowler et al., 2022; Li et al., 2023; B.-X. Wang et al., 2023; Wen et al., 2022b). In recent years, the exceptional benefits of metamaterials have been further harnessed in the development of meta-devices in both three-dimensional and planar configurations. The most recent advancements in meta-devices encompass a broad spectrum of applications including infrared plasmonic photodetectors, ultraviolet beam sources, functional modulators, and highly sensitive meta-sensors (Garg & Jain, 2023; Zheludev & Kivshar, 2012).

The concept of metamaterials was initially introduced by Victor Veselago in 1968 through a theoretical examination of materials exhibiting simultaneous negative permittivity (ε) and permeability (μ) (Veselago, 1968). These materials can be categorized into four groups based on the signs of their ε and μ values: double-positive, epsilon-negative, double-negative, and mu-negative materials. Mu-negative materials, also referred to as magnetic metamaterials (MMs), focus solely on the magnetic component, operating primarily at low MHz frequencies(Son et al., 2022). While most metamaterial structures designed for GHz and THz frequency ranges exhibit both electric and magnetic responses, MM stands out for its emphasis on magnetic fields. Leveraging its negative permeability properties, MM has the capability to amplify evanescent waves in the near-field region. This unique characteristic enables MM to enhance the performance of wireless power transfer (WPT) systems and facilitate near-field data transmission, showcasing its potential in improving efficiency across these applications (Pham et al., 2019; B.-X. Wang et al., 2023).

Following the groundbreaking contributions of Tesla in the early 20th century (Tesla, 1914), a resurgence of interest in WPT was sparked by a cohort of researchers at the Massachusetts Institute of Technology (MIT) in 2007 (Kurs et al., 2007). Over the course of a decade of dedicated experimentation and exploration, WPT has emerged as a technology with diverse applications spanning from implantable medical devices to electric vehicles, capable of delivering power outputs ranging from microwatts to several hundred watts. Moreover, researchers have delved into the integration of metamaterials into the realm of WPT, particularly focusing on evanescent wave amplification (Lu et al., 2020). This innovative approach has yielded promising results, showcasing a notable enhancement in the efficiency of WPT over extended distances. By incorporating metamaterials into the design and implementation of WPT systems, researchers have paved the way for significant advancements in the field, offering new possibilities for enhancing power transfer efficiency and expanding the scope of applications for this modern technology (Adepoju et al., 2022; Hui et al., 2014).

Currently, many WPT systems utilize metamaterial or metasurface to enhance performance. These systems can employ a two-dimensional (2-D) or three-dimensional (3-D) metamaterial structure to enhance the evanescent field, thereby increasing the transmission efficiency of the system. Some studies use a cavity configuration to focus the magnetic field into a desired area, reducing leakage magnetic fields and optimizing the transmission performance of the system. Configurations using metasurfaces are also used to transmit energy and information in a planar configuration. Magneto-inductive waves (MIWs) resulting from the propagation of a varying magnetic field across a continuous array of elementary cells of the metasurface provide an efficient method to simultaneously transmit energy and information (Hiep et al., 2022; Son et al., 2022; Stevens, 2015).

DESIGN AND ANALYSIS OF METASUFACE FOR WIRELESS POWER TRANSMISSION

Figure 1. The categorization of materials based on permittivity (ε) and permeability (μ) values

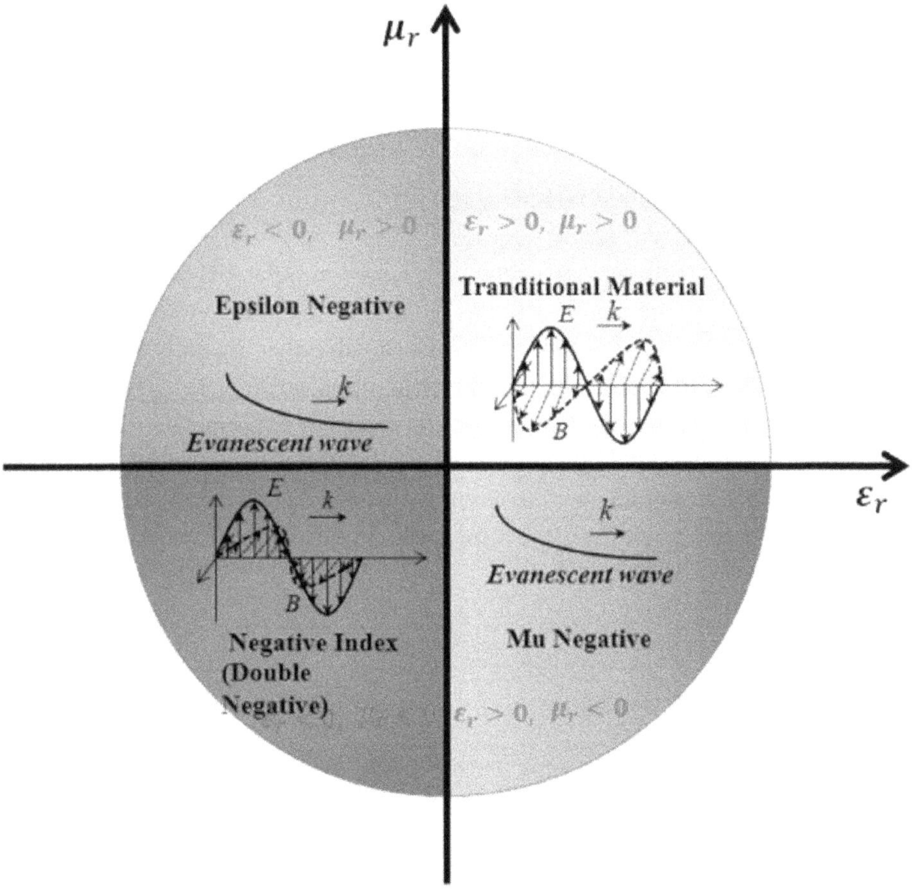

(Rong et al., 2021)

Based on the electromagnetic wave response of materials, the classification of materials can be categorized into four classes as shown in Fig. 1. The first class comprises common materials with positive permittivity (ε) and positive permeability (μ), which are naturally occurring materials. The second class consists of materials with negative permittivity (ε) and positive permeability (μ), commonly referred to as epsilon negative materials. The third class includes materials with negative permittivity (ε) and negative permeability (μ), known as double negative or negative index materials. Lastly, the fourth group encompasses materials with positive permittivity (ε) and negative permeability (μ), known as mu negative materials. Mu negative materials find various applications in the MHz frequency range, particularly in enhancing the transmission efficiency of magnetic resonant WPT systems, where the focus is solely on the magnetic field component.

Transmission performance is an important factor in the WPT system. This efficiency depends on the internal values of the system such as the quality-factor (Q-factor) of the coils and the distance between the coils and is expressed in the following formula:

$$\eta = \frac{k^2 Q_1 Q_2}{\left(1 + \sqrt{1 + k^2 Q_1 Q_2}\right)^2} = \frac{U^2}{\left(1 + \sqrt{1 + U^2}\right)^2} \qquad (1)$$

where k is the coupling coefficient, Q_1, Q_2 are the Q-factors of resonators 1 and 2, respectively. U is the figure-of-merit (FOM) of the WPT system, $U = k\sqrt{Q_1 Q_2}$. In which the Q-factor of the resonator is a value that depends on the material and configuration of the resonators, which is often difficult to change, however, the coupling coefficient (k) depends on the distance between the resonators and the environment surrounding the resonators. With the negative refraction index metasurface, the amplitude of evanescent waves can be enhanced and the coupling coefficient of two resonators can be improved. Finally, the efficiency of WPT system can be improved.

Figure 2. Classification of the metamaterial slab structure for WPT

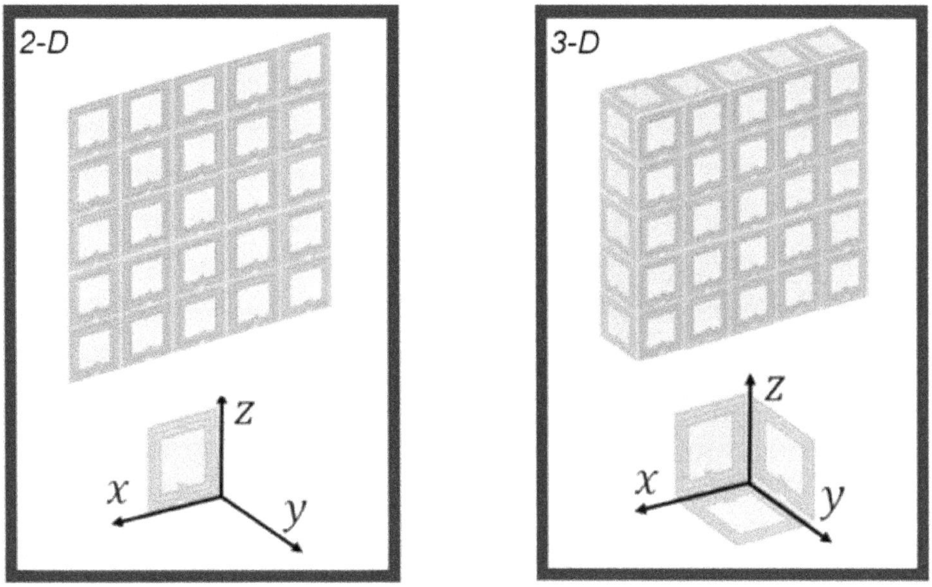

(Lee & Yoon, 2020)

There are many approaches to using metamaterials for WPT, of which there are two basic configurations of metamaterials that can be used: slab (2-D) and bulk (3-D), as shown in Fig. 2. Both of these configurations have their own advantages and disadvantages. However, in terms of structural simplicity and manufacturing cost, 2-D configurations are more popular. 2-D metamaterial configurations in the form of slab or metasurfaces are widely used to improve the performance of WPT systems (Lee & Yoon, 2023; Ranaweera et al., 2014; B. Wang et al., 2011; Zhang et al., 2023).

Figure 3. (a) An active metasurface with 9×9 metamaterial unit cell, (b) an equivalent circuit of a unit cell.

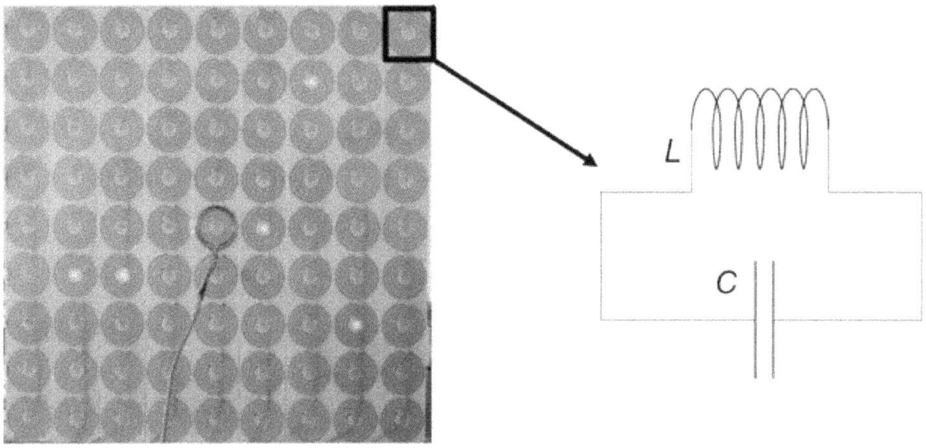

Metamaterials represent a class of artificial materials characterized by a periodic arrangement that endows them with unique electromagnetic properties. These metamaterial structures can manifest in one-dimensional (1D), two-dimensional (2D), or three-dimensional (3D) configurations. Within the scope of this discourse, our focus will be directed towards an in-depth examination of two-dimensional metamaterial structures, commonly referred to as metasurfaces. The pervasive interest in metasurfaces stems from their immense potential for diverse applications and their relative ease of fabrication compared to bulk metamaterials. Metasurfaces have garnered significant attention for their exceptional capabilities in absolute absorption of electromagnetic waves (Zheng et al., 2024), as well as their utilization in various information-related applications such as coding and programmable metasurfaces (Ke et al., 2022; S. R. Wang et al., 2023). Notably, these applications predominantly operate within the GHz or THz frequency ranges, presenting the versatility of metasurfaces in high-frequency domains.

In contrast, the exploration of metasurfaces at lower frequencies, particularly within the MHz range, has been a focal point of extensive research endeavors. In this regime, metasurfaces have been extensively investigated for their potential applications in WPT and information transfer through the near field. By delving into the realm of lower frequencies, researchers aim to harness the unique properties of metasurfaces to optimize the efficiency and performance of systems operating in this frequency band, thereby expanding the horizons of metamaterial applications in the domain of energy and information transmission.

Figure 3 depicts a visual representation of a metasurface configuration comprising a grid of 9×9 individual unit cells. Each unit cell measures 5×5 cm, resulting in a collective metasurface size of 45×45 cm. This metasurface structure has been specifically tailored for investigations pertaining to near-field WPT, with a designated operating frequency falling within the MHz range, around 13.56 MHz. The unit cell design features a spiral structure consisting of five turn, fabricated from a thin layer of copper with a thickness of 0.035 mm. Positioned atop an FR-4 substrate measuring 1.6 mm in thickness and characterized by a dielectric constant of 4.3, the unit cells are equipped with a control circuit affixed to the rear. This control circuit facilitates the adjustment of capacitor values at both terminals of the unit cell, thereby enabling precise modulation of the resonant frequency of each individual unit and consequently, the operational frequency of the metasurface.

Each unit cell within this intricate metamaterial structure can be effectively modeled using an *LC* circuit, as illustrated in the accompanying subfigure. The primary focus lies in elucidating the resonant properties of the metamaterial; thus, it is viable to simplify the analysis by conceptualizing each unit cell as an ideal *LC* circuit devoid of resistive components. This streamlined approach allows for a comprehensive exploration of the resonant characteristics exhibited by the metamaterial configuration, shedding light on its intricate behavior and performance attributes in the context of near-field wireless power transmission applications at MHz frequencies.

Figure 4. Coupling between two metamaterials unit cells

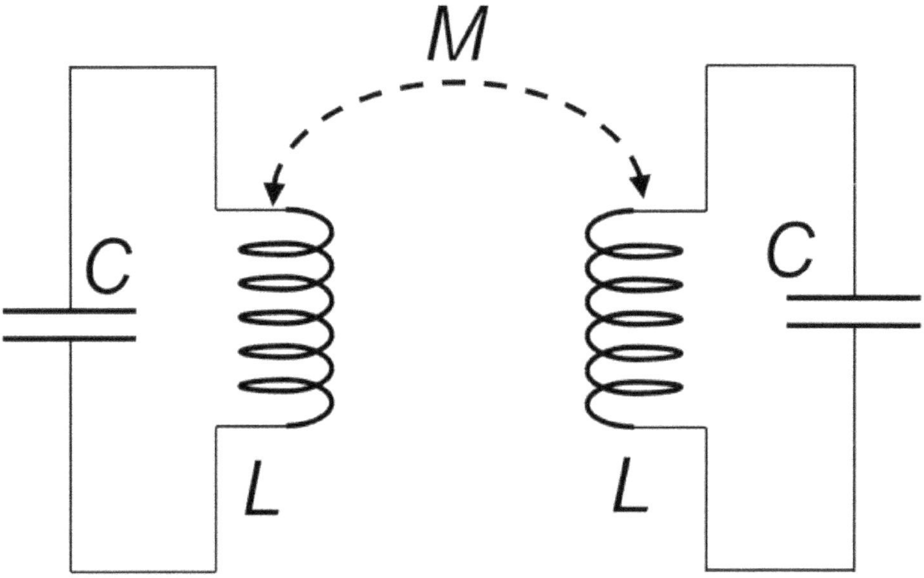

Figure 4 illustrates a schematic representation two metamaterial unit cells positioned adjacently, poised to establish interconnection through the integration of two inductors characterized by mutual inductance denoted as M. The magnitude of M is contingent upon the spatial separation between the paired unit cells. The quantification of the coupling coefficient governing the interaction between these two unit cells can be computed utilizing the following formula:

$$k = \frac{M}{L} \tag{2}$$

Figure 5. Schematic of a metasurface composed by 2D metamaterials structure

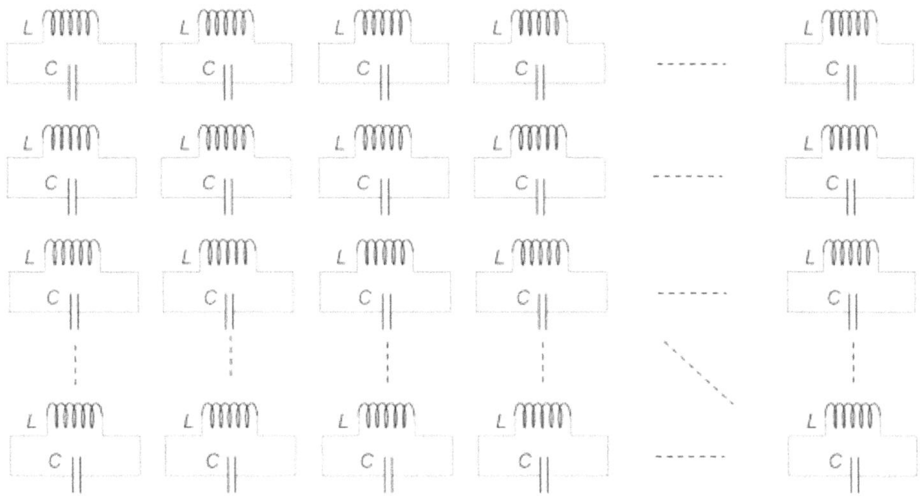

Given that each unit cell within the metasurface configuration can be effectively represented by an *LC* circuit model, the schematic depiction of the metasurface, illustrated in Fig. 5, encapsulates the interconnected network of these individual *LC* circuits. The interconnection between these *LC* circuits is facilitated through the self-inductance inherent in the inductor, derived from the spiral configuration comprising five turns of wire.

In the coplanar model delineated in Fig. 3, the substantial spatial separation between non-adjacent base cells necessitates a simplified analysis focusing solely on the coupling dynamics between neighboring base cells along both the vertical and horizontal axes. This pragmatic approach allows for a streamlined examination of the intercellular interactions within the metasurface structure, emphasizing the interplay between proximal unit cells while disregarding the influence of distantly positioned base cells for computational expediency and analytical clarity.

Figure 6. (a) Magnetic field distribution for the case of defect cavity metasurface, (b) magnetic field intensity as a function of length/wavelength ratio

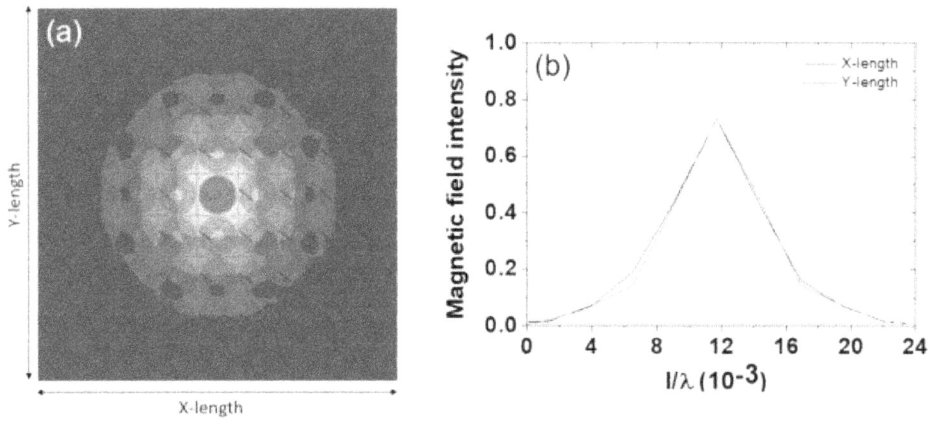

Subsequently, we delve into an exploration of the metasurface architecture featuring the introduction of a defected cavity. Building upon the premise established in the preceding section, wherein each unit cell within the metasurface structure wields the capability to manipulate its resonant frequency through the modulation of an external capacitor, thanks to that, we can form a unit cell harboring a distinct resonant frequency compared to its counterparts dispersed across the metasurface. By aligning with the requisite frequency conditions conducive to inducing Fano interference, as referenced in (Pham et al., 2017), we engender a scenario wherein the magnetic field converges towards the cavity within the metasurface. This orchestrated convergence results in the confinement of magnetic energy within the confines of the defected cavity, while concurrently extinguishing its presence in the surrounding regions of the metasurface.

Figure 6(a) shows the outcomes derived from simulations elucidating the distribution of magnetic fields across the metasurface. The simulation were conducted utilizing the CST Studio Suite simulation software, with the application of an "open" boundary condition. Within this configuration, strategic adjustments were made to the central unit cell of the metasurface, thereby configuring it to manifest as a defected cavity. Noteworthy observations reveal the localization of magnetic field intensities within this central unit cell, accentuating a stark contrast against the markedly subdued magnetic field strengths permeating through the remaining segments of the metasurface. This configuration underscores the intricate interplay between localized magnetic field phenomena and the dynamic spatial distribution of magnetic energy within the metasurface, showing the capabilities and potential applications of tailored metasurface designs.

Figure 6(b) provides an insightful depiction of the magnetic field intensity distribution along the central cross-section of the metasurface, along both the vertical and horizontal axes. The spatial values are represented in relation to the wavelength, offering a comprehensive perspective on the magnetic field dynamics within the metasurface. Notably, the magnetic field values are quantified in reference to S_{21} measurements, unveiling intriguing insights into the localization and confinement of magnetic fields within distinct spatial domains. Evident from the observations, the central unit cell emerges as the focal point where the magnetic field attains its peak intensity, as discerned through S_{21} measurements. Remarkably, the spatial region encapsulating this heightened magnetic field intensity manifests with dimensions significantly smaller than the prevailing wavelength. In fact, the confinement space housing the localized magnetic field spans a mere fraction of 8×10^{-3} wavelengths, underscoring the remarkable capability of the metasurface in confining and concentrating magnetic fields within spatial regions that transcend conventional wavelength scales. This unique property heralds a myriad of potential applications for the metasurface technology, particularly in the realms of field focusing within spatial domains smaller than the wavelength, wave guiding across sub-wavelength regions, and in-depth investigations into the intricate interplay between electromagnetic waves and matter across spatial scales that defy conventional diffraction limit.

Figure 7. Transmission of uniform metasurface mode and cavity mode

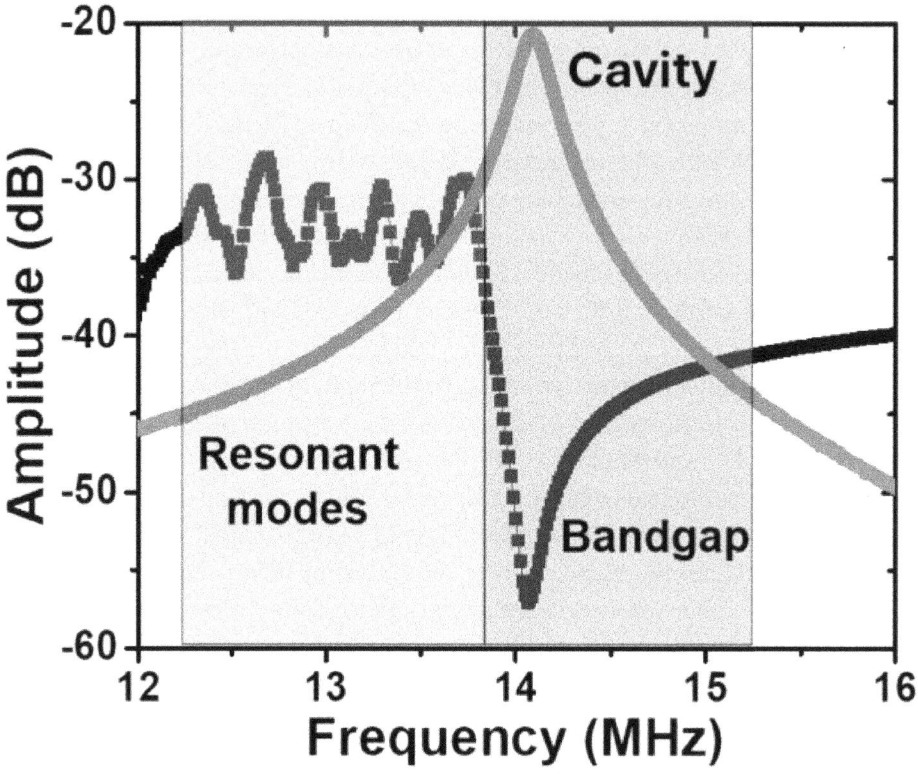

The phenomenon of magnetic confinement within the defected cavity can be elucidated through the lens of hybridization bandgap theory, as expounded in (Ranaweera et al., 2019). Figure 7 encapsulates the frequency response profiles of both the metasurface and the defected cavity, shedding light on the intricate resonance dynamics governing these structures. Notably, the frequency response analysis unveils compelling insights into the interplay between unit cells within the metasurface and the emergence of resonance phenomena within the defected cavity. Upon closer examination, it becomes apparent that within the metasurface configuration, the intercellular interactions engender a broadened resonance region spanning from 12.2 to 13.8 MHz. An intriguing observation arises in the form of a distinct no-transmission region situated immediately following the resonant band characterized by heightened waveguide amplitudes, commonly referred to as the hybridization bandgap. Remarkably, the resonance frequency of the defected cavity aligns with this frequency range, positioning it within the confines of the bandgap region. Consequently, at the resonance frequency specific to the cavity, a fascinating phenomenon unfolds wherein the propagation of magnetic fields within the metasurface is effectively inhibited, confining the magnetic energy exclusively within the confines of the defected cavity. This distinctive property not only underscores the unique capabilities of the defected cavity and metasurface architectures but also paves the way for a diverse array of applications and functionalities that leverage this intrinsic characteristic. Subsequent sections will delve deeper into elucidating the manifold applications of metasurface.

METASURFACE FOR WIRELESS POWER TRANSFER AND MAGNETO-INDUCTIVE WAVEGUIDE

In the preceding sections, we have investigated into the intricacies of metasurface and defected cavity configurations, unraveling their underlying principles and resonant behaviors. Building upon this foundational knowledge, let us now explore of the diverse applications facilitated by the utilization of defected cavities. The unique capability of these cavities to confine and localize magnetic fields within spatial domains smaller than the prevailing wavelength engenders a host of innovative applications, with one notable example being the utilization of 1-dimensional cavities for guiding MIW.

Figure 8 shows a line cavity configuration tailored for MIW transmission, exemplifying the practical implementation of this concept. By initiating energy excitation within the unit cell positioned at the midpoint of the left edge of the metasurface, magnetic field energy propagates along the defined cavity line from left to right. A discernible trend emerges wherein the magnetic field undergoes attenuation as it traverses cavities distanced from the energy source. Despite this attenuation effect, it is noteworthy that the magnetic field remains steadfastly localized within the cavity line, refraining from leaking into adjacent regions across the metasurface. A closer examination along the X and Y-axes unveils the distinctive characteristic of magnetic field localization and propagation within spatial confines significantly smaller than the prevailing wavelength. This spatial confinement underscores the efficacy of the cavity line in guiding and manipulating MIW, offering the potential for such configurations to enable precise control and propagation of magnetic fields within confined spatial domains. The magnetic field localization and controlled propagation herald a new frontier in waveguiding technologies, showcasing the capacity of defected cavities to apply in diverse applications spanning fields such as telecommunications and sensing technologies.

Figure 8. Field intensity follow cavity position in metasurface

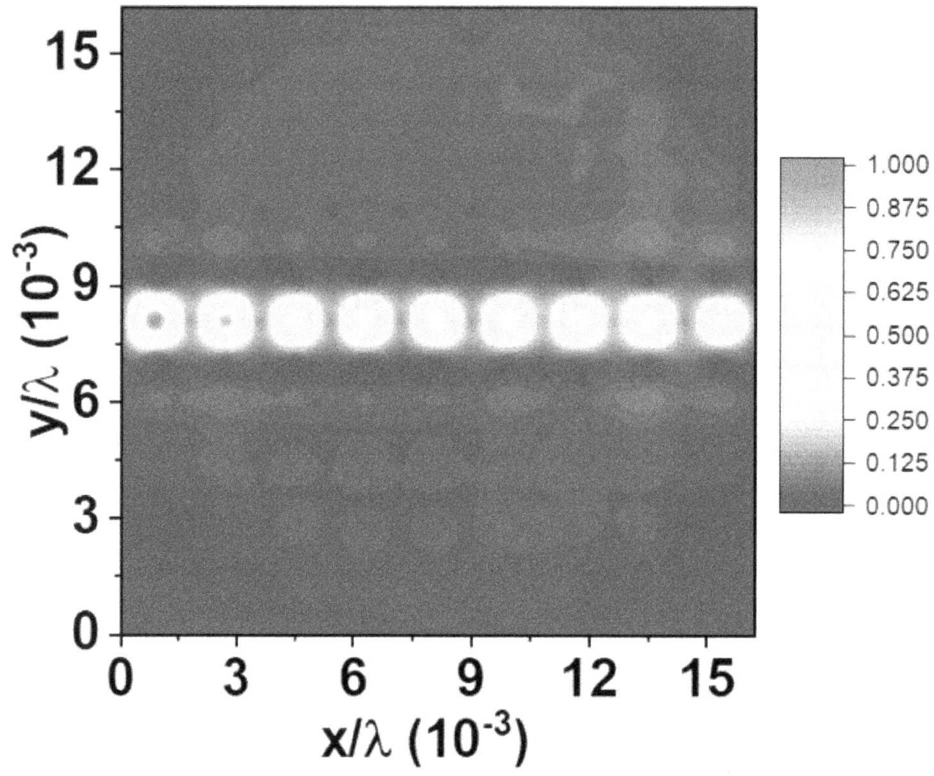

(Ranaweera et al., 2019)

In the next, an analysis of the wave guiding and WPT capabilities inherent in the uniform metasurface configuration, as depicted in Fig. 9. Analogous to our exploration of the cavity line, we initiate energy excitation within the unit cell situated at the midpoint of the left edge to unravel the propagation dynamics across the metasurface. In this scenario, a distinct pattern emerges as the magnetic field emanating from the excitation source permeates throughout the entirety of the metasurface, resulting in a rapid dissipation of energy as one moves away from the point of excitation. A assessment of the transmittance characteristics along the X-axis, corresponding to a fixed Y-axis location of 5, unveils a notable trend wherein the transmittance coefficient experiences a swift attenuation from -5 to -21 dB as one traverses away from the excitation source. Notably, unit cells positioned at considerable distances from the source exhibit low energy transmission values hovering around -20 dB. Shifting our focus to a vertical analysis at location 5 along the X-axis, we observe a relatively consistent transmittance coefficient profile characterized by values approximating -20 dB. The transmittance coefficient values derived from the uniform metasurface configuration underscore a pronounced diminishment in energy transmission efficacy, with values exhibiting substantial attenuation. While these modest transmittance coefficients hold promise for applications in information transmission, their inadequacy for facilitating WPT becomes apparent. This discrepancy in energy transmission efficiency highlights the nuanced interplay between metasurface design and functionality, underscoring the imperative of tailored configurations to optimize energy transfer capabilities for diverse applications in wireless power transmission and communication technologies.

Figure 9. Magnetic field distribution of cavity line in uniform metasurface

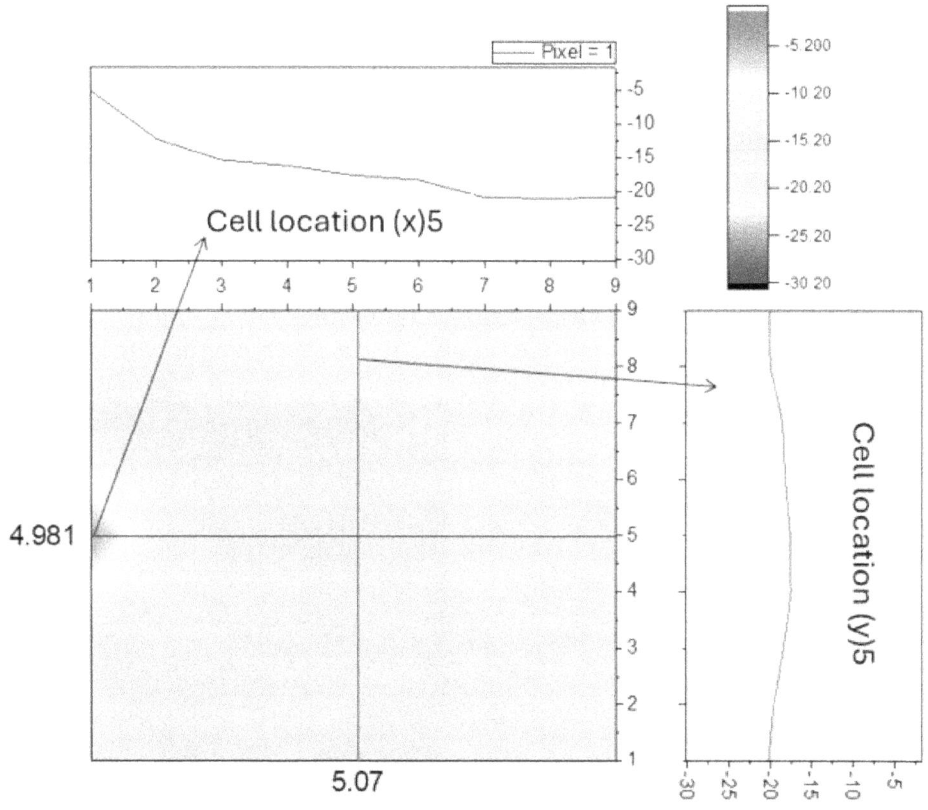

In contrast to the uniform metasurface scenario, the cavity line configuration in the non-uniform metasurface exhibits significantly enhanced transmittance coefficients, underscoring its superior wave guiding and energy transfer capabilities. Figure 10

encapsulates the spatial transmittance values across the metasurface, illuminating the distinct performance disparities between the cavity line and the uniform metasurface configurations. Notably, the transmittance coefficient within the cavity line retains a notably elevated value in comparison to other regions within the metasurface. The magnetic field energy distribution along the X-axis at a fixed Y-axis location of 5 shows a substantial amplification in energy levels when compared to the uniform metasurface counterpart. In contrast to the uniform metasurface, regions outside the cavity line exhibit transmittance coefficients hovering around -27 dB, a marked reduction from the -20 dB levels observed in the uniform metasurface case. Conversely, within the cavity line itself, the transmittance coefficient surges to -5 dB, signifying a pronounced concentration of magnetic field energy within this localized region. This distinctive property underscores the efficacy of the cavity line configuration in confining and channeling magnetic field energy along specific pathways, as opposed to the dispersion observed in uniform metasurface designs. The localization of magnetic field energy within the cavity line not only enhances energy transfer efficiency but also the precision and control afforded by tailored cavity structures in optimizing wave guiding capabilities.

EXPERIMENT RESULTS

Figure 11. Experiment setup of metasurface for wireless power transfer and waveguide (a) metasurface with controlled circuit, (b) computer interface for controlling

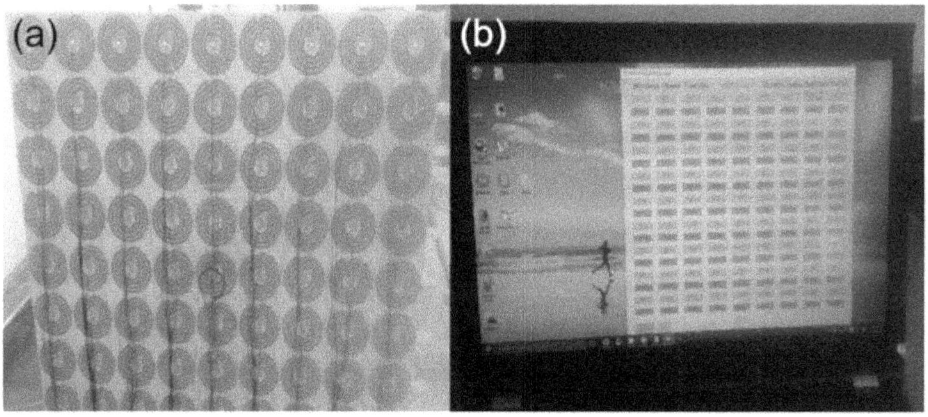

Within this section, we aim to delve into the experimental configurations of metasurfaces tailored for applications in wireless energy transfer and the guidance of magnetic induction waves. Figure 11(a) unveils a customized material sheet comprising a grid of 9×9 unit cells, each equipped with frequency control circuits affixed to the rear surface of the metasurface. Notably, the orientation of the metasurface, whether vertical or horizontal, does not impinge upon its inherent capabilities or operational dynamics, underscoring the versatility and adaptability of this innovative technology. The resonance frequency tuning and cavity formation processes are seamlessly facilitated through the utilization of computer-aided tools endowed with graphical visualization capabilities, as exemplified in Fig. 11(b). This sophisticated approach streamlines the configuration and optimization of resonance frequencies within the metasurface structure, offering a user-friendly interface for manipulating and fine-tuning the metasurface properties to suit specific application requirements. By leveraging advanced computational tools and visualization techniques, researchers can meticulously tailor the metasurface properties to achieve desired resonance frequencies and cavity formations, thereby unlocking a spectrum of possibilities for enhancing WPT efficiency and guiding magnetic induction waves with precision and efficacy.

Figure 12. Measured transmission efficiency of cavity line follow unit cell position

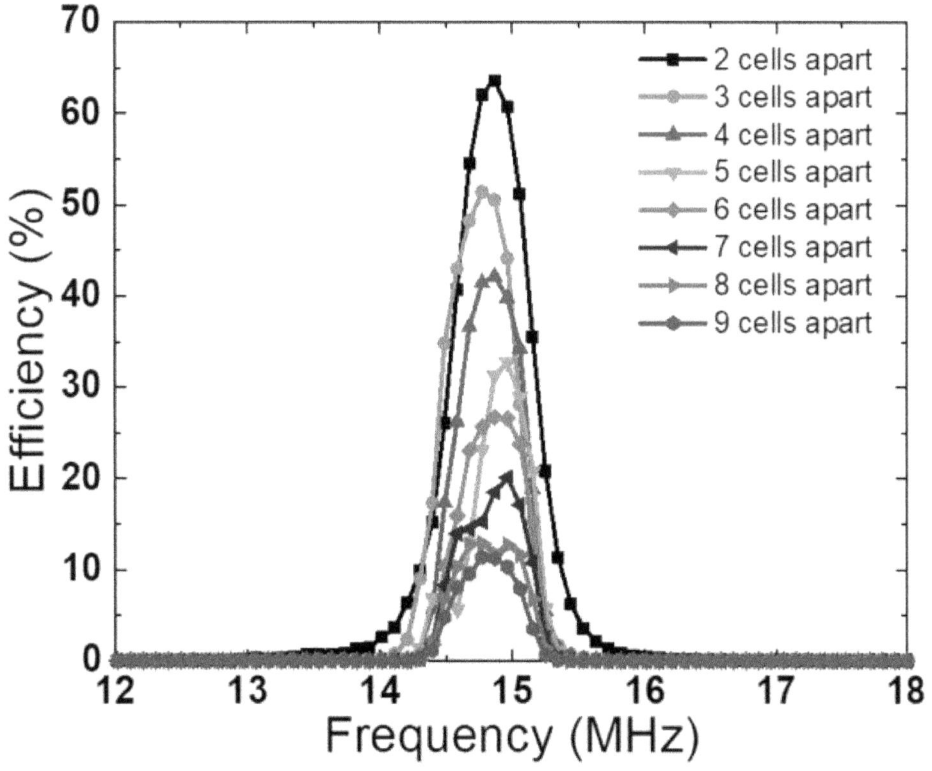

Figure 12 present an in-depth analysis of the transmission efficiency along the unit cells of the cavity line configuration. The transmission efficiency within the cavity line exhibits a discernible trend, showcasing a gradual decline as one progresses from the 2nd to the 9th unit cell. Interestingly, a notable consistency in performance is observed between the 8th and 9th unit cells, attributed to the phenomenon of reflection occurring at the terminal position of the cavity line. This reflection mechanism serves to maintain transmission efficiency levels between these final unit cells. Upon closer examination, the transmission efficiency values at the 2nd unit cell register a commendable 65%, progressively diminishing to 13% by the time we reach the 9th unit cell. Notably, these transmission efficiency values significantly surpass those observed in the uniform metasurface scenario, underscoring the efficacy of the cavity line configuration in facilitating robust energy transfer capabilities. The substantial transmission efficiency levels exhibited by the cavity line configuration render it eminently suitable for applications in wireless energy transfer, magnetic induction wave transmission, and information transmission.

Table 1. Comparison of the absorption performance among different metamaterial absorber configurations

Ref.	**Metamaterial configurations**	**Operating frequency (MHz)**	**Diameter of resonator (mm)**	**Tranfer distance (mm)**	**Efficiency (%)**
(B. Wang et al., 2011)	3D	27	400	500	35
(Ranaweera et al., 2014)	3D	6.78	600	1000	53
(Zhang et al., 2023)	2D Broadband	8.7-10.9	51	100	11.51
(B. Wang et al., 2011)	2D	27	400	500	47
(M. Wang et al., 2023)	13.56	13.56	90	110	76.5
(Lee & Yoon, 2023)	2D	6.78	180	600	46
This chapter	2D planar WPT	14.2	50	50-450	65-13

Table 1 presents some comparisons between different metamaterial configurations for applications in WPT. Configurations vary from 2D to 3D, single-peak or broadband. Meanwhile, the results presented in this chapter demonstrate a 2D metamaterial or metasurface configuration applied in planar WPT that achieves transmission performance that is comparable to previous work.

These elevated transmission efficiency values not only underscore the utility and practicality of cavity line configurations in optimizing WPT processes but also highlight their potential in enabling efficient and reliable transmission of MIW and information across diverse spatial domains. The pronounced performance enhancements in the cavity line configuration herald a new era in metasurface applications, offering a promising avenue for advancing WPT technologies and enhancing communication systems with heightened efficiency and reliability.

CONCLUSION

Within this chapter, we have delved into the realm of 2D metamaterials, commonly called metasurfaces, and their application in WPT and the propagation of MIW. A distinctive characteristic of these metasurfaces lies in their operational frequency, which predominantly resides within the low MHz region—a departure from the conventional GHz or THz frequency ranges typically associated with traditional metamaterials. This unique frequency domain affords us the opportunity to focus solely on the interactions between the magnetic field and the metasurface, disregarding the electric field dynamics.

By manipulating external capacitance values, we can engineer defected cavity configurations within the metasurface, offering a controlled environment for energy confinement and transmission. The introduction of cavity line configurations further enhances energy transmission efficiency compared to uniform metasurface designs, underscoring the pivotal role of tailored structures in optimizing energy transfer processes. These metasurfaces, operating within the MHz frequency spectrum, exhibit immense potential for diverse applications in WPT and near-field data transfer.

REFERENCES

Adepoju, W., Bhattacharya, I., Sanyaolu, M., Bima, M. E., Banik, T., Esfahani, E. N., & Abiodun, O. (2022). Critical review of recent advancement in metamaterial design for wireless power transfer. *IEEE Access : Practical Innovations, Open Solutions*, 10, 42699–42726. 10.1109/ACCESS.2022.3167443

Fowler, C., Silva, S., Thapa, G., & Zhou, J. (2022). High efficiency ambient RF energy harvesting by a metamaterial perfect absorber. *Optical Materials Express*, 12(3), 1242–1250. 10.1364/OME.449494

Garg, P., & Jain, P. (2023). A review of metamaterial absorbers and their application in sensors and radar cross-section reduction. *Microwave and Optical Technology Letters*, 65(2), 387–411. 10.1002/mop.33496

Hiep, L. T. H., Pham, T. S., Khuyen, B. X., Tung, B. S., Ngo, Q. M., Hien, N. T., Minh, N. T., & Lam, V. D. (2022). Enhanced transmission efficiency of magneto-inductive wave propagating in non-homogeneous 2-D magnetic metamaterial array. *Physica Scripta*, 97(2), 025504. 10.1088/1402-4896/ac4a3a

Hui, S. Y. R., Zhong, W., & Lee, C. K. (2014). A critical review of recent progress in mid-range wireless power transfer. *IEEE Transactions on Power Electronics*, 29(9), 4500–4511. 10.1109/TPEL.2013.2249670

Ke, J. C., Dai, J. Y., Zhang, J. W., Chen, Z., Chen, M. Z., Lu, Y., Zhang, L., Wang, L., Zhou, Q. Y., Li, L., Ding, J. S., Cheng, Q., & Cui, T. J. (2022). Frequency-modulated continuous waves controlled by space-time-coding metasurface with nonlinearly periodic phases. *Light, Science & Applications*, 11(1), 1. 10.1038/s41377-022-00973-836104318

Kurs, A., Karalis, A., Moffatt, R., Joannopoulos, J. D., Fisher, P., & Soljačić, M. (2007). Wireless power transfer via strongly coupled magnetic resonances. *Science*, 317(5834), 83–86. 10.1126/science.114325417556549

Lee, W., & Yoon, Y.-K. (2020). Wireless power transfer systems using metamaterials: A review. *IEEE Access : Practical Innovations, Open Solutions*, 8, 147930–147947. 10.1109/ACCESS.2020.3015176

Lee, W., & Yoon, Y.-K. (2023). High-efficiency wireless-power-transfer system using fully rollable Tx/Rx coils and metasurface screen. *Sensors (Basel)*, 23(4), 4. 10.3390/s2304197236850570

Li, L., Wen, J., Wang, Y., Jin, Y., Wen, Y., Sun, J., Zhao, Q., Li, B., & Zhou, J. (2023). A Transparent broadband all-dielectric water-based metamaterial absorber based on laser cutting. *Physica Scripta*, 98(5), 055516. 10.1088/1402-4896/accc15

Lu, C., Huang, X., Tao, X., Rong, C., & Liu, M. (2020). Comprehensive analysis of side-placed metamaterials in wireless power transfer system. *IEEE Access : Practical Innovations, Open Solutions*, 8, 152900–152908. 10.1109/ACCESS.2020.3017492

Pham, T. S., Bui, H. N., & Lee, J.-W. (2019). Wave propagation control and switching for wireless power transfer using tunable 2-D magnetic metamaterials. *Journal of Magnetism and Magnetic Materials*, 485, 126–135. 10.1016/j.jmmm.2019.04.034

Pham, T. S., Ranaweera, A. K., Ngo, D. V., & Lee, J. W. (2017). Analysis and experiments on Fano interference using a 2D metamaterial cavity for field localized wireless power transfer. *Journal of Physics. D, Applied Physics*, 50(305102), 1–10. 10.1088/1361-6463/aa7988

Ranaweera, A. L. A. K., Duong, T. P., & Lee, J.-W. (2014). Experimental investigation of compact metamaterial for high efficiency mid-range wireless power transfer applications. *Journal of Applied Physics*, 116(4), 043914. 10.1063/1.4891715

Ranaweera, A. L. A. K., Pham, T. S., Bui, H. N., Ngo, V., & Lee, J.-W. (2019). An active metasurface for field-localizing wireless power transfer using dynamically reconfigurable cavities. *Scientific Reports*, 9(1), 1. 10.1038/s41598-019-48253-731409834

Rong, C., Lu, C., Zeng, Y., Tao, X., Liu, X., Liu, R., He, X., & Liu, M. (2021). A critical review of metamaterial in wireless power transfer system. *IET Power Electronics*, 14(9), 1541–1559. 10.1049/pel2.12099

Son, P. T., Khuyen, B. X., Tung, B. S., Hiep, L. T. H., & Lam, V. D. (2022). A critical review on wireless power transfer systems using metamaterials. *Vietnam Journal of Science and Technology*, 60(4), 4. 10.15625/2525-2518/16954

Stevens, C. J. (2015). Magnetoinductive waves and wireless power transfer. *IEEE Transactions on Power Electronics*, 30(11), 6182–6190. 10.1109/TPEL.2014.2369811

Tesla, N. (1914). *Apparatus for transmitting electrical energy* (United States Patent US1119732A). https://patents.google.com/patent/US1119732A/en

Veselago, V. G. (1968). The electrodynamics of substances with simultaneously negative values of ε and μ. *Soviet Physics - Uspekhi*, 10(4), 509. 10.1070/PU1968v010n04ABEH003699

Wang, B., Teo, K. H., Nishino, T., Yerazunis, W., Barnwell, J., & Zhang, J. (2011). Experiments on wireless power transfer with metamaterials. *Applied Physics Letters*, 98(25), 254101. 10.1063/1.3601927

Wang, B.-X., Xu, C., Duan, G., Xu, W., & Pi, F. (2023). Review of broadband metamaterial absorbers: From principles, design strategies, and tunable properties to functional applications. *Advanced Functional Materials*, 33(14), 2213818. 10.1002/adfm.202213818

Wang, M., Guo, J., Shi, Y., Wang, M., Song, G., & Yin, R. (2023). A metamaterial-incorporated wireless power transmission system for efficiency enhancement. *International Journal of Circuit Theory and Applications*, 51(7), 3051–3065. 10.1002/cta.3587

Wang, S. R., Dai, J. Y., Zhou, Q. Y., Ke, J. C., Cheng, Q., & Cui, T. J. (2023). Manipulations of multi-frequency waves and signals via multi-partition asynchronous space-time-coding digital metasurface. *Nature Communications*, 14(1), 1. 10.1038/s41467-023-41031-037666804

Wen, J., Ren, Q., Peng, R., & Zhao, Q. (2022a). Multi-functional tunable ultra-broadband water-based metasurface absorber with high reconfigurability. *Journal of Physics. D, Applied Physics*, 55(28), 285103. 10.1088/1361-6463/ac683e

Wen, J., Ren, Q., Peng, R., & Zhao, Q. (2022b). Ultrabroadband Saline-Based Metamaterial Absorber With Near Theoretical Absorption Bandwidth Limit. *IEEE Antennas and Wireless Propagation Letters*, 21(7), 1388–1392. 10.1109/LAWP.2022.3169467

Zhang, R., Wang, K., Wang, X., Luo, X., & Zhao, C. (2023). Broadband and precise reconfiguration of megahertz electromagnetic metamaterials for wireless power transfer. *Physica Scripta*, 98(11), 115529. 10.1088/1402-4896/acfea9

Zheludev, N. I., & Kivshar, Y. S. (2012). From metamaterials to metadevices. *Nature Materials*, 11(11), 11. Advance online publication. 10.1038/nmat343123089997

Zheng, H., Pham, T. S., Chen, L., & Lee, Y. (2024). Metamaterial Perfect Absorbers for Controlling Bandwidth: Single-Peak/Multiple-Peaks/Tailored-Band/Broadband. *Crystals*, 14(1), 1. 10.3390/cryst14010019

Chapter 6
Planar Metamaterial Microwave Sensors for Characterization of Dielectric Materials

Man Seng Sim
http://orcid.org/0000-0001-7776-2239
Universiti Teknologi Malaysia, Malaysia

Kok Yeow You
http://orcid.org/0000-0001-5214-7571
Universiti Teknologi Malaysia, Malaysia

Stephanie Yen Nee Kew
http://orcid.org/0009-0001-2957-8848
Universiti Teknologi Malaysia, Malaysia

Raimi Dewan
http://orcid.org/0000-0002-2023-132X
Universiti Teknologi Malaysia, Malaysia

Fahmiruddin Esa
Universiti Tun Hussein Onn Malaysia, Malaysia

DiviyaDevi Paramasivam
http://orcid.org/0009-0005-8901-295X
Universiti Teknologi Malaysia, Malaysia

Fandi Hamid
Universiti Teknologi Malaysia, Malaysia

ABSTRACT

Metamaterials can be integrated into planar microwave sensors for field localization and enhancement as well as sensitivity improvement in permittivity-based sensing measurements. This chapter reviews metamaterial-based planar microwave sensors

DOI: 10.4018/979-8-3693-2599-5.ch006

for characterizing dielectric materials. It begins by introducing planar microwave sensors. The subsequent section focuses on the sensing principles of planar microwave sensors loaded with metamaterial-inspired resonators, namely frequency shift, frequency splitting, and amplitude or phase variation. Furthermore, recent advances in metamaterial-integrated sensors are discussed, focusing on the types of samples under test, including solid samples in bulk and powder form, and liquid samples in fluid tubes or microfluidic channels. Furthermore, planar antennas loaded with metamaterial elements are also explored. To enhance understanding of frequency-variation sensors, a microstrip transmission line loaded with complementary split ring resonators is simulated. Finally, the design strategies of planar metamaterial-based sensors are reviewed.

INTRODUCTION

In recent days, there has been a growing demand for microwave sensors, largely due to the proliferation of Internet of Things (IoT) technology (Costanzo et al., 2023). Microwave sensors are sensitive, non-destructive to materials under test (MUTs), robust, and offer remote wireless sensing capabilities (Dai et al., 2020; Salim & Lim, 2018; Velez et al., 2017). They are among the extensively utilized sensors employed for material characterization in various industries such as biomedical, agriculture, food, construction, and manufacturing (Prakash & Gupta, 2022; You et al., 2020). These sensors utilize microwave signals ranging from 300 MHz up to approximately 300 GHz to interact with the MUTs, probing their electromagnetic responses and providing a comprehensive understanding of the properties. Microwave sensors encompass two main categories: planar and non-planar designs. Planar microwave sensors use planar technology and are implemented on a substrate (Rahman et al., 2017), while non-planar sensors employ microwave devices such as waveguides (Iftimie et al., 2018; Seng et al., 2020; You & Abbas, 2011), coaxial probes (You & Sim, 2018), and antennas (Sim et al., 2023; You, Sim, et al., 2017).

In terms of operation mode, microwave sensors function in either reflection mode and/or transmission mode (You, 2017; You, Esa, et al., 2017). For single-port reflection sensors, they operate by sending a signal and measuring the reflected signal. The dielectric properties of the MUT can be measured by analyzing the changes in amplitude or phase of the reflected signals. This reflection mode technique is commonly used in applications such as moisture content measurement, non-destructive testing for defect detection, and density or concentration determination. For two-port transmission sensors, they send an electromagnetic signal and measure the transmitted signal at another port. Typically, *S*-parameters are the measured parameters for microwave sensing devices, in which S_{11} represents the reflection coefficient (the ratio of the reflected signal power to the incident signal power at port 1), and S_{21} represents the transmission coefficient (the amount of power transferred from port 1 to port 2). They are expressed as complex numbers, and therefore variation of their magnitude and/or phase, for different MUTs introduced to the measurement system, can be calibrated with the parameters to be measured.

Microwave sensors can be integrated with metamaterial-based resonant elements for performance enhancement (Bogue, 2017; Sim et al., 2021; Q. Wang et al., 2023). Metamaterials are engineered materials designed with unique structures to manipulate electromagnetic waves (Puentes Vargas, 2014). Microwave metamaterials typically comprise of subwavelength unit cells which are arranged in periodic manner to provide a collective behavior for interactions with electromagnetic waves (Sim et al., 2020). They can take various forms, such as split ring resonators (SRRs) and complementary split ring resonators (CSRRs), and resonate at specific frequencies (RoyChoudhury et al., 2016; Sim et al., 2018). These resonators can be integrated into microwave devices such as microstrip transmission line, waveguides, and antennas. In recent publication, the terms "metamaterial-inspired" or "metamaterial-based" are commonly used to describe these resonator-loaded microwave sensors to highlight the concept of metamaterials, where properties mainly arise from geometrical design rather than the properties of the constituent materials (Govind & Akhtar, 2019; Harnsoongnoen, 2021; Kayal et al., 2020).

This book chapter aims to review planar metamaterial-loaded microwave sensors for dielectric material characterization applications. The chapter is organized as follows. Firstly, planar microwave sensors are introduced. Next, the sensing principles of metamaterial-integrated planar microwave sensors are explained, followed by their recent advances. Subsequently, a planar microstrip transmission line-based sensor integrated with CSRRs is designed and evaluated based on its performance in characterizing the properties of thin planar solid dielectric samples. Finally, the design strategies of planar metamaterial-inspired microwave sensors are discussed.

PLANAR MICROWAVE SENSORS FOR MATERIAL CHARACTERIZATION

Planar microwave sensors are microwave sensors implemented using planar technology (Muñoz-Enano et al., 2020; Rahman et al., 2017). They can utilize various substrates, including rigid, flexible, organic, and biodegradable substrates, depending on their applications. The two main common fabrication techniques for planar microwave sensors are conventional Printed Circuit Board (PCB) printing and Computer Numerical Control (CNC) engraving. Both techniques are subtractive processes in which unwanted copper is removed from the substrate. PCB printing employs chemicals to etch away the unmasked copper on the substrate, resulting in the desired circuit traces. One of the benefits of employing this technique is achieving mass production at a more economical price. On the other hand, CNC engraving uses engraving machines to mechanically remove the unwanted copper, allowing for fast execution of parts with precisely defined geometry (Dalgac et al., 2020; Dalgaç, Furat, et al., 2020).

Some common types of planar transmission lines include microstrip transmission lines, strip-lines, and coplanar waveguides (CPWs) (Rahman et al., 2017). Generally, these transmission lines are used for signal transmission and reception, power transfer, and impedance matching between various RF and microwave components. They typically consist of flat metallic strips structures, with the top surface being air and the bottom surface being a dielectric substrate. The electric and magnetic fields are distributed in a quasi-TEM mode for these transmission lines. For sensing applications, these planar transmission lines can be used to characterize the dielectric properties of many materials by monitoring changes in the magnitude and phase of S-parameters. Figure 1 shows an example of a microstrip transmission line integrated with multiple SRRs, loaded with a solid MUT. The S-parameters responses are measured using a two-port vector network analyzer (VNA). In addition to transmission lines, planar antennas are antennas fabricated using planar fabrication techniques (Esmail et al., 2022; Milias et al., 2021). The metallic structures of planar antennas can take different forms, such as patches, slots, or arrays on a substrate. While commonly used in communication applications, planar antennas can also be applied in material characterization for contactless free-space or remote sensing (Elwi, 2018).

Figure 1. (a) Perspective view of planar split-ring-resonator-based microwave sensor loaded with material under test, (b) Top view of the fabricated microwave sensor, (c) Measurement using vector network analyzer

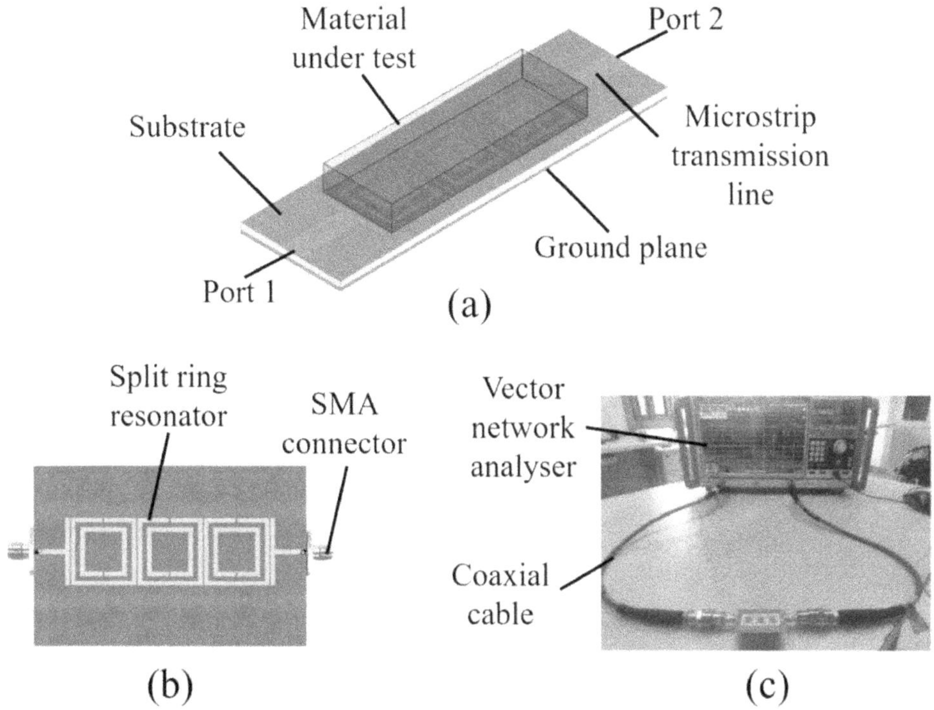

Planar microwave sensors are low-cost, low-profile, and compact. Constructed from a series of transmission line segments on a substrate, the design of these sensors is fairly simple and easy to fabricate. Additionally, it is easy to integrate these sensors with other microwave circuitry for communication and post-processing. The measured data can be easily transferred to a microcontroller or computer for analysis and converted to comprehensible data for display to users. Furthermore, advancements in technologies such as microfluidic and lab-on-a-chip have expanded the range of possible sensing applications (Zhang et al., 2018). In summary, planar microwave sensors have become a preferred sensing approach for specific sensing applications due to factors such as cost-effectiveness, low-profile, robustness, design simplicity, and compatibility with associated technologies.

PLANAR MICROWAVE SENSORS INTEGRATED WITH METAMATERIAL-INSPIRED RESONATORS

Planar microwave sensors can be enhanced by integrating them with metamaterial-inspired resonators (Vivek et al., 2019). As illustrated in Figure 2, the microstrip line is loaded and coupled with split ring resonators. Typically, the MUTs are positioned in close proximity to these resonators. The variation in the reflection responses (S_{11} for one-port systems), or transmission responses (S_{21} for two-port systems) is used to determine the properties of MUTs. This section discusses the sensing principles and recent developments of planar metamaterial-based sensors.

Figure 2. (a) One-port; (b) Two-port microstrip transmission line integrated with metamaterial elements

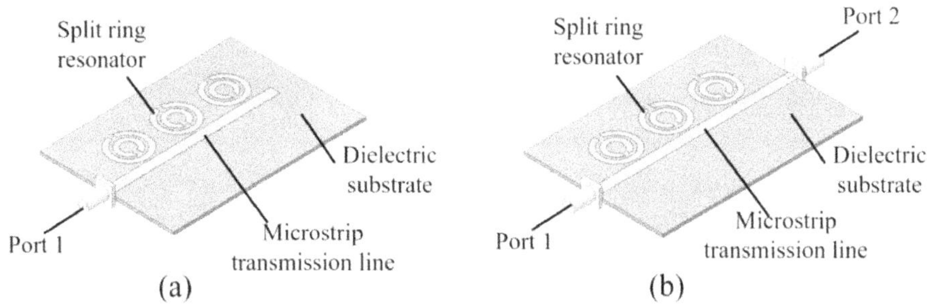

Sensing Principle of Planar Metamaterial Microwave Sensors

For planar metamaterial-based sensors designed for material characterization applications, three main sensing principles are utilized: frequency shift, frequency splitting, and amplitude or phase variation (Mayani et al., 2021; Su et al., 2017). The simplified S-parameter responses for each principle are illustrated in Figure 3. In a resonant-based sensor, a distinct resonant response occurs at a specific frequency, denoted as the 'REF' curve, serving as the reference. Upon loading the sensor with MUTs, the resonant responses alter, as depicted by the red dotted curve labeled 'MUT' in the graphs.

Figure 3. S-parameter against frequency curves for planar metamaterial-inspired sensors based on (a) Frequency Shift, (b) Frequency Splitting, and (c) Amplitude Variation. 'REF' denotes measurements without MUTs or with standard reference samples, while 'MUT' represents measurements with MUTs loaded

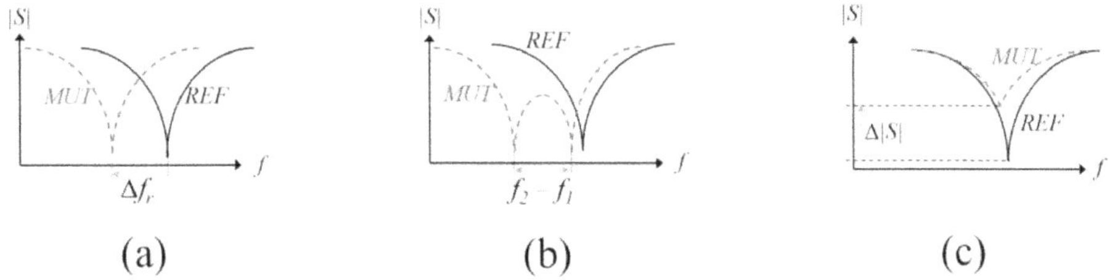

Resonant Frequency Shift

Most planar metamaterial sensors are resonant-based, and the basic working principle involves measuring the shift of the resonant frequency (Chen et al., 2012). A planar transmission line loaded with metamaterial elements exhibits a transmission zero at the resonant frequency, f_r. When a MUT is loaded on the metamaterial sensing elements, the overall effective permittivity around the metamaterial changes. The total capacitance increases due to the addition of the loaded MUT. This addition shifts the resonant frequency to a new resonant frequency, f_r'. The shift in resonant frequency, Δf_r is expressed as (Javed et al., 2020):

$$\Delta f_r = f_r' - f_r \quad (1)$$

These types of sensors are very simple, and monitoring the frequency shift, Δf_r can be used to determine many different variables, not only specifically for the determination of the dielectric constant, ε_r'. This is because ε_r' is dependent on material characteristics [including composition (Dalgaç, Karada , et al., 2020), concentration (Shen et al., 2024), position (Memon et al., 2020), thickness (Haq & Koziel, 2023), and moisture content (Oguzhan Akgol & Unal, 2018)] and environmental conditions [such as temperature (Lu et al., 2018)]. Frequency-variation sensors suffer from cross-sensitivity issues, in which their accuracy is affected by environmental conditions (Ebrahimi et al., 2018). Therefore, they need to be properly calibrated for accurate measurements.

Frequency Splitting

For sensors based on frequency-splitting working principle, the transmission line is loaded with a pair of identical resonators in a symmetrical configuration (Mata-Contreras et al., 2017). When the sensing structure is unloaded (without MUT), a transmission zero appears at the resonant frequency, f_r. When the symmetry is disrupted (*e.g.* one of the resonators is loaded with MUT), the transmission zero splits into two, at two resonant frequencies, f_1 and f_2 (Ebrahimi et al., 2018). The difference in the frequency, f between both frequencies at the transmission zeros is calculated as:

$$f = f_2 - f_1 \tag{2}$$

This sensing strategy is useful for the sensor to work as a comparator. This type of sensor is also commonly known as differential sensors. One of the resonators is loaded with the reference sample, whereas another resonator is loaded with MUT. The value of f indicates the difference between the permittivity values of the MUT and the reference sample. Besides permittivity determination, the sensors can be used to detect defects or abnormalities in the MUTs. One limitation of differential sensors is the degradation of sensitivity due to coupling between resonators.

Amplitude/ Phase Variation

Amplitude or phase-variation sensors can operate at a single frequency, which helps reduce the cost of the electronics required for signal processing and analysis (Castillo et al., 2018). In these sensors, changes in material properties are detected through variations in either the amplitude or phase of the transmitted or reflected signals (Abdulkarim, Deng, Karaaslan, Dalgaç, et al., 2020; S. Mohammadi & Zarifi, 2021; Tefek et al., 2023; Velez et al., 2023; Viswanathan et al., 2020). For instance, when MUTs with high loss are introduced, the amplitude of the signals usually decreases due to energy absorption or dissipation. Meanwhile, the phase of a signal may also shifts depending on the dielectric properties of the MUTs. By monitoring these changes, amplitude- or phase-variation sensors can effectively characterize the properties of the MUTs.

Recent Advance of Planar Metamaterial Microwave Sensors

In this subsection, various recent planar metamaterial-based microwave sensors for dielectric material characterization applications are discussed. Subsequently, metamaterial-integrated planar antennas are reviewed. Additionally, a table comparing recent metamaterial-based microwave sensors is provided.

Metamaterial-Loaded Planar Transmission Line Sensors

The planar transmission line sensors can be loaded with metamaterial-inspired resonators for field localization and enhancement. The region with enhanced fields can be loaded with MUTs for dielectric characterization. There are various types of MUTs, including thin solid, powdered solid, bulk solid, liquid in sample holder or fluidic channels, and gas.

i. Thin Solid MUTs

Thin solid MUTs can be directly placed in close proximity to metamaterial-inspired resonators. Figure 4 shows a two-port planar microwave sensor integrated with a "52"-shaped complementary resonator for characterization of low-permittivity materials (Haq et al., 2020). The sensor comprises a microstrip transmission line on top of an FR-4 dielectric substrate. The complementary resonator is etched out from the ground plane directly below the microstrip line. For the unloaded sensing structure (without MUT), the structure resonates at 15.12 GHz with a notch depth, $|S_{21}|$ of –44.33 dB. The thin dielectric MUT is then loaded on top of the complementary resonator structure [refer Figure 4 (b)]. Based on Figure 4 (c), the resonant frequency shifts to lower frequency when the dielectric constant, ε_r increases. This shift of resonant frequency allows the sensing structure to perform permittivity measurement for thin dielectric materials. For this type of configuration, it is important to ensure the surface of the MUTs is flat to avoid the presence of an air gap between the MUTs and the resonators.

Figure 4. (a) Microstrip transmission line loaded with complementary "52"-shaped resonator, (b) Complementary resonator loaded with thin solid MUT, (c) Transmission coefficient, $|S_{21}|$ against frequency, f for MUTs with different dielectric constant, ε_r

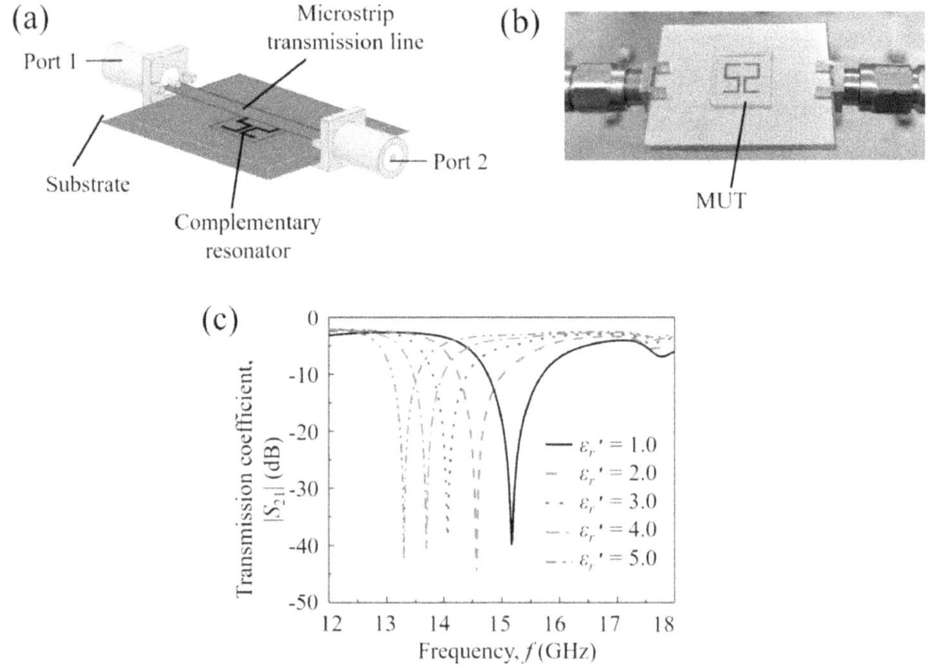

ii. Powdered Solid MUTs

Solid MUTs in powder form can be directly introduced into the sensing region (Kurland et al., 2023). Figure 5 shows a two-port planar microwave sensor integrated with a square CSRR for the characterization of coal powder (Shahzad et al., 2022). The sensor comprises a microstrip transmission line on top of an FR-4 dielectric substrate. The CSRR resonator is etched out from the ground plane directly below the microstrip line. For the unloaded sensing structure (without MUT), the structure resonates at 4.75 GHz with a notch depth, $|S_{21}|$ of –41 dB. The coal powder is then loaded on top of the CSRR structure [refer Figure 5 (b)]. As shown in Figure 5 (c), the resonant frequencies for black powder (Anthracite) and grey powder (Bituminous) are at 4.139 GHz and 4.305 GHz, respectively. This distinctive resonant behaviour allows the sensor to distinguish between coal powders. To improve the accuracy and repeatability, it is crucial to cover the sensing region with the same amount of powder. Therefore, a sample holder is suggested to ensure the powder is compact, covering the sensing region to a particular thickness.

Figure 5. (a) Microstrip transmission line on FR-4 substrate (top view) and CSRR on the ground plane (bottom view), (b) CSRR loaded with coal powder, (c) Transmission coefficient, $|S_{21}|$ against frequency, f for air (without sample), black coal powder (Anthracite), and grey coal powder (Bituminous)

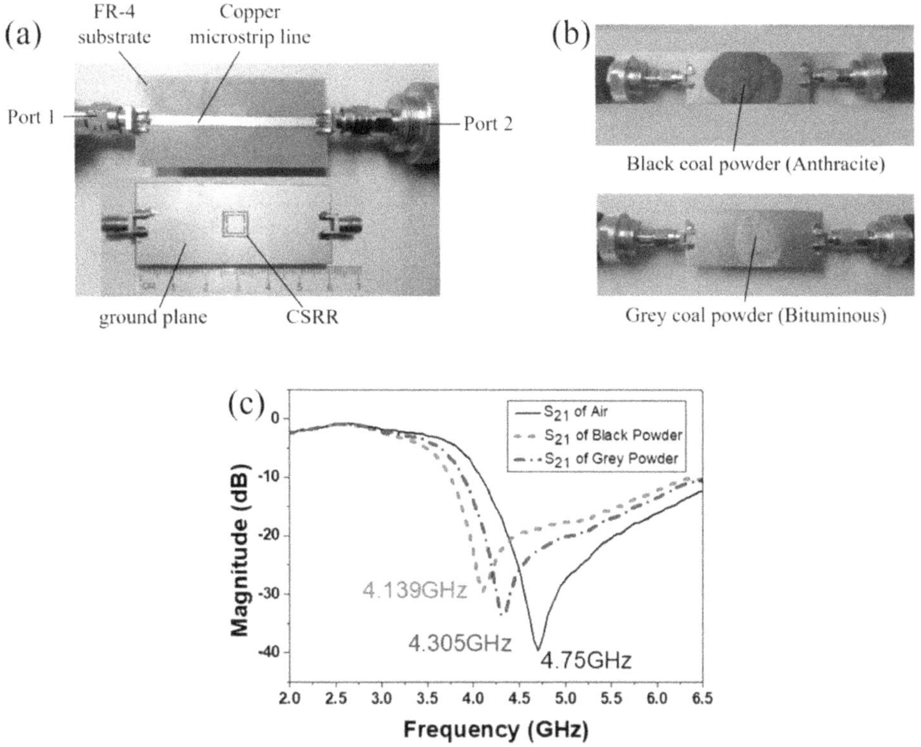

iii. Liquid MUTs in Sample Holders

A small amount of liquid sample can be placed on the resonator using the dropping method (Ali et al., 2023; Alrayes & Hussein, 2021; Jafari & Rad Malekshahi, 2023). The liquid sample is typically placed in the gap structure of the resonator. However, controlling the amount of liquid and the position of the drop using this technique can be challenging. Therefore, a sample holder is preferred if the amount of liquid MUTs is adequate. A hole can be drilled near to the resonator, acting as a reservoir, for the placement of liquid MUTs (Abdulkarim et al., 2021). Figure 6 shows a two-port planar microwave sensor integrated with an Omega-shaped split ring resonator (SRR) for characterization of liquid chemicals (Abdulkarim, Deng, Karaaslan, Altıntaş, et al., 2020). The sensor comprises a microstrip transmission line bent around the Omega-shaped SRR. For the unloaded sensing structure (without MUT), the structure resonates at 1.9 GHz with a notch depth, $|S_{21}|$ of –7 dB. The liquid methanol is then loaded into the sensor layer, which is a cylindrical hole in the substrate layer [refer Figure 6 (b)]. Based on Figure 6 (c), the resonant frequency for methanol shifts to a higher frequency when the concentration increases. This distinctive resonant behaviour allows the sensor to measure the concentration of liquid chemicals.

Figure 6. (a) Microstrip transmission line on FR-4 substrate integrated with an Omega-shaped split ring resonator, (b) Liquid methanol sample loaded in the sensor layer, (c) Transmission coefficient, $|S_{21}|$ against frequency, f for methanol solution of different concentration

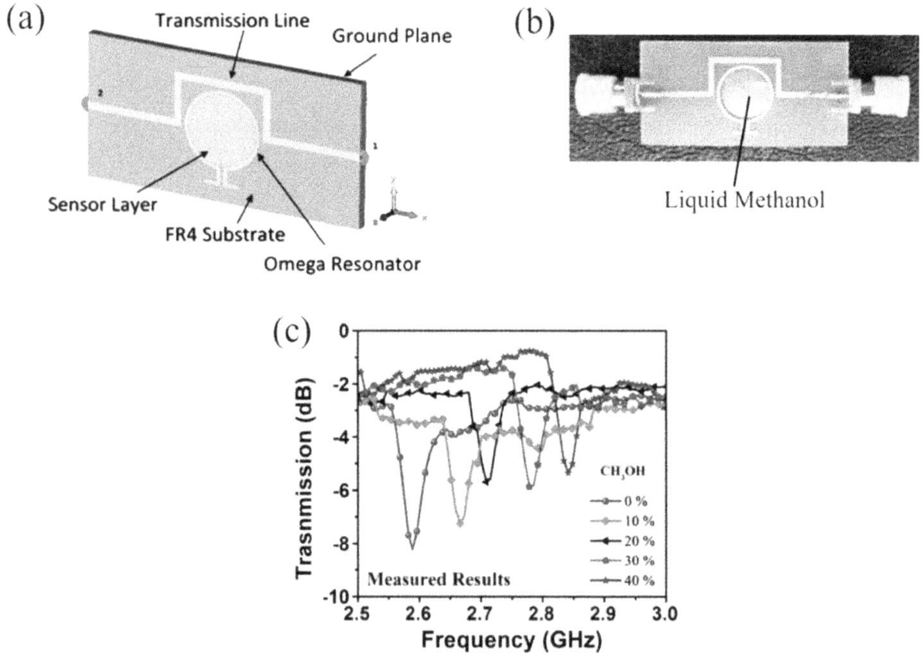

Adapted from Abdulkarim et al. (2020) licensed under Creative Commons Attribution (CC BY 4.0 Deed). Copyright 2020, MDPI.

iv. Liquid MUTs through Fluidic Channels

Liquid chemicals can be introduced onto microwave sensors through capillary tubes (Yasin et al., 2023) or microfluidic channels (Song & Huang, 2023). The fluidic channels can be placed horizontally on top of the resonators or below the resonators (through the substrate). Moreover, the fluidic channels can also pass through the substrate vertically through a drilled hole (Gulsu et al., 2021; Kumar et al., 2021). The microfluidic channels can be meandered and bent around to achieve a better coupling configuration and improved sensitivity. Figure 7 shows a substrate-integrated waveguide (SIW) filter with a capillary tube placed horizontally on top of the structure. The capillary tube is loaded with different liquid chemical MUTs having different dielectric constant, ε_r . The transmission coefficient, $|S_{21}|$ responses are shown in Figure 7 (b). Both peaks in the curve shift to a lower frequency when the liquid MUTs are loaded compared to the empty capillary tube. Continuous monitoring of liquid MUTs is possible by integrating a pumping system to pump the liquid MUTs though the capillary tubes.

Figure 7. (a) Substrate-integrated waveguide (SIW) filter with loosely-coupled feed lines and capillary tubes containing liquid samples, (b) Transmission coefficient, $|S_{21}|$ against frequency for different liquid chemicals

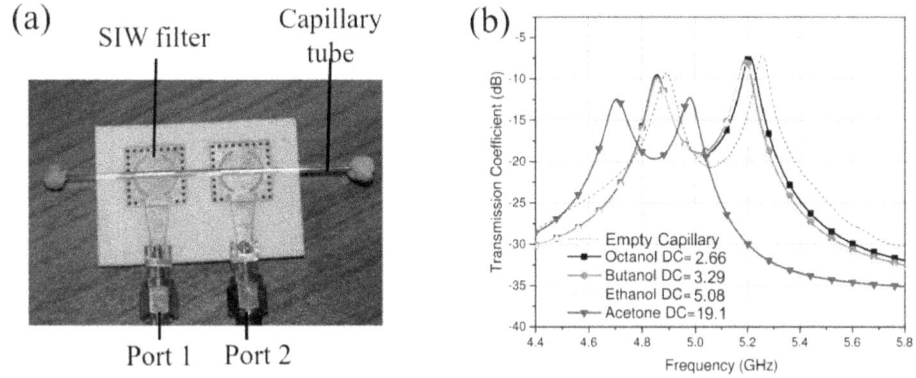

v. Bulk Solid or Liquid MUTs

Planar metamaterial sensors can be used to characterize bulk solid MUTs such as concretes and wood products by properly design their topology. For example, a V-shaped resonator is designed to measure the properties of construction materials at the 90° corners (O. Akgol et al., 2019). For MUTs such as soils, fertilizers, and liquids, submersible sensing structures can be designed to immerse the sensing structure into the MUTs. For submersible sensors, the transmission line can be designed so that both ports are at the same side of the substrate (Dalgac et al., 2021; Galindo-Romera et al., 2016; Tiwari et al., 2020). Figure 8 shows an example of submersible sensor. The planar sensor is based on a rectangular SRR placed between a pair of monopole antennas. The resonant frequency, f_r shifts to a lower frequency range for liquid MUTs of higher dielectric constant, ε_r .

Figure 8. (a) Submersible sensor with a rectangular split ring resonator (SRR) in between a pair of monopole antennas, submerged into a liquid material under test (MUT), (b) The transmission coefficient, $|S_{21}|$ against frequency for liquid MUT of different dielectric constant, ε_r.

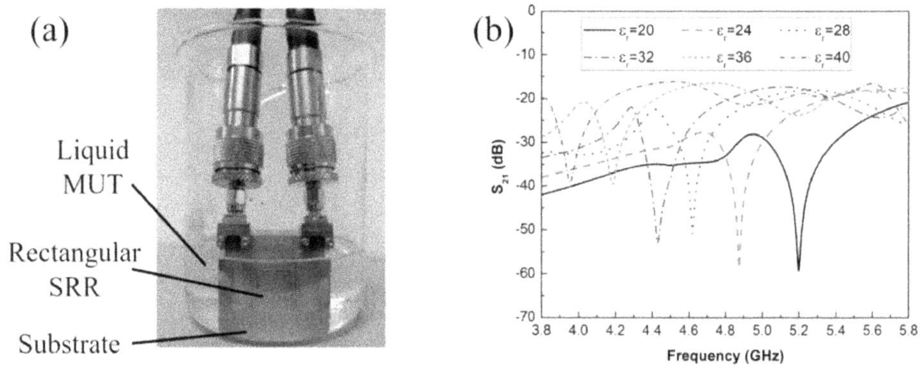

vi. Gas MUTs

Sensing measurement of volatile vapours such as ethanol, methanol, acetone vapours can be conducted with metamaterial-based sensors aided by porous substrate materials such as the Rogers series. The tested gas sample would be absorbed into the substrate, thereby changing the permittivity of the substrate (Wu et al., 2021). The substrate permittivity increases according to the concentration of the gas MUTs, which affects the rate of absorption of the substrate. The resonance responses can be monitored to estimate the concentration of the volatile gas. It is worth mentioning that while gas sensing is possible using metamaterials in the microwave regime, much research has been focused on utilizing mid-infrared terahertz metasurfaces for gas detection (Tabassum et al., 2022).

Metamaterial-Loaded Planar Antennas

Metamaterials can be integrated into planar antennas for gain enhancement and size reduction (Dewan et al., 2017; Esmail et al., 2022; Islam et al., 2018; Milias et al., 2021). For example, the Vivaldi antenna is loaded with metamaterial elements [refer to Figure 9 (a)] for breast tumour detection (Slimi et al., 2022). The Vivaldi antenna achieves a return loss of 25 dB and a gain of 6.2 dB at 5 GHz. The incorporation of metamaterial elements enhances and increases the distribution of magnetic fields. The peak values of magnetic fields are higher for the Vivaldi antenna with metamaterial elements compared to the Vivaldi antenna without metamaterial elements. As depicted in Figure 9 (b), an array of 8 antennas is arranged around the breast phantom in a circular and symmetrical configuration. The result shows that the return loss for the antenna closest to the tumour is the lowest. This conceptual design proves its possible application in breast tumour detection and positioning.

Figure 9. (a) Vivaldi antenna loaded with metamaterial elements, (b) Array Vivaldi antennas for breast tumour detection, (c) Magnetic field distribution of Vivaldi antenna without and with metamaterials at 5 GHz

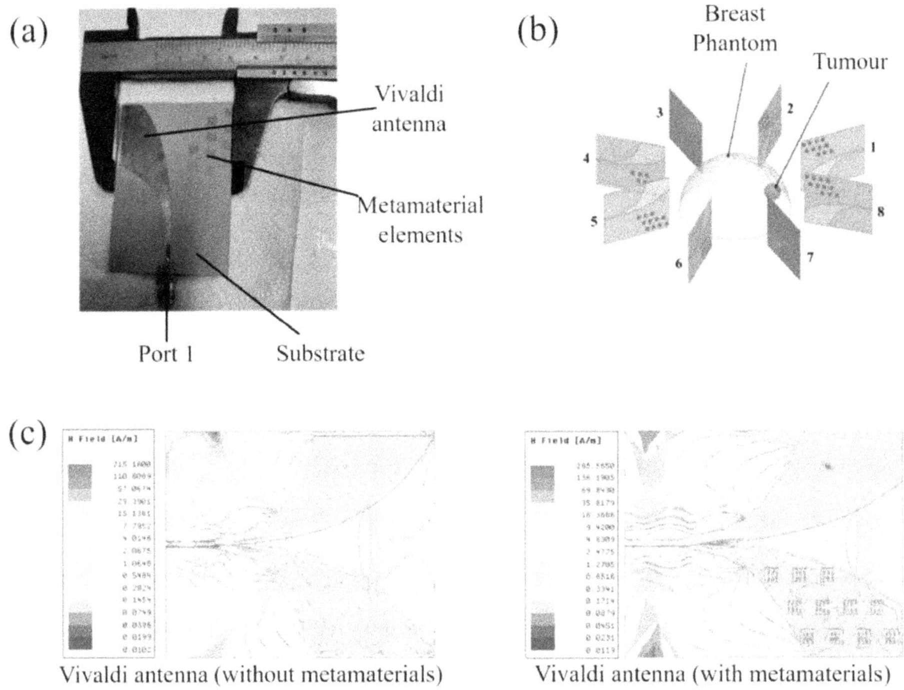

Adapted from Slimi et al., (2022) licensed under Creative Commons Attribution (CC BY 4.0 Deed). Copyright 2022, MDPI.

Comparison of Planar Metamaterial-based Sensors

Table 1 summarizes selected recent planar metamaterial-based sensors for material characterization and compares their types of resonators, topology, applications, operating frequencies, sensing principles, and sensitivity or performance metrics. For sensors measuring the variation of dielectric constants, ε_r, the sensitivity, S can be calculated as (Ali et al., 2023; Rabbani et al., 2024):

$$S(\%) = \frac{\Delta f_r}{f_r(\Delta \varepsilon_r')} \times 100 \quad (3)$$

where Δf_r represents the shift of resonant frequency, f_r is the unloaded or reference resonant frequency, $\Delta\varepsilon_r$ is the difference in the dielectric constants of MUT ($\Delta\varepsilon_r = \varepsilon_r - 1$, for the case in which the sensor compares the dielectric constants of MUT and air). This sensitivity, S (in %) is useful for comparing the sensitivity of different dielectric sensors operating at different resonant frequencies, f_r. It is worth mentioning that sensitivity can be expressed according to their applications and sensing principles (Chowdhury et al., 2024). For example, for a temperature sensor based on amplitude-variation sensing principle, the sensitivity, S can be expressed as the ratio of the change in magnitude of the transmission coefficient, $|S_{21}|$ to the change in temperature, and therefore its unit becomes dB/°C.

Recent advancements in planar metamaterial sensors have introduced various design configurations. For instance, power divider/combiner sections (Kiani et al., 2020), rat-race couplers (Velez et al., 2023), and defected ground structures (Alrayes & Hussein, 2021) are being explored. Different coupling configurations between transmission lines, resonators, and MUTs are being investigated. The state-of-the-art planar metamaterial-inspired sensors has broadened their applications beyond focusing solely on dielectric material properties. Utilizing similar sensing principles, they are now capable of applications such as measuring liquid flow rates (Niksan et al., 2022), monitoring bacterial growth (Jain et al., 2020), and imaging breast tissue for tumour detection (Islam et al., 2018; Slimi et al., 2022). In addition, machine learning integration has also been observed in automating the design of metamaterial-inspired resonator topology (Saadat-Safa et al., 2019; B. X. Wang et al., 2021) and microfluidic channels routing (Zhao et al., 2022). Furthermore, machine learning algorithms are employed to predict the properties of the MUTs (Harrsion et al., 2020; Prakash & Gupta, 2023), and to mitigate cross-sensitivity to ambient factors (Kazemi et al., 2021).

Table 1. List of recent planar metamaterial microwave sensors with their applications and performances

Reference		Sensor Topology	Applications	Unloaded f_r (GHz)	Sensing Principle	Sensitivity, S
S. Mohammadi & Zarifi (2021)	SRR	•Power divider/ combiner section •Differential	Determine the concentration of volatile organic compound (VOC) in real-time	5.12 and 5.65	Frequency splitting	71 MHz/ ε_r
P. Mohammadi et al. (2023)	Modified SRR	•Two-port microstrip transmission line (TL)	Measure glucose concentration in aqueous solutions	2.32	Frequency shift	•0.78 MHz/(mg/dL) based on the transmission poles •0.95 MHz/(mg/dL) based on the transmission zero

continued on following page

Table 1. Continued

Reference		Sensor Topology	Applications	Unloaded f_r (GHz)	Sensing Principle	Sensitivity, S
Javed et al. (2020)	Multiple CSRR	•Two-port microstrip TL •Capillary tube vertically through the substrate	Characterize liquid chemicals (butanol, ethanol, methanol)	2.45	Frequency shift	•Maximum resonant frequency shift = 400 MHz (for distilled water) •Maximum S = 0.214%
Jain et al. (2020)	Single SRR	•Power divider •Differential	Monitor the growth of *Escherichia coli* under various nutritional conditions	1.514 and 1.797	Amplitude variation	0.236 dB/hour (for bacteria growth in 2% glucose concentration)
Gan et al. (2020)	CSRR	•Two-port microstrip TL •Differential •Microfluidic channel	Determine the relative permittivity and loss tangent of the liquid sample	1.618	Frequency shift	0.626%
Niksan et al. (2022)	SRR	•Two-port microstrip TL •3D-printed liquid channel on top of the resonator	Measure the flow rates of uniform liquids	4.852 (Bare sensor); 4.612 (Sensor with 3D-printed channel)	Amplitude variation	0.02 dB/mL/h (Resonant amplitude to flow rate variations for water)

CASE STUDY: COMPLEMENTARY SPLIT-RING-RESONATOR-BASED PLANAR METAMATERIAL MICROWAVE SENSORS

In this section, a planar microstrip transmission line loaded with a CSRR is discussed. Firstly, the theory and analytical calculation of the CSRR are presented. Furthermore, the structure is simulated using Computer Simulation Technology (CST) Microwave Studio software. Finally, the performance of the sensing structure for loading of MUT with varying dielectric constants, ε_r is discussed.

Theory

A resonant circuit, also known as an *LC* circuit and tank circuit, is an electrical circuit consisting of an inductor and a capacitor that store and exchange energy at a resonant frequency, f_r. When a voltage source applies a voltage across the circuit, current flows through the inductor, which opposes the change in current. The inductor builds up a magnetic field and stores energy as the current flows through it. Meanwhile, the capacitor begins to charge due to the applied voltage, storing energy as an electric field when voltage is applied across it. Energy oscillates between the electric field of the capacitor and the magnetic field of the inductor. The resonant frequency, f_r can be calculated as: -

$$f_r = \frac{1}{2\pi\sqrt{L_C C_C}}$$

. (4)

where C_C and L_C represents the capacitance and inductance associated with the resonator, respectively.

A CSRR can be etched out from the ground plane and coupled to a microstrip transmission line on the other side of the substrate, as shown in Figure 10 (a). The CSRR is a complementary structure of the SRR. By applying the Babinet principle to the SRR and CSRR, the electromagnetic response of the CSRR is inversely related to that of the SRR. The SRR behaves as a resonant magnetic dipole that can be excited by an axial magnetic field, whereas the CSRR behaves as an electric dipole that can be excited by an axial electric field. The inductances in the SRR become capacitances in the CSRR, whereas the capacitances in the SRR become inductances in the CSRR. The equivalent circuit model for the microstrip transmission line loaded with a CSRR structure is depicted in Figure 10 (b) (Baena et al., 2005). L and C represent the per-section inductance and capacitance of the microstrip transmission line.

Figure 10. (a) Topology of the microstrip transmission line integrated with a complementary split ring resonator (CSRR) and its (b) equivalent circuit model

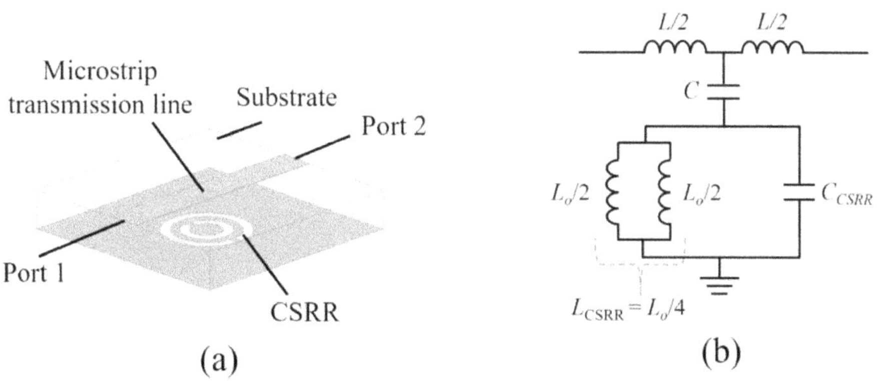

The values of C_{CSRR} and L_{CSRR} are determined by equation (5) and equation (6), respectively (Baena et al., 2005; Marqués et al., 2003) as:

$$C_{CSRR} = \frac{\pi^3 \varepsilon_o}{c^2} \int_0^\infty \frac{[bB(kb) - aB(ka)]^2}{k^2} \left[\frac{1}{2} \left(1 + \frac{1 + \varepsilon/\varepsilon_o \tanh(kh)}{1 + \varepsilon_o/\varepsilon \tanh(kh)} \right) \right] dk \quad (5)$$

$$L_{CSRR} = \frac{\pi^3 \mu_o}{4c^2} \int_0^\infty \frac{[bB(kb) - aB(ka)]^2}{k^2} dk \quad (6)$$

where a and b are the geometrical parameters, h is the thickness of the substrate, B is a function involving Struve-Bessel function integral.

The geometrical parameter, a can be calculated as:

$$a = r_o - \frac{c}{2} \quad (7)$$

where r_o is the average radius of the C_{CSRR}, c is the width of the ring.

The geometrical parameter, b can be calculated as:

$$b = a + c \quad (8)$$

The function B is defined as:

$$B(x) = S_o(x) J_1(x) - S_1(x) J_o(x) \quad (9)$$

where S_n and J_n is the nth-order Struve and Bessel functions, respectively. The S_n and J_n follow equation (10) and equation (11), respectively, which can be computed using standard integration routines (Goswami et al., 2019).

$$S_n(x) = \sum_{m=0}^{\infty} \frac{(-1)^m}{\Gamma(m + 3/2)\Gamma(m + n + 3/2)} \left(\frac{x}{2} \right)^{m+1} \quad (10)$$

$$J_n(x) = \left(\frac{x}{2} \right)^n \sum_{m=0}^{\infty} \frac{(-x^2/4)^m}{m!\Gamma(n + m + 1)} \quad (11)$$

From these expressions, analytical computation of the resonant frequency of the CSRR structure can be determined.

Design and Simulation

The substrate is FR-4 (dielectric constant, $\varepsilon_{FR4} = 4.3$; thickness, $h = 1.6$ mm). The width of the microstrip transmission line, w is 3.11 mm. The dimensions of the CSRR [refer to Figure 12 (b)] are: radius of the outer ring, $r = 4.55$ mm; gap of the split, $g = 0.5$ mm; gap distance between inner and outer ring, $s = 0.5$ mm; width of the ring, $w = 0.5$ mm). The analytically determined inductance, $L_{CSRR} = 17.232$ nH; capacitance, $C_{CSRR} = 214.199$ fF; resonant frequency, $f_r = 2.62$ GHz. The structure is modelled and simulated using CST Microwave Studio software. Initially, the structure is modelled as depicted in Figure 11, with two waveguide ports added for both ends of the microstrip line, labelled as port 1 and port 2.

Figure 11. Modelling of the microstrip line loaded with CSRR in CST microwave studio software

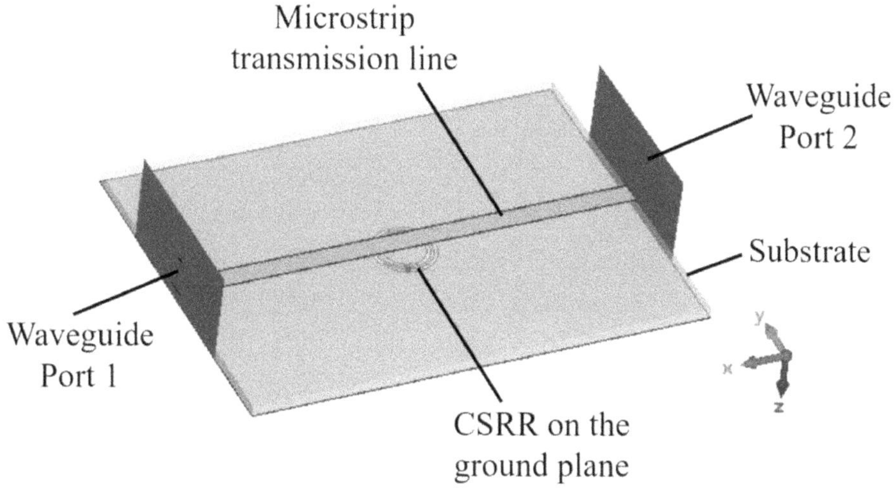

RESULTS AND DISCUSSION

In this subsection, the S-parameter responses for the microstrip line coupled with a CSRR and multiple CSRRs are discussed. Furthermore, the shift in resonant frequencies when the sensing structure is loaded with MUTs of different dielectric properties is investigated.

Microstrip Transmission Line Coupled with a CSRR

The *S*-parameter response as a function of frequency for the microstrip transmission line coupled with a CSRR is depicted in Figure 12. The transmission coefficient, $|S_{21}|$ exhibits a transmission zero at 2.45 GHz. The notch depth, $|S_{21}|$, at the resonant frequency achieves –19 dB. Based on Figure 13, a parametric study is carried out to investigate the effect of the outer radius of outer ring, *r*; width of the ring, *w*; gap of the split, *g*; and gap distance between the inner and outer ring, *s*, on the $|S_{21}|$ responses. The increment of the outer radius of the outer ring, *r*, shifts the resonant frequency, f_r, to a lower frequency as the electrical length of the ring increases.

Figure 12. (a) Perspective view of the 50Ω microstrip transmission line coupled with a complementary split ring resonator (CSRR) (b) Enlarged view of the CSRR (r = 4.55 mm, g = s = w = 0.5 mm), (c) Simulated magnitude of reflection coefficient, $|S_{11}|$ and magnitude of transmission coefficient, $|S_{21}|$ as a function of frequency for TL-1CSRR

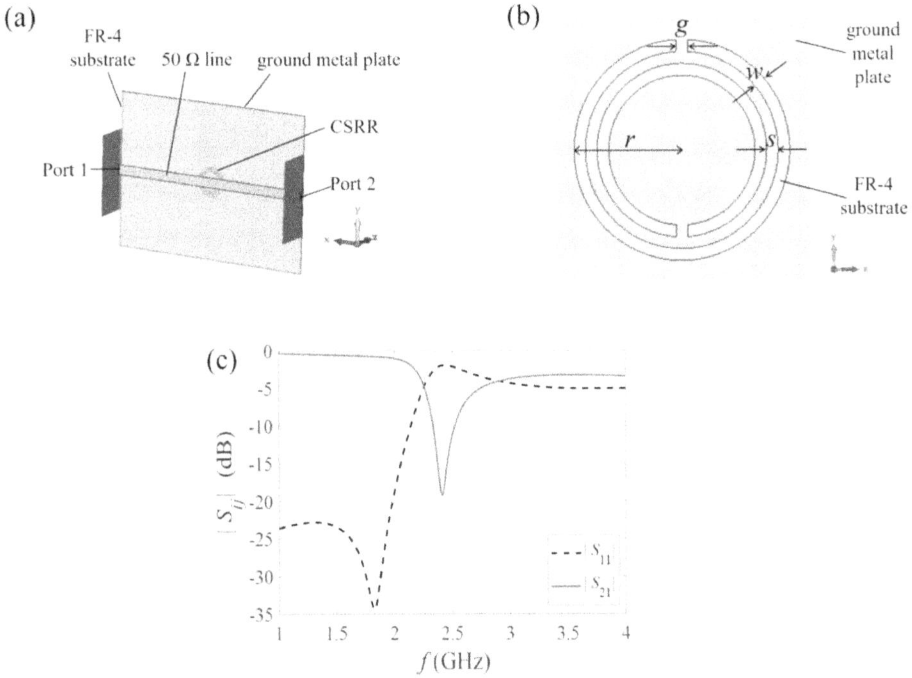

Figure 13. Parametric study of the microstrip transmission line coupled with single-CSRR with varying (a) outer radius of outer ring, r (b) width of ring, w, (c) gap of split, g, (d) gap distance between inner and outer ring, s

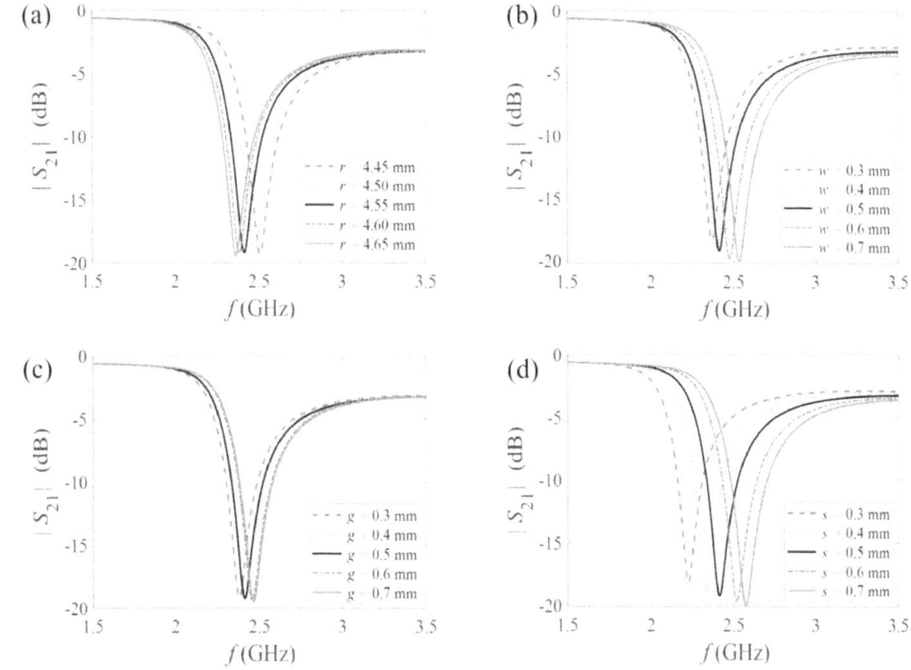

Microstrip Transmission Line Coupled with Multiple CSRRs

The microstrip transmission line is loaded with three CSRRs with a periodicity, *p* of 14.1 mm (gap distance between CSRRs, $g_{CSRR} = 5.0$ mm), and the *S*-parameter response is shown in Figure 14. The notch depth of the transmission coefficient, $|S_{21}|$, at resonant frequency achieves –42 dB. The notch depth improves compared to the line loaded with a single CSRR. With reference to Figure 15, a parametric study is carried out to investigate the effect of periodicity, *p*, on the transmission coefficient, $|S_{21}|$, response. When the periodicity increases (the gap between CSRRs increases from $g_{CSRR} = 1.0$ mm to 9.0 mm, with a step of 2.0 mm), the quality factor slightly improves.

Figure 14. (a) Perspective view of the 50Ω microstrip transmission line coupled with multiple-CSRRs (b) CSRRs with periodicity, p, (c) Simulated magnitude of reflection coefficient, $|S_{11}|$ and magnitude of transmission coefficient, $|S_{21}|$ as a function of frequency for TL-3CSRR

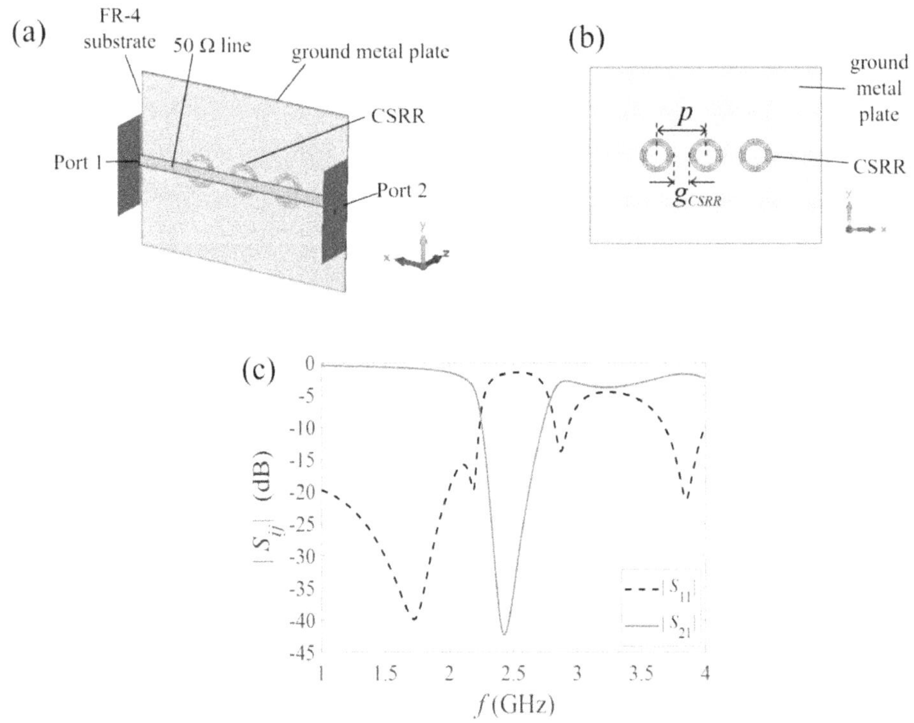

Figure 15. Parametric study of the microstrip transmission line coupled with 3-CSRR with varying periodicity, p

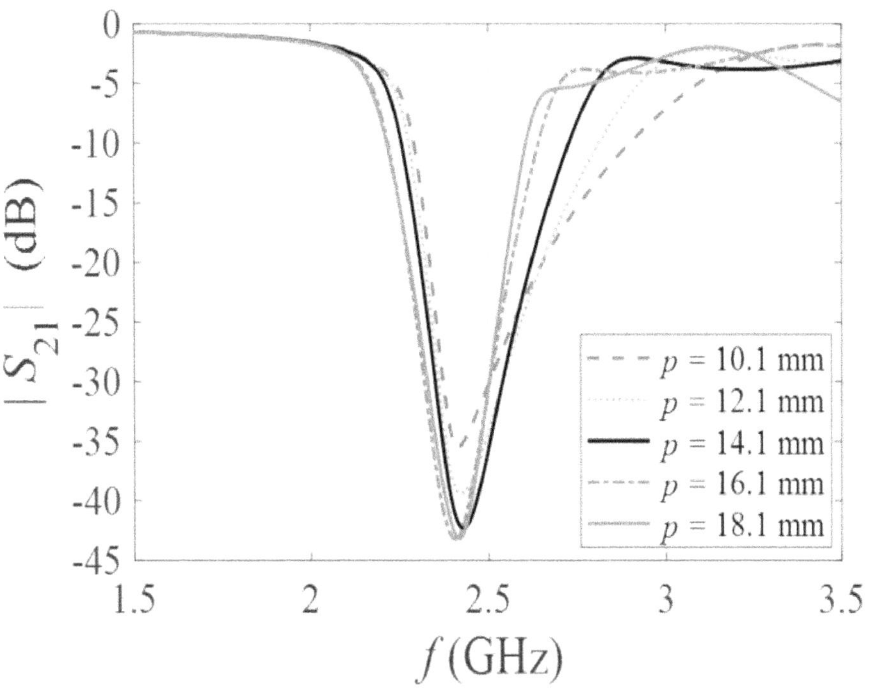

Frequency Shift Analyses for TL-CSRR Loaded With Dielectric MUTs

As depicted in Figure 16, a dielectric MUT with a thickness, t_{MUT}, of 10 mm is loaded in close proximity to the ground metal plane covering the CSRR structures. Figure 17 shows the transmission coefficient response as a function of frequency for the sensing structure loaded with MUTs of different dielectric constants, ε_r, ranging from 1 to 10. When the ε_r of the MUT increases, the resonant frequency shifts to a lower frequency range. These data can be used to estimate the ε_r of MUTs of the same type with unknown ε_r in future measurements.

Figure 16. The sensing structure loaded with flat MUT of thickness, t_{MUT}

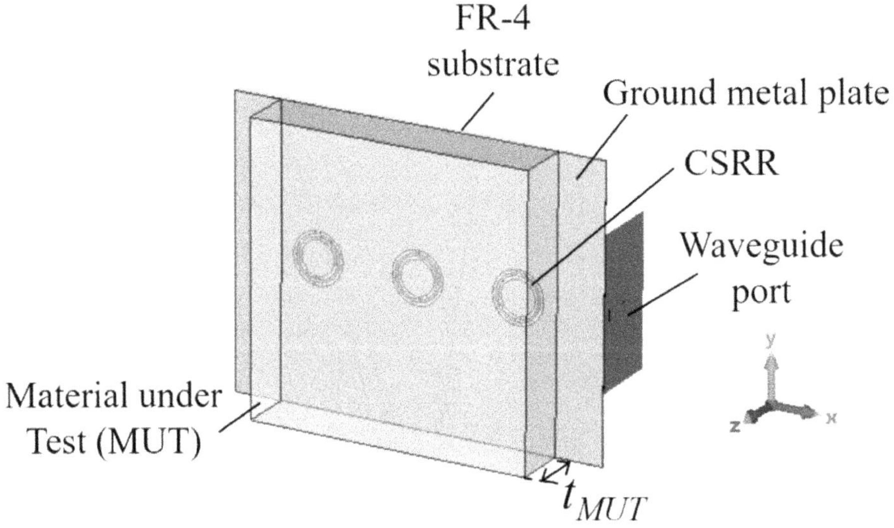

Figure 17. Shift of resonant frequencies for TL-3CSRR loaded with MUTs of different dielectric constant, ε_r

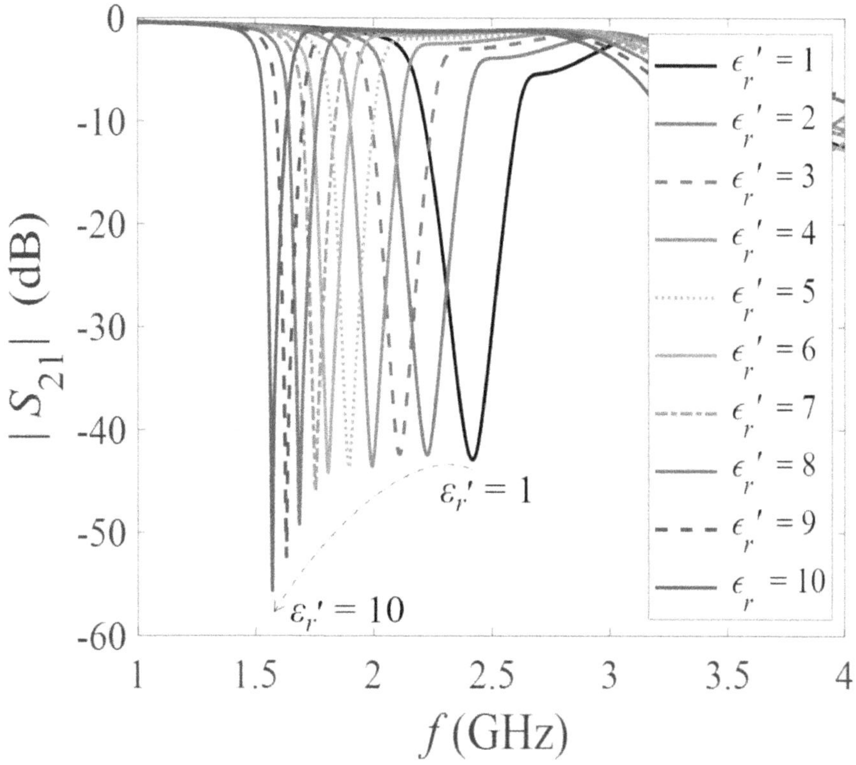

DESIGN STRATEGIES OF PLANAR METAMATERIAL MICROWAVE SENSORS

In this section, the design strategies to improve the performance and sensitivity of metamaterial-based microwave sensors are discussed.

Optimization of Resonator Pattern

The optimization of microwave sensor resonators is an essential factor that greatly contributes to improving the overall sensitivity and performance of these sensors. With careful design and optimization, the resonator geometrical pattern can be tailored to maximize sensitivity. In this regard, the resonator pattern may heavily affect the sensor's response to dielectric changes in its surrounding environment. There are several techniques that can be used to optimize resonators, including computer-aided design (CAD) and advanced algorithmic tools. Simulation software is used to carry out the optimization process, which involves adjusting the shapes, sizes, and arrangements of the resonators. However, this method can be time-consuming and arbitrary. Deep reinforcement learning, which combines deep learning with the decision-making abilities of reinforcement learning, can be applied to automatically optimize resonator patterns (B. X. Wang et al., 2021).

Swarm intelligence algorithms can also be implemented to optimize resonator patterns. Figure 18 shows an example of a microstrip line loaded with a complementary resonator designed using a binary particle swarm optimization algorithm for maximum sensitivity (Saadat-Safa et al., 2019). The sensing area is first pixelated [refer to Figure 18(b)]. Each pixel is then determined to be metalized or not. The final outcome is for the resonator pattern to yield the highest possible sensitivity. As shown in Figure 18(c), the fabricated sensor is used to measure the dielectric constants of chemical liquid materials. The results show that the optimized sensor has higher sensitivity compared to that of a simple CSRR sensor.

Figure 18. (a) Microstrip transmission line loaded with a complementary resonator, (b) Pixelated resonator optimized using binary optimization algorithm, (c) Fabricated sensor with the finalized optimized resonator pattern for characterization of dielectric liquid materials

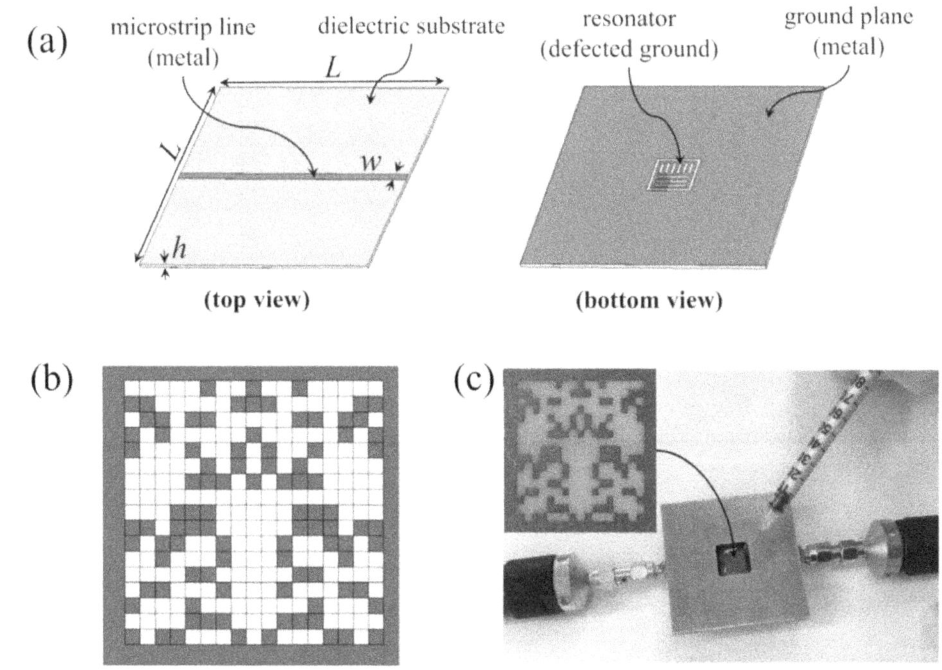

Coupling Configuration Enhancement

The coupling configuration between resonators and samples plays an important role in the design, functionality, and the ability of planar microwave sensors to provide accurate data in a variety of material characterization applications. The proper arrangement and positioning of MUTs are responsible for ensuring that the electromagnetic fields produced by the resonators interact with the MUTs effectively. For example, the capillary tube for the liquid sample can pass through the substrate via a drilled hole [in the direction normal to the ground plane as illustrated in Figure 19 (a)], so that the liquid sample is positioned in the region with the highest fields around the resonators (Chuma et al., 2018; Javed et al., 2020). As illustrated in Figure 19 (b), the liquid channel can also be embedded in the substrate right below the resonator (Zhou et al., 2018). Microfluidic channels are also used to properly position the dielectric liquids in the appropriate region for sensitivity enhancement (Abdelwahab et al., 2021; Salim & Lim, 2018).

Figure 19. Planar sensing structure for liquid characterization with liquid channel through the substrate (a) vertically (normal to the ground plane), (b) horizontally

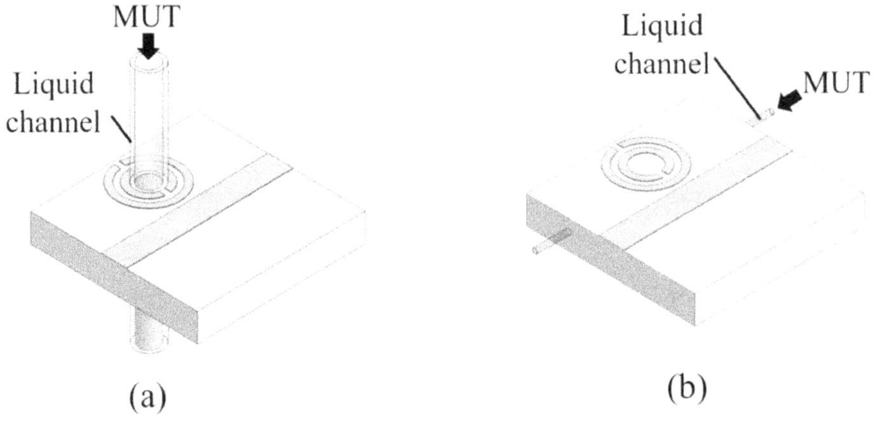

Neutralizing Environmental Factors for Cross-Sensitivities Mitigation

Most planar microwave sensors suffer from cross-sensitivity to ambient environmental factors such as temperature and humidity, leading to inaccurate measurements. Differential sensors can be designed, which load the microstrip line with two identical resonators (Ebrahimi et al., 2018; Velez et al., 2017). One resonator serves as a reference, while the other is loaded with MUT. Typically, this type of measurement follows the frequency splitting sensing principle. Using this approach, the robustness against environmental influences can be improved (Xie et al., 2023). Besides loading identical resonators onto a single microstrip transmission line, a power divider can also be used to realize differential measurement by delivering input power equally to two resonators (Jain et al., 2020). The isolation between the resonators is improved as the coupling between them is minimized.

The use of machine learning in neutralizing environmental factors is also an innovative approach. By using machine learning algorithms, it can potentially aid in interpreting and mitigating the impact of the environmental influences on sensor data. For instance, a sensor guided by machine learning can precisely measure the concentration of the tested material in the mixture without being affected by the surrounding environment (Kazemi et al., 2021). For machine learning to have the capacity to identify patterns related to environmental influences, an extensive dataset covering a variety of environmental conditions is required during system training. This ensures that the sensor is able to adjust and compensate for environment changes, leading to more reliable and stable functioning. It is undoubtedly recognized that machine learning allows the sensor to distinguish between alterations caused by the desired material and those induced by external factors, thereby increasing the overall accuracy of planar microwave sensors in real-world situations.

CONCLUSION

Metamaterial-inspired microwave sensors find extensive applications in material characterization. This chapter discusses the recent progress of metamaterial-based microwave sensors for measuring various types of samples. In brief, metamaterials enhance these sensing devices by localizing and amplifying the fields. By loading samples onto the appropriate sensing region, the sensors can measure changes according to their respective sensing principles, such as frequency shift, frequency splitting, and amplitude or phase variation. Additionally, this chapter demonstrates the simulation of a complementary-split-ring-loaded microstrip line and the shifting of resonant frequencies for loading of thin dielectric samples with different dielectric constants. Furthermore, design strategies for metamaterial-based microwave sensors, including resonator pattern optimization, coupling enhancement, and cross-sensitivity mitigation, are provided.

ACKNOWLEDGMENT

This work was supported in part by Universiti Teknologi Malaysia under UTM Fundamental Research Grant (Q.J130000.3851.22H13).

REFERENCES

Abdelwahab, H., Ebrahimi, A., Tovar-Lopez, F. J., Beziuk, G., & Ghorbani, K. (2021). Extremely sensitive microwave microfluidic dielectric sensor using a transmission line loaded with shunt LC resonators. *Sensors, 21*(20), 6811-1-6811–6813. 10.3390/s21206811

Abdulkarim, Y. I., Deng, L., Karaaslan, M., Altıntaş, O., Awl, H. N., Muhammadsharif, F. F., Liao, C., Unal, E., & Luo, H. (2020). Novel metamaterials-based hypersensitized liquid sensor integrating omega-shaped resonator with microstrip transmission line. *Sensors (Switzerland), 20*(3), 943-1-943–18. 10.3390/s20030943

Abdulkarim, Y. I., Deng, L., Karaaslan, M., Dalgaç, Ş., Mahmud, R. H., Alkurt, F. O., Muhammadsharif, F. F., Awl, H. N., Huang, S., & Luo, H. (2020). The detection of chemical materials with a metamaterial-based sensor incorporating oval wing resonators. *Electronics, 9*(5), 825-1-825–15. 10.3390/electronics9050825

Abdulkarim, Y. I., Muhammadsharif, F. F., Bakır, M., Awl, H. N., Karaaslan, M., Deng, L., & Huang, S. (2021). Hypersensitized metamaterials based on a corona-shaped resonator for efficient detection of glucose. *Applied Sciences (Basel, Switzerland)*, 11(1), 1–19. 10.3390/app11010103

Akgol, Oguzhan, & Unal, H. (2018). Metamaterial-based multifunctional sensor design for moisture, concrete aging and ethanol density sensing applications. *Modern Physics Letters B, 32*(23). 10.1142/S0217984918502718

Akgol, O., Unal, E., Bağmancı, M., Karaaslan, M., Sevim, U. K., Öztürk, M., & Bhadauria, A. (2019). A nondestructive method for determining fiber content and fiber ratio in concretes using a metamaterial sensor based on a V-shaped resonator. *Journal of Electronic Materials*, 48(4), 2469–2481. 10.1007/s11664-019-06937-w

Ali, U., Jabbar, A., Yi, X., Naveed, M. A., Mehmood, M. Q., Zubair, M., & Massoud, Y. (2023). A novel fractal Hilbert curve-based low-cost and highly sensitive microwave sensor for dielectric characterization of liquid materials. *IEEE Sensors Journal*, 23(20), 23950–23957. 10.1109/JSEN.2023.3312309

Alrayes, N., & Hussein, M. I. (2021). Metamaterial-based sensor design using split ring resonator and Hilbert fractal for biomedical application. *Sensing and Bio-Sensing Research*, 31, 100395-1, 100395–10. 10.1016/j.sbsr.2020.100395

Baena, J. D., Bonache, J., Martín, F., Sillero, R. M., Falcone, F., Lopetegi, T., Laso, M. A. G., García-García, J., Gil, I., Portillo, M. F., & Sorolla, M. (2005). Equivalent-circuit models for split-ring resonators and complementary split-ring resonators coupled to planar transmission lines. *IEEE Transactions on Microwave Theory and Techniques*, 53(4), 1451–1460. 10.1109/TMTT.2005.845211

Bogue, R. (2017). Sensing with metamaterials: A review of recent developments. *Sensor Review*, 37(3), 305–311. 10.1108/SR-12-2016-0281

Castillo, E. S. R., Fernandez, E., Aranibar, P. C., & Vargas, D. S. (2018). Metamaterial inspired multiband planar array to detect glyphosate in water by real-time electromagnetic wave sensor. *2018 IEEE MTT-S Latin America Microwave Conference, LAMC 2018 - Proceedings*, (pp. 1–3). IEEE. 10.1109/LAMC.2018.8699077

Chen, T., Li, S., & Sun, H. (2012). Metamaterials application in sensing. *Sensors (Basel)*, 12(3), 2742–2765. 10.3390/s120302742 22736975

Chowdhury, M. Z. B., Islam, M. T., Alzamil, A., Soliman, M. S., & Samsuzzaman, M. (2024). A tunable star-shaped highly sensitive microwave sensor for solid and liquid sensing. *Alexandria Engineering Journal, 86*(July 2023), 644–662. 10.1016/j.aej.2023.12.001

Chuma, E. L., Iano, Y., Fontgalland, G., & Bravo Roger, L. L. (2018). Microwave sensor for liquid dielectric characterization based on metamaterial complementary split ring resonator. *IEEE Sensors Journal*, 18(24), 9978–9983. 10.1109/JSEN.2018.2872859

Costanzo, A., Augello, E., Battistini, G., Benassi, F., Masotti, D., & Paolini, G. (2023). Microwave devices for wearable sensors and IoT. *Sensors (Basel)*, 23(9), 4356. 10.3390/s23094356 37177569

Dai, L., Zhao, X., Guo, J., Feng, S., Fu, Y., Kang, Y., & Guo, J. (2020). Microfluidics-based microwave sensor. *Sensors and Actuators. A, Physical*, 309, 111910-1, 111910–111916. 10.1016/j.sna.2020.111910

Dalgac, S., Akdogan, V., Kiris, S., Incesu, A., Akgol, O., Unal, E., Basar, M. T., & Karaaslan, M. (2021). Investigation of methanol contaminated local spirit using metamaterial based transmission line sensor. *Measurement*, 178(April), 109360-1, 109360–109369. 10.1016/j.measurement.2021.109360

Dalgac, S., Baklr, M., Karadag, F., Karaaslan, M., Akgol, O., Unal, E., & Sabah, C. (2020). Microfluidic sensor applications by using chiral metamaterial. *Modern Physics Letters B, 34*(5), 2050031-1-2050031–14. 10.1142/S0217984920500311

Dalgaç, Ş., Furat, M., Karaaslan, M., Akgöl, O., Karadağ, F., Zile, M., & Bakir, M. (2020). Grease oil humidity sensor by using metamaterial. *Journal of Electromagnetic Waves and Applications*, 34(18), 2488–2498. 10.1080/09205071.2020.1824690

Dalgaç, Ş., Karada , F., Ünal, E., Özkaner, V., Baklr, M., Akgöl, O., Sevim, U. K., Delihaclo lu, K., Öztürk, M., Karaaslan, M., & Sabah, C. (2020). Metamaterial sensor application concrete material reinforced with carbon steel fiber. *Modern Physics Letters B, 34*(10). 10.1142/S0217984920500979

Dewan, R., Rahim, M. K. A., Hamid, M. R., Yusoff, M. F. M., Samsuri, N. A., Murad, N. A., & Kamardin, K. (2017). Artificial magnetic conductor for various antenna applications: An overview. *International Journal of RF and Microwave Computer-Aided Engineering*, 27(6), 1–18. 10.1002/mmce.21105

Ebrahimi, A., Scott, J., & Ghorbani, K. (2018). Differential sensors using microstrip lines loaded with two split-ring resonators. *IEEE Sensors Journal*, 18(14), 5786–5793. 10.1109/JSEN.2018.2840691

Elwi, T. A. (2018). Metamaterial based a printed monopole antenna for sensing applications. *International Journal of RF and Microwave Computer-Aided Engineering*, 28(7), 1–10. 10.1002/mmce.21470

Esmail, B. A. F., Koziel, S., & Szczepanski, S. (2022). Overview of planar antenna loading metamaterials for gain performance enhancement: The two decades of progress. *IEEE Access : Practical Innovations, Open Solutions*, 10, 27381–27403. 10.1109/ACCESS.2022.3157634

Galindo-Romera, G., Javier Herraiz-Martínez, F., Gil, M., Martínez-Martínez, J. J., & Segovia-Vargas, D. (2016). Submersible printed split-ring resonator-based sensor for thin-film detection and permittivity characterization. *IEEE Sensors Journal*, 16(10), 3587–3596. 10.1109/JSEN.2016.2538086

Gan, H. Y., Zhao, W. S., Liu, Q., Wang, D. W., Dong, L., Wang, G., & Yin, W. Y. (2020). Differential microwave microfluidic sensor based on microstrip complementary split-ring resonator (MCSRR) structure. *IEEE Sensors Journal*, 20(11), 5876–5884. 10.1109/JSEN.2020.2973196

Goswami, S., Sarmah, K., Sarma, A., Sarma, K. K., & Baruah, S. (2019). Design considerations pertaining to the application of complementary split ring resonators in microstrip antennas. In *Emerging Innovations in Microwave and Antenna Engineering* (pp. 25–56). Springer. 10.4018/978-1-5225-7539-9.ch002

Govind, G., & Akhtar, M. J. (2019). Metamaterial-inspired microwave microfluidic sensor for glucose monitoring in aqueous solutions. *IEEE Sensors Journal*, 19(24), 11900–11907. 10.1109/JSEN.2019.2938853

Gulsu, M. S., Bagci, F., Can, S., Yilmaz, A. E., & Akaoglu, B. (2021). Minkowski-like fractal resonator-based dielectric sensor for estimating the complex permittivity of binary mixtures of ethanol, methanol and water. *Sensors and Actuators. A, Physical*, 330, 112841-1, 112841–10. 10.1016/j.sna.2021.112841

Haq, T., & Koziel, S. (2023). New complementary resonator for permittivity- and thickness-based dielectric characterization. *Sensors (Basel)*, 23(22), 1–14. 10.3390/s2322913838005525

Haq, T., Ruan, C., Zhang, X., Ullah, S., Fahad, A. K., & He, W. (2020). Extremely sensitive microwave sensor for evaluation of dielectric characteristics of low-permittivity materials. *Sensors, 20*(7), 1916-1-1916–1917. https://doi.org/10.3390/s2007191

Harnsoongnoen, S. (2021). Metamaterial-inspired microwave sensor for detecting the concentration of mixed phosphate and nitrate in water. *IEEE Transactions on Instrumentation and Measurement*, 70, 9509906. 10.1109/TIM.2021.3086901

Harrsion, L., Ravan, M., Tandel, D., Zhang, K., Patel, T., & Amineh, R. K. (2020). Material identification using a microwave sensor array and machine learning. *Electronics (Switzerland)*, 9(2). Advance online publication. 10.3390/electronics9020288

Iftimie, N., Faktorová, D., Fabo, P., Savin, A., & Steigmann, R. (2018). Evaluation of dielectric materials properties using microwave enhanced metamaterials sensor. *IOP Conference Series: Materials Science and Engineering, 444*(2), 022007-1-022007–022009. 10.1088/1757-899X/444/2/022007

Islam, M. T., Samsuzzaman, M., Islam, M. T., & Kibria, S. (2018). Experimental breast phantom imaging with metamaterial-inspired nine-antenna sensor array. *Sensors, 18*(12), 4427-1-4427–19. 10.3390/s18124427

Jafari, F., & Rad Malekshahi, M. (2023). A low-cost microwave microfluidic sensor based on planar ring resonator. *IEEE Sensors Journal*, 23(18), 21070–21077. 10.1109/JSEN.2023.3301813

Jain, M. C., Nadaraja, A. V., Vizcaino, B. M., Roberts, D. J., & Zarifi, M. H. (2020). Differential microwave resonator sensor reveals glucose-dependent growth profile of *E. coli* on solid agar. *IEEE Microwave and Wireless Components Letters*, 30(5), 531–534. 10.1109/LMWC.2020.2980756

Javed, A., Arif, A., Zubair, M., Mehmood, M. Q., & Riaz, K. (2020). A low-cost multiple complementary split-ring resonator-based microwave sensor for contactless dielectric characterization of liquids. *IEEE Sensors Journal*, 20(19), 11326–11334. 10.1109/JSEN.2020.2998004

Kayal, S., Shaw, T., & Mitra, D. (2020). Design of metamaterial-based compact and highly sensitive microwave liquid sensor. *Applied Physics. A, Materials Science & Processing*, 126(1), 1–9. 10.1007/s00339-019-3186-4

Kazemi, N., Abdolrazzaghi, M., & Musilek, P. (2021). Comparative analysis of machine learning techniques for temperature compensation in microwave sensors. *IEEE Transactions on Microwave Theory and Techniques*, 69(9), 4223–4236. 10.1109/TMTT.2021.3081119

Kiani, S., Rezaei, P., & Navaei, M. (2020). Dual-sensing and dual-frequency microwave SRR sensor for liquid samples permittivity detection. *Measurement*, 160, 107805-1, 107805–107808. 10.1016/j.measurement.2020.107805

Kumar, A., Rajawat, M. S., Mahto, S. K., & Sinha, R. (2021). Metamaterial-inspired complementary split ring resonator sensor and second-order approximation for dielectric characterization of fluid. *Journal of Electronic Materials*, 50(10), 5925–5932. 10.1007/s11664-021-09099-w

Kurland, Z., Goyette, T., & Gatesman, A. (2023). A novel technique for ultrathin inhomogeneous dielectric powder layer sensing using a W-Band metasurface. *Sensors (Basel)*, 23(2), 842. Advance online publication. 10.3390/s2302084236679638

Lu, F., Tan, Q., Ji, Y., Guo, Q., Guo, Y., & Xiong, J. (2018). A novel metamaterial inspired high-temperature microwave sensor in harsh environments. *Sensors, 18*(9), 2879-1-2879–12. 10.3390/s18092879

Marqués, R., Mesa, F., Martel, J., & Medina, F. (2003). Comparative analysis of edge- and broadside-coupled split ring resonators for metamaterial design - theory and experiments. *IEEE Transactions on Antennas and Propagation*, 51(10), 2572–2581. 10.1109/TAP.2003.817562

Mata-Contreras, J., Su, L., & Martín, F. (2017). Microwave sensors based on symmetry properties and metamaterial concepts: a review of some recent developments (invited paper). *2017 IEEE 18th Wireless and Microwave Technology Conference, WAMICON 2017*, (pp. 1–6). IEEE. 10.1109/WAMICON.2017.7930278

Mayani, M. G., Herraiz-Martinez, F. J., Domingo, J. M., & Giannetti, R. (2021). Resonator-based microwave metamaterial sensors for instrumentation: Survey, classification, and performance comparison. *IEEE Transactions on Instrumentation and Measurement*, 70, 1–14. 10.1109/TIM.2020.3040484

Memon, M. U., Salim, A., Jeong, H., & Lim, S. (2020). Metamaterial inspired radio frequency-based touchpad sensor system. *IEEE Transactions on Instrumentation and Measurement*, 69(4), 1344–1352. 10.1109/TIM.2019.2908507

Milias, C., Andersen, R. B., Lazaridis, P. I., Zaharis, Z. D., Muhammad, B., Kristensen, J. T. B., Mihovska, A., & Hermansen, D. D. S. (2021). Metamaterial-inspired antennas: A review of the state of the art and future design challenges. *IEEE Access : Practical Innovations, Open Solutions*, 9, 89846–89865. 10.1109/ACCESS.2021.3091479

Mohammadi, P., Mohammadi, A., & Kara, A. (2023). Dual-frequency microwave resonator for noninvasive detection of aqueous glucose. *IEEE Sensors Journal*, 23(18), 21246–21253. 10.1109/JSEN.2023.3303170

Mohammadi, S., & Zarifi, M. H. (2021). Differential microwave resonator sensor for real-time monitoring of volatile organic compounds. *IEEE Sensors Journal*, 21(5), 6105–6114. 10.1109/JSEN.2020.3041810

Muñoz-Enano, J., Vélez, P., Gil, M., & Martín, F. (2020). Planar microwave resonant sensors: A review and recent developments. *Applied Sciences, 10*(7), 2615-1-2615–2630. 10.3390/app10072615

Niksan, O., Jain, M. C., Shah, A., & Zarifi, M. H. (2022). A nonintrusive flow rate sensor based on microwave split-ring resonators and thermal modulation. *IEEE Transactions on Microwave Theory and Techniques*, 70(3), 1954–1963. 10.1109/TMTT.2022.3142038

Prakash, D., & Gupta, N. (2022). Applications of metamaterial sensors: A review. *International Journal of Microwave and Wireless Technologies*, 14(1), 19–33. 10.1017/S1759078721000039

Prakash, D., & Gupta, N. (2023). CSRR based metamaterial inspired sensor for liquid concentration detection using machine learning. *Progress in Electromagnetics Research C. Pier C*, 130(February), 255–267. 10.2528/PIERC22110101

Puentes Vargas, M. (2014). Fundamentals of metamaterial structures. In *Planar Metamaterial Based Microwave Sensor Arrays for Biomedical Analysis and Treatment* (pp. 7–31). Springer. 10.1007/978-3-319-06041-5_2

Rabbani, M. G., Islam, M. T., Hoque, A., Bais, B., Albadran, S., Islam, M. S., & Soliman, M. S. (2024). Orthogonal centre ring field optimization triple-band metamaterial absorber with sensing application. *Engineering Science and Technology, an International Journal, 49*(June 2023), 101588. 10.1016/j.jestch.2023.101588

Rahman, N. A., Zakaria, Z., Rahim, R. A., Dasril, Y., & Mohd Bahar, A. A. (2017). Planar microwave sensors for accurate measurement of material characterization: A review. [Telecommunication Computing Electronics and Control]. *Telkomnika*, 15(3), 1108–1118. 10.12928/telkomnika.v15i3.6684

Reyes-Vera, E., Acevedo-Osorio, G., Arias-Correa, M., & Senior, D. E. (2019). A submersible printed sensor based on a monopole-coupled split ring resonator for permittivity characterization. *Sensors (Switzerland), 19*(8), 1936-1-1936–12. 10.3390/s19081936

RoyChoudhury, S., Rawat, V., Jalal, A. H., Kale, S. N., & Bhansali, S.RoyChoudhury. (2016). Recent advances in metamaterial split-ring-resonator circuits as biosensors and therapeutic agents. *Biosensors & Bioelectronics*, 86, 595–608. 10.1016/j.bios.2016.07.02027453988

Saadat-Safa, M., Nayyeri, V., Ghadimi, A., Soleimani, M., & Ramahi, O. M. (2019). A pixelated microwave near-field sensor for precise characterization of dielectric materials. *Scientific Reports*, 9(1), 1–12. 10.1038/s41598-019-49767-w31527610

Salim, A., & Lim, S. (2018). Review of recent metamaterial microfluidic sensors. *Sensors, 18*(1), 232-1-232–25. 10.3390/s18010232

Seng, S. M., You, K. Y., Esa, F., & Mayzan, M. Z. H. (2020). Dielectric and magnetic properties of epoxy with dispersed iron phosphate glass particles by microwave measurement. *Journal of Microwaves, Optoelectronics and Electromagnetic Applications*, 19(2), 165–1676. 10.1590/2179-10742020v19i2824

Shahzad, W., Hu, W., Ali, Q., Raza, H., Abbas, S. M., & Ligthart, L. P. (2022). A low-cost metamaterial sensor based on DS-CSRR for material characterization applications. *Sensors (Basel)*, 22(5), 1–11. 10.3390/s2205200035271147

Shen, Z., Zhang, H., & Zhang, J. (2024). A high Q-factor metamaterial sensor based on electromagnetically induced transparency-like. *Materials Research Express*, 11(1), 015801. 10.1088/2053-1591/ad1a61

Sim, M. S., You, K. Y., Dewan, R., Esa, F., Salim, M. R., Kew, S. Y. N., & Hamid, F. (2023). Dual-band metamaterial microwave absorber using ring and circular patch with slits. *Advanced Electromagnetics*, 12(4), 36–44. 10.7716/aem.v12i4.2324

Sim, M. S., You, K. Y., & Esa, F. (2020). Electromagnetic metamaterials in microwave regime. In *Handbook of Research on Recent Developments in Electrical and Mechanical Engineering* (pp. 64–86). Springer. 10.4018/978-1-7998-0117-7.ch002

Sim, M. S., You, K. Y., Esa, F., & Chan, Y. L. (2021). Nanostructured electromagnetic metamaterials for sensing applications. In *Applications of Nanomaterials in Agriculture* (pp. 141–164). Food Science, and Medicine. 10.4018/978-1-7998-5563-7.ch009

Sim, M. S., You, K. Y., Esa, F., Dimon, M. N., & Khamis, N. H. (2018). Multiband metamaterial microwave absorbers using split ring and multiwidth slot structure. *International Journal of RF and Microwave Computer-Aided Engineering*, 28(7), 1–13. 10.1002/mmce.21473

Slimi, M., Jmai, B., Dinis, H., Gharsallah, A., & Mendes, P. M. (2022). Metamaterial Vivaldi antenna array for breast cancer detection. *Sensors (Basel)*, 22(10), 3945. 10.3390/s2210394535632355

Song, J., & Huang, J. (2023). A microfluidic antenna-sensor for contactless characterization of complex permittivity of liquids. *IEEE Sensors Journal*, 23(22), 27251–27261. 10.1109/JSEN.2023.3318213

Su, L., Mata-contreras, J., Vélez, P., & Martín, F. (2017). A review of sensing strategies for microwave sensors based on metamaterial-inspired resonators: Dielectric characterization, displacement, and angular velocity measurements for health diagnosis, telecommunication, and space applications. *International Journal of Antennas and Propagation*, 2017, 1–13. 10.1155/2017/5619728

Tabassum, S., Nayemuzzaman, S. K., Kala, M., Kumar Mishra, A., & Mishra, S. K. (2022). Metasurfaces for sensing applications: Gas, bio and chemical. *Sensors, 22*(18), 6896-1-6896–29. 10.3390/s22186896

Tefek, U., Sari, B., Alhmoud, H. Z., & Hanay, M. S. (2023). Permittivity-based microparticle classification by the integration of impedance cytometry and microwave resonators. *Advanced Materials*, 35(46), 2304072. 10.1002/adma.20230407237498158

Tiwari, N. K., Singh, S. P., & Akhtar, M. J. (2020). Adulteration detection in petroleum products using directly loaded coupled-line-based metamaterial-inspired submersible microwave sensor. *IET Science, Measurement & Technology*, 14(3), 376–385. 10.1049/iet-smt.2018.5687

Velez, P., Paredes, F., Casacuberta, P., Elgeziry, M., Su, L., Munoz-Enano, J., Costa, F., Genovesi, S., & Martin, F. (2023). Portable reflective-mode phase-variation microwave sensor based on a rat-race coupler pair and gain/phase detector for dielectric characterization. *IEEE Sensors Journal*, 23(6), 5745–5756. 10.1109/JSEN.2023.3240771

Velez, P., Su, L., Grenier, K., Mata-Contreras, J., Dubuc, D., & Martin, F. (2017). Microwave microfluidic sensor based on a microstrip splitter/combiner configuration and split ring resonators (SRRs) for dielectric characterization of liquids. *IEEE Sensors Journal*, 17(20), 6589–6598. 10.1109/JSEN.2017.2747764

Viswanathan, A. P., Moolat, R., Mani, M., Shameena, V. A., & Pezholil, M. (2020). A simple electrically small microwave sensor based on complementary asymmetric single split resonator for dielectric characterization of solids and liquids. *International Journal of RF and Microwave Computer-Aided Engineering*, 30(12), 1–13. 10.1002/mmce.22462

Vivek, A., Shambavi, K., & Alex, Z. C. (2019). A review: Metamaterial sensors for material characterization. *Sensor Review*, 39(3), 417–432. 10.1108/SR-06-2018-0152

Wang, B. X., Zhao, W. S., Wang, D. W., Wang, J., Li, W., & Liu, J. (2021). Optimal design of planar microwave microfluidic sensors based on deep reinforcement learning. *IEEE Sensors Journal*, 21(24), 27441–27449. 10.1109/JSEN.2021.3124294

Wang, Q., Chen, Y., Mao, J., Yang, F., & Wang, N. (2023). Metasurface-assisted terahertz sensing. *Sensors (Basel)*, 23(13), 1–17. 10.3390/s2313590237447747

Wu, J., Yang, D., Huang, X., Li, Y., & Xia, Y. (2021). The design and experiment of a novel microwave gas sensor loaded with metamaterials. *Physics Letters, Section A: General. Physics Letters. [Part A]*, 389, 127080-1, 127080–127086. 10.1016/j.physleta.2020.127080

Xie, B., Gao, Z., Wang, C., Ali, L., Muhammad, A., Meng, F., Qian, C., Ding, X., Adhikari, K. K., & Wu, Q. (2023). High-sensitivity liquid dielectric characterization differential sensor by 1-bit coding DGS. *Sensors, 23*(1), 372-1-372–12. 10.3390/s23010372

Yasin, A., Gogosh, N., Sohail, S. I., Abbas, S. M., Shafique, M. F., & Mahmoud, A. (2023). Relative permittivity measurement of microliter volume liquid samples through microwave filters. *Sensors, 23*(6), 2884-1-2884–13. 10.3390/s23062884

You, K. Y. (2017). Materials characterization using microwave waveguide system. In *Microwave Systems and Applications* (pp. 341–358). 10.5772/66230

You, K. Y., & Abbas, Z. (2011). Waveguide techniques for determination of moisture content in materials. In *Antenna and Applied Electromagnetic Application* (pp. 113–130).

You, K. Y., & Esa, F. Bin, & Abbas, Z. (2017). Macroscopic characterization of materials using microwave measurement methods - A survey. *2017 Progress in Electromagnetics Research Symposium-Fall (PIERS-FALL)*, pp. 194–204. 10.1109/PIERS-FALL.2017.8293135

You, K. Y., & Sim, M. S. (2018). Precision permittivity measurement for low-loss thin planar materials using large coaxial probe from 1 to 400 MHz. *Journal of Manufacturing and Materials Processing, 2*(4), 81-1-81–15. 10.3390/jmmp2040081

You, K. Y., Sim, M. S., & Abdullah, S. N. (2020). Emerging microwave technologies for agricultural and food processing. In *Precision Agriculture Technologies for Food Security and Sustainability* (pp. 94–148). 10.4018/978-1-7998-5000-7.ch005

You, K. Y., Sim, M. S., Mutadza, H., Esa, F., & Chan, Y. L. (2017). Free-space measurement using explicit, reference-plane and thickness-invariant method for permittivity determination of planar materials. *2017 Progress in Electromagnetics Research Symposium-Fall (PIERS-FALL)*, (pp. 222–228). IEEE. 10.1109/PIERS-FALL.2017.8293139

Zhang, W., Song, Q., Zhu, W., Shen, Z., Chong, P., Tsai, D. P., Qiu, C., & Liu, A. Q. (2018). Metafluidic metamaterial: A review. *Advances in Physics: X*, 3(1), 165–184. 10.1080/23746149.2017.1417055

Zhao, W. S., Wang, B. X., Wang, D. W., You, B., Liu, Q., & Wang, G. (2022). Swarm intelligence algorithm-based optimal design of microwave microfluidic sensors. *IEEE Transactions on Industrial Electronics*, 69(2), 2077–2087. 10.1109/TIE.2021.3063873

Zhou, H., Hu, D., Yang, C., Chen, C., Ji, J., Chen, M., Chen, Y., Yang, Y., & Mu, X. (2018). Multi-band sensing for dielectric property of chemicals using metamaterial integrated microfluidic sensor. *Scientific Reports*, 8(1), 1–11. 10.1038/s41598-018-32827-y30287826

ADDITIONAL READING

Ahmed, S., & Lim, S. (2018). Review of recent metamaterial microfluidic sensors. *Sensors (Basel)*, 18(1), 232. 10.3390/s1801023229342953

Chen, T., Li, S., & Sun, H. (2012). Metamaterials application in sensing. *Sensors (Basel)*, 12(3), 2742–2765. 10.3390/s12030274222736975

Lee, Y., Kim, S. J., Park, H., & Lee, B. (2017). Metamaterials and metasurfaces for sensor applications. *Sensors (Basel)*, 17(8), 1726. 10.3390/s1708172628749422

Prakash, D., & Gupta, N. (2022). Applications of metamaterial sensors: A review. *International Journal of Microwave and Wireless Technologies*, 14(1), 19–33. 10.1017/S1759078721000039

Salim, A., & Lim, S. (2018). Review of recent metamaterial microfluidic sensors. *Sensors (Basel)*, 18(1), 232. 10.3390/s1801023229342953

Su, L., Mata-Contreras, J., Vélez, P., & Martín, F. (2017). A review of sensing strategies for microwave sensors based on metamaterial-inspired resonators: Dielectric characterization, displacement, and angular velocity measurements for health diagnosis, telecommunication, and space applications. *International Journal of Antennas and Propagation*, 2017, 2017. 10.1155/2017/5619728

Tabassum, S., Nayemuzzaman, S. K., Kala, M., Kumar Mishra, A., & Mishra, S. K. (2022). Metasurfaces for sensing applications: Gas, bio and chemical. *Sensors (Basel)*, 22(18), 6896. 10.3390/s2218689636146243

Vivek, A., Shambavi, K., & Alex, Z. C. (2018). A review: Metamaterial sensors for material characterization. *Sensor Review*, 39(3), 417–432. 10.1108/SR-06-2018-0152

You, K. Y., & Sotirios, K. G. (2017). Materials characterization using microwave waveguide system. *Microwave systems and applications*, pp.341-358.

Zhang, W., Song, Q., Zhu, W., Shen, Z., Chong, P., Tsai, D. P., Qiu, C., & Liu, A. Q. (2018). Metafluidic metamaterial: A review. *Advances in Physics: X*, 3(1), 1417055. 10.1080/23746149.2017.1417055

KEY TERMS AND DEFINITIONS

Frequency-Variation Sensor: A sensing device which measures a physical property by measuring the variation or shift of resonant frequency.

Metamaterial: An array of engineered subwavelength unit cell structures which collectively exhibit properties not usually found in natural materials.

Microfluidic Channel: A confined tunnel or chamber with micro-meters size for the flow of small amounts of fluids.

Microwave Sensor: A sensing device which uses microwave signals to interact with samples under test to measure a particular physical property.

Planar Sensor: A sensing device which is implemented on a substrate using planar technology.

Chapter 7
Performance Analysis of Dual-Band Microstrip Patch Antenna With and Without Electromagnetic Band Gap Structures:
A Comparative Study

Mehaboob Mujawar
http://orcid.org/0000-0002-8260-7062
Bearys Institute of Technology, India

Subuh Pramono
http://orcid.org/0000-0001-9505-1588
Universitas Sebelas Maret, Surakarta, Indonesia

ABSTRACT

This chapter investigates the optimization of dual-band MPA antennas in modern wireless communication systems. By integrating Electromagnetic Bandgap (EBG) structures, performance metrics such as bandwidth, gain, and radiation efficiency are enhanced. The chapter presents a thorough literature review, design methodology, and simulation results analyzing key metrics. Insights gained from this comparative study contribute to antenna design advancements. Additionally, the impact of antenna bending on wearable applications is explored, along with Specific Absorption Rate (SAR) analysis ensuring safety within regulatory limits.

DOI: 10.4018/979-8-3693-2599-5.ch007

INTRODUCTION

In the domain of contemporary wireless communication systems, the demand for compact, efficient, and multi-functional antennas is ever-growing. Microstrip patch antennas have become a preferred option because of their flat design, lightweight nature, simple manufacturing process, and suitability for integrated circuit technologies. However, to meet the evolving requirements of wireless communication systems, there is a continual need for antennas with enhanced performance characteristics, including multi-band operation and improved radiation properties. Dual-band have attracted considerable capability to function across two separate frequency bands, providing enhanced flexibility and adaptability for a range of application. Nonetheless, challenges persist in achieving optimal performance. One promising approach to address these challenges is the integration of periodic artificial materials designed to manipulate electromagnetic waves by creating stop bands or band gaps within the frequency spectrum. By incorporating EBG (Ramesh & Rajya Lakshmi, 2013) structures into microstrip (Jaget Singh & Badhan, 2018) patch antennas, it becomes feasible to mitigate surface wave excitation, suppress mutual coupling between antenna elements, and improve radiation characteristics, thereby augmenting the overall antenna performance. The chapter aims to provide an in-depth analysis and evaluation of various performance metrics, including impedance bandwidth, radiation patterns, gain, and efficiency, to assess the effectiveness of EBG structures in enhancing antenna performance. The chapter begins with a review of relevant literature concerning dual-band microstrip patch antennas and Electromagnetic Band Gap structures. It then outlines the design methodology and configuration of the proposed antennas. Following this, the chapter presents the simulation setup and results obtained from numerical simulations. A detailed analysis and discussion of the obtained results are provided, highlighting the advantages and limitations of integrating EBG structures into dual-band microstrip patch antennas. Through this comparative study, the chapter seeks to offer insights into the efficacy of integrating EBG (Alam et al., 2011) structures with dual-band microstrip patch antennas, thereby contributing to the advancement of antenna design techniques for next-generation wireless communication systems.

DEVELOPMENT PROCESS OF THE ANTENNA DESIGN UNDER CONSIDERATION

Traditional Methodology for Antenna Design

The investigation employed Computer Simulation Technology MICROWAVE STUDIO software for all design iterations and simulations starting from 2 GHz and varying till 5.5 GHz. Figure 1 illustrates the proposed antenna. The substrate material utilized was polyethylene foam, with a thickness of 3 millimeters, compact dimensions of 68mm×73mm. The radiating patch's overall dimensions were 44.75mm×47mm. In the fabrication of the recommended dual-band antenna, a 50-ohm microstrip feed line was integrated. The dimensions were established by applying transmission line principles to the design challenge.

Figure 1. Geometric configuration of the antenna under consideration

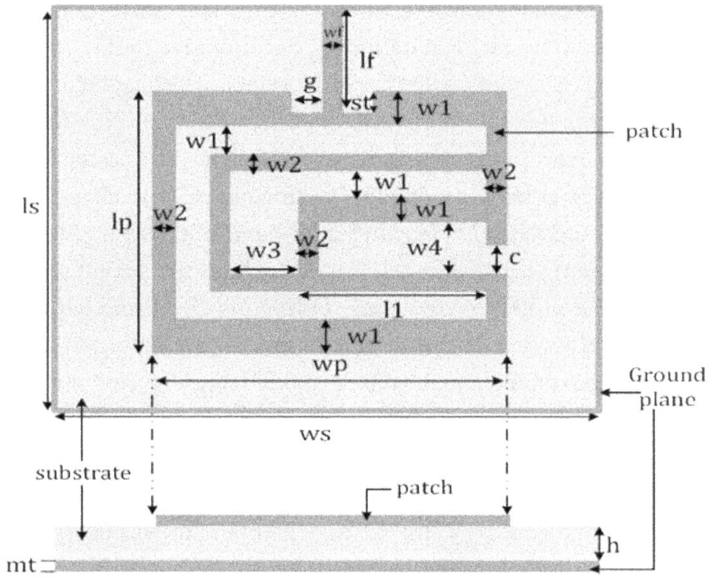

Dual-Band Electromagnetic Band Gap Unit Cells

This section details the creation and assessment of two unique unit cells derived from separate electromagnetic bandgap (EBG) structures tailored to accommodate specific frequency bands (2.5 GHz and 5.196 GHz). Each of these EBG structures, serving as the substrate material, has a thickness of 3 millimeters and consists of polyethylene. The resonant frequency of the unit cell is dictated by its effective capacitance (C) and inductance (L), which are influenced by the permeability and permittivity of free space. The first unit cell adopts a traditional mushroom-like shape, with dimensions of 32.5 millimeters by 32.5 millimeters and a square form, featuring a 1.5 millimeter radius through. In contrast, the second unit cell is modified to resemble a mushroom with a plus-shaped slot. The inclusion of slots enhances the capacitance, resulting in a more compact design. The unit cell with a plus-shaped slot measures 26.5 millimeters by 26.5 millimeters, making it roughly 18.46% smaller than the standard mushroom-shaped unit cell in size. Figure 3 illustrates the resonance behavior of the dual-band Electromagnetic Band Gap (EBG) unit cells through their in-phase reflection. These unit cells feature a plus-shaped slot within the second unit cell, with dimensions of 3 millimeters in width and 13 millimeters in length. At specified frequencies of 2.5 GHz and 5.2 GHz, each EBG unit cell exhibits a distinctive in-phase response, indicating resonance. This resonance behavior is crucial for their effectiveness in manipulating electromagnetic waves within the specified frequency bands.

Figure 2. Two types of EBG unit cells: (a) A mushroom-shaped EBG unit cell. (b) A unit cell with slots arranged in a plus-shaped configuration

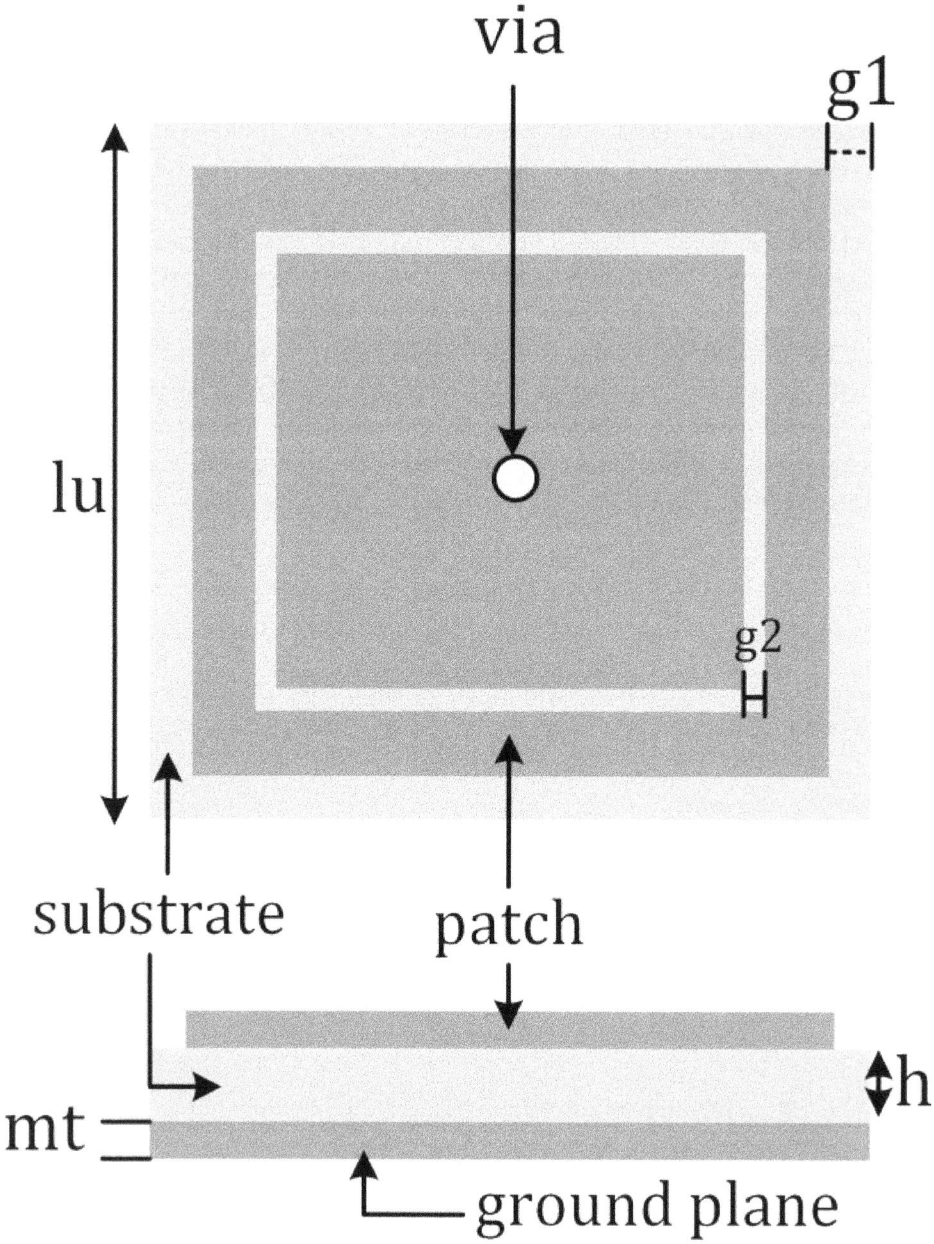

Figure 3. In-phase reflection from the EBG unit cells: (a) Mushroom-shaped unit cell. (b) Unit cell with slots arranged in a plus-shaped configuration

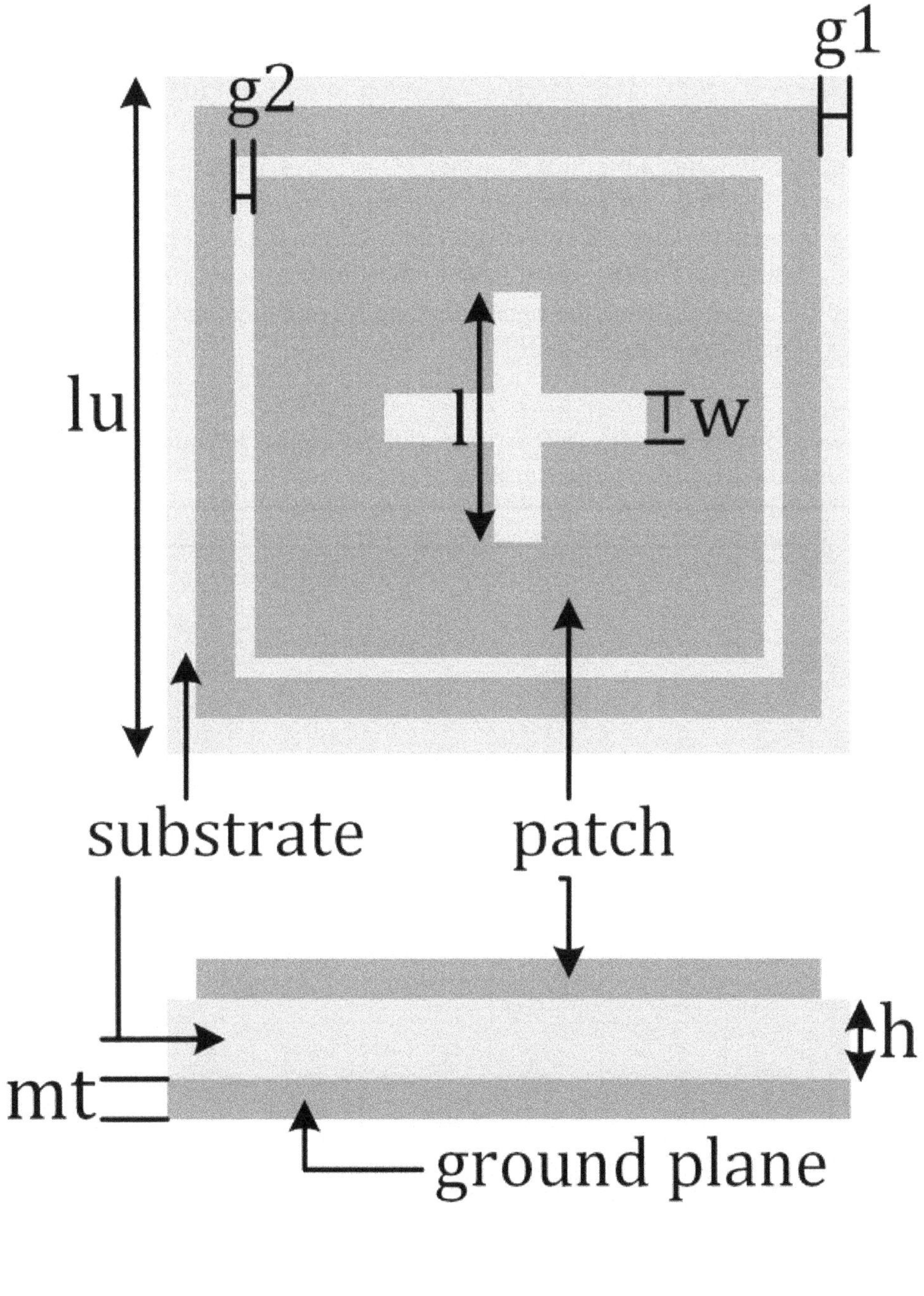

The suppression of waves stands as a crucial aspect of EBG structures, as it effectively prevents their propagation. This suppression not only reduces the size of antenna side and back lobes but also enhances attributes like gain and directivity (Mujawar et al., 2021). The scattering properties of the two EBG surfaces can be acquired as illustrated in Figure 4. Here, one of the ports serves as the excitation source, while the other port, situated on the right, acts as a matched load. The second EBG array, featuring plus-shaped slots, occupies a total size of 136 square millimeters, whereas the mushroom-like EBG array encompasses a total size of 176.5 square millimeters. The transmission and reflection coefficients are illustrated in Figure 5. It is evident from the figures that the S_{21} value remains below 15 dB at both frequencies for each EBG unit.

Figure 4. A 5×5 array composed of: (a) Mushroom-shaped EBG units and (b) EBG units with plus-shaped slots and suspended strips

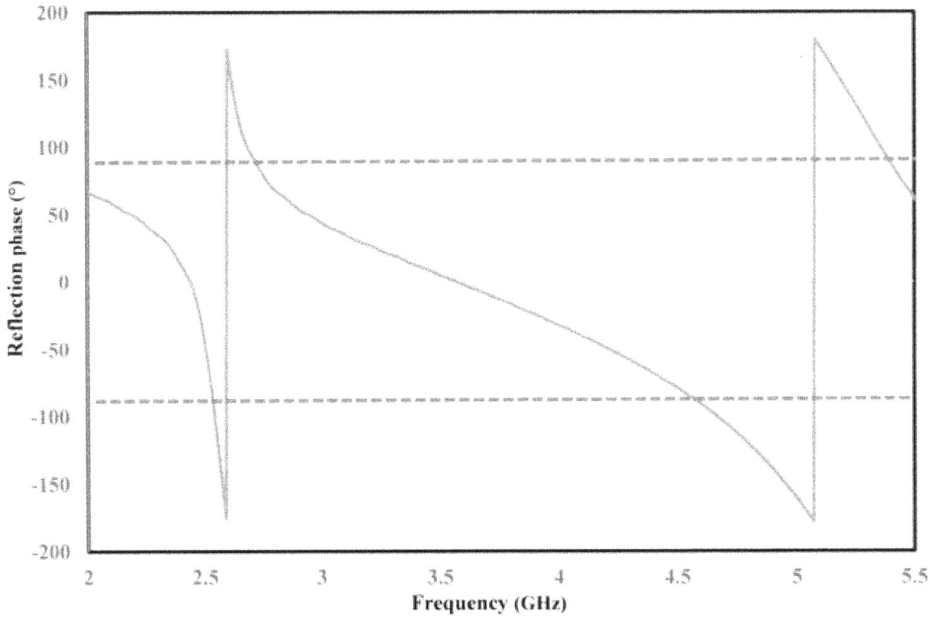

(a)

Figure 5. Reflection and transmission coefficients of a 5×5 array of EBG units: (a) Array using mushroom-like EBG units, (b) Array utilizing EBG units with plus-shaped slots

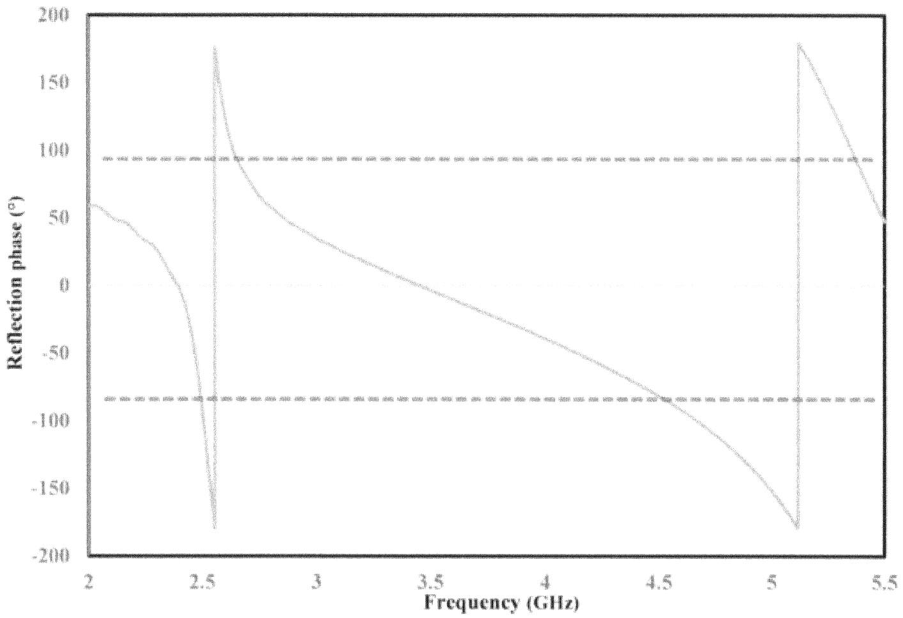

(b)

Antenna Utilizing Electromagnetic Band Gap (EBG) Technology

As depicted in Figure 6, the dual-band antenna design was situated atop two separate 5×5 EBG ground planes. The dimensions of the antenna on the EBG ground plane with mushroom-like slots and plus-shaped slots measured 176.5 mm by 176.5 mm, and 136 mm by 136 mm respectively.

Figure 6. Antenna positioned above EBG ground planes: (a) EBG ground plane with mushroom-like slots, (b) EBG ground plane with plus-shaped slots

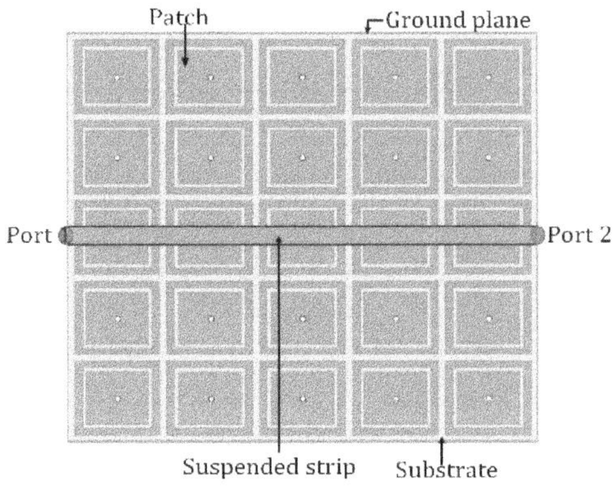

RESULTS

Off-Body Analysis: Conventional Antenna

Figure 1 illustrates the construction of the recommended antenna designed to operate at ISM band frequencies using software. In Figure 7, the return loss of the planned antenna is depicted. The antenna exhibits input impedance bandwidths of 23.8 MHz and 60.4 MHz at resonance frequencies of 2.5 GHz and 5.2 GHz, respectively. At 2.5 GHz and 5.5 GHz, the return losses are measured at 24.975 and 28.234, respectively. Figure 8 showcases the excellent matching of the proposed antenna. Figures 9 and 10 illustrate the directivity and gain patterns. At these frequencies, the directivities are measured at 8.39 dB and 9.01 dB, respectively, representing the highest directivities achieved. The 3 dB angular widths are 83.4 degrees at 2.5 GHz and 76.9 degrees at 5.2 GHz. The antenna achieves gains of 8.74 dBi and the highest gains at 2.5 GHz and 5.2 GHz, respectively. However, it is noted that at the higher resonance frequency, the patterns are significantly disrupted. This disruption can be attributed to the small size of the proposed antenna. The procedure of diminishing the dimensions of an antenna inevitably involves a trade-off between antenna size and radiation characteristics. To achieve the compact dimensions of 68×73 mm^2, the intended dual-band antenna incorporates numerous slots.

Figure 7. Reflection characteristics of the traditional dual-band antenna

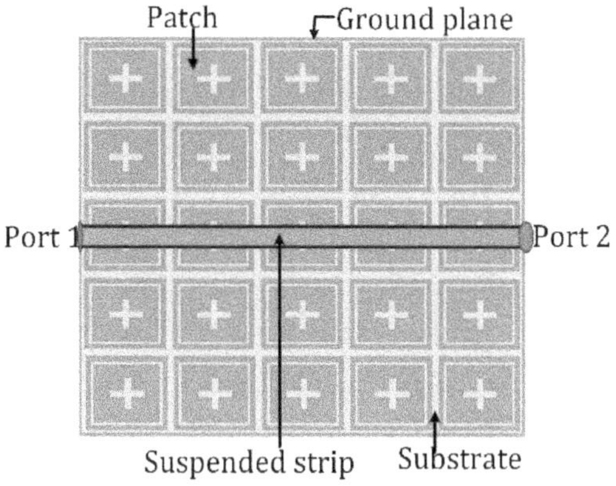

Figure 8. Voltage standing wave ratio (VSWR) of the traditional dual-band antenna

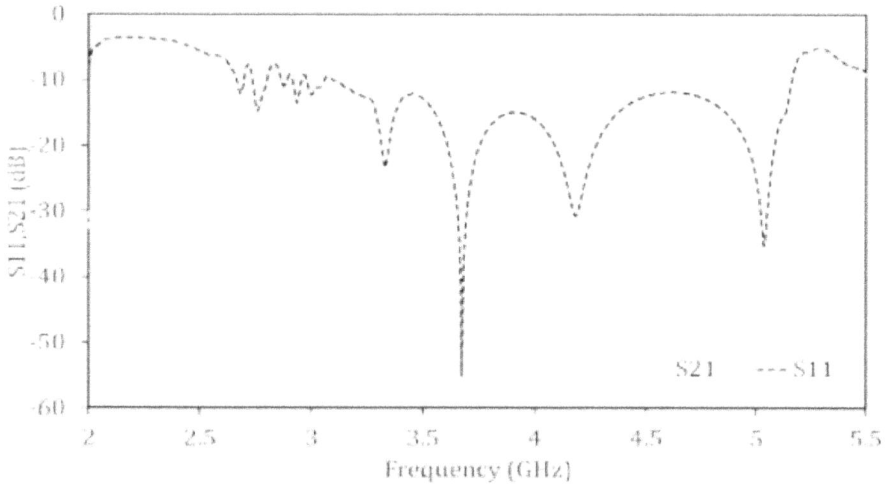

Figure 9 depicts the surface current distribution in the conventional wearable antenna at 2.5 and 5.2 GHz, providing clear indications of the resonant lengths.

Figure 9. Description of the surface current distribution

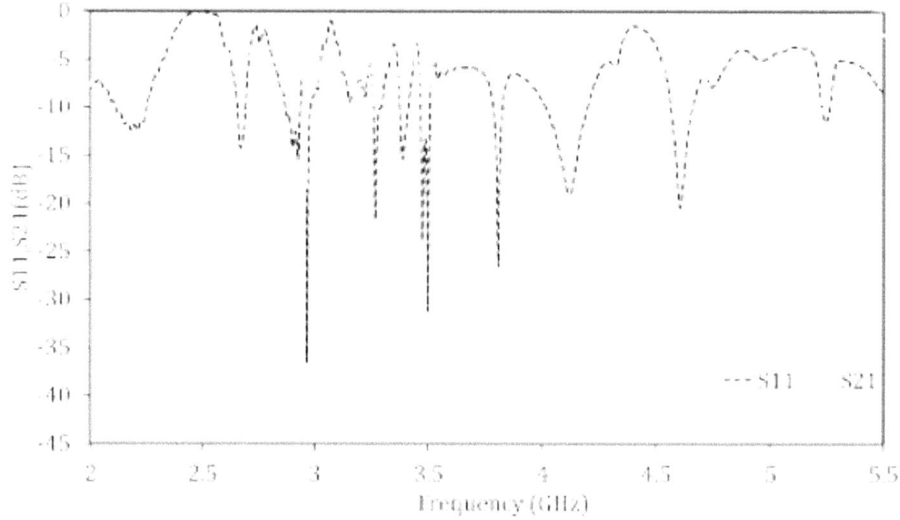

Figure 10 illustrates the Return loss for antennas without electromagnetic bandgap (EBG) ground planes compared to those with EBG ground planes (Tu et al., 2017).

Figure 10. Return loss for antennas without electromagnetic bandgap (EBG) ground planes compared to those with EBG ground planes

The utilization of mushroom-shaped and plus-shaped electromagnetic bandgap (EBG) ground planes resulted in notable enhancements in antenna bandwidth and performance. More precisely, employing the mushroom-shaped EBG ground plane led to a bandwidth increase of 5.2 MHz and 33.6 MHz for the lower and higher frequency bands, respectively. Meanwhile, employing a plus-shaped slot EBG ground plane extended bandwidth by 7.9 MHz and 16.7 MHz for the 2.5 GHz and 5.2 GHz bands, respectively. These enhancements were accompanied by a reduction in return loss across both EBG configurations. Moreover, a slight shift in resonance frequencies was observed, with the plus-shaped slots EBG causing a shift to the right for the lower frequency band and the mushroom-shaped EBG leading to a shift to the left. Similar resonance frequency shifts were noted when operating the antenna on metamaterial-based ground planes.

Table 1 presents a summary of comparison of the performance between the conventional dual-band antenna, both with and without electromagnetic bandgap.

Table 1. A comparison of the performance between proposed antenna, with EBG and without EBG

Parameter	Traditional dual-band antenna		Antenna utilizing a mushroom-shaped electromagnetic bandgap (EBG) design		Antenna based on electromagnetic bandgap (EBG) with plus-shaped slots	
Frequency (GHz)	**2**	**5**	**2**	**4.2**	**2.4**	**5.4**
Bandwidth (MHz)	22.8	60	28	93	31	77
Return loss (dB)	-23.5	-27.4	-18.2	-23.1	-23.6	-18.5
Directivity (dB)	8	9	7.55	10.6	8.6	8.3
Gain (dBi)	8	7.74	7.4	10.4	8.43	8.1
Radiation efficiency (%)	92.06	92.4	96.3	95.7	93.9	94.1
Total efficiency (%)	91.76	92.8	95.41	94.2	93.2	93.8
Angular width (Deg)	82.4	75	80	97.4	61.8	71.2

Conventional Antenna Under Varying Free Space Bending Conditions

The application currently available includes fitness, sports, and security devices designed to be worn on the body. Given the nature of these applications, maintaining a flat antenna pose becomes challenging as users engage in various activities while wearing these gadgets. Consequently, antenna bending occurs during normal usage, This could potentially impact the performance of the connected device system if the antenna is incorporated within it. To address this concern, we conducted a comprehensive investigation into the behavior of a wearable antenna under different bending conditions, aiming to enhance designs for future wearable devices. We subjected the antenna to bending configurations shown in subsequent figures, simulating its wrapping around various body locations. Through this investigation, we analyzed under different degrees of bending. In this section, we present an overview of our findings. Notably, both the healthy arm and the child's arm exhibit smaller radii compared to the leg. It's important to note that the return loss resulting from antenna bending on specific body parts remains relatively consistent regardless of the bending radius. For instance, at a frequency of 2.5 GHz, the return loss decreases by approximately 15.16 dB. Similarly, the gain increases by about 5.68 decibels at 5.2 GHz when the antenna is bent around a leg with a radius of 70 millimeters. Figure 15 illustrates that the S_{11} is minimally impacted by varying bending radii.

Figure 11. Reflection coefficient of the conventional dual-band antenna when placed off the body

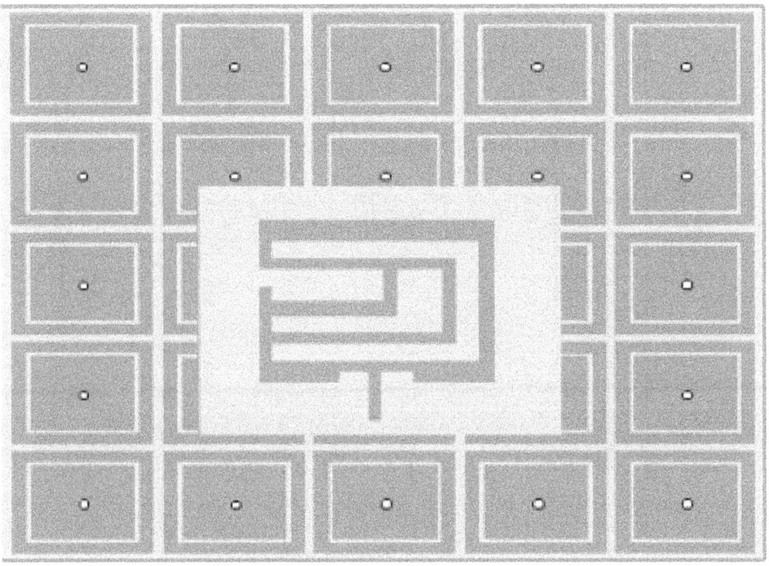

ANALYSIS CONDUCTED ON THE BODY

In this section, we evaluate performance when positioned on the human body. Our analysis considers different wearable antenna factors.

Return Loss

Figure 12 compares S_{11} across various scenarios: on a flat body phantom, bending around more rounded body sections, and without a body (in scenarios involving the leg and arms). The resonance frequencies of the antenna tend to increase when positioned over different body sections. Furthermore, as the bending radius decreases, the resonance frequencies shift further to the right. Smaller bending radii lead to a greater reduction in the effective length of the antenna, causing a shift in the resonant frequency toward higher frequencies. However, it can be demonstrated that the proposed antenna remains consistently tuned, despite various environmental factors that may be present.

Figure 12. Return loss analysis conducted while the conventional dual-band antenna is positioned on the body

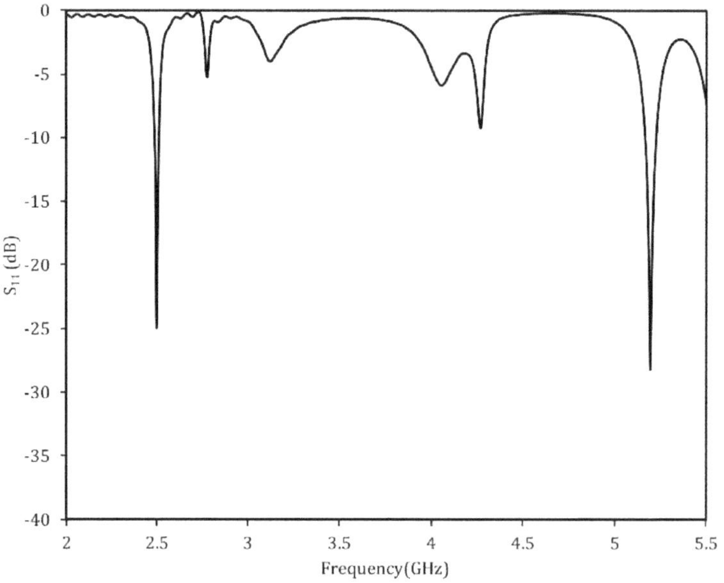

Analysis of Specific Absorption Rate (SAR)

The Analysis of Specific Absorption Rate (SAR) in antenna design is a critical evaluation process to assess the potential health risks associated with the electromagnetic fields emitted by antennas, particularly when they are placed close to the human body, such as in wearable applications. SAR measures the rate at which energy is absorbed by biological tissues exposed to electromagnetic fields, typically expressed in watts per kilogram (W/kg). The Specific Absorption Rate (SAR) refers to the rate at which energy from radiofrequency electromagnetic fields is absorbed by the human body. It is a measure of the amount of electromagnetic energy absorbed per unit of mass of biological tissue, typically measured in watts per kilogram (W/kg). SAR is used to evaluate the potential health risks associated with exposure to electromagnetic radiation from devices such as mobile phones, wireless routers, and other wireless communication devices. Regulatory bodies often establish SAR limits to ensure that exposure levels remain within safe thresholds for human health.

Figure 13. Specific Absorption Rate (SAR) values of the conventional antenna on a flat body phantom at (a) 2.5 GHz and (b) 5.2 GHz

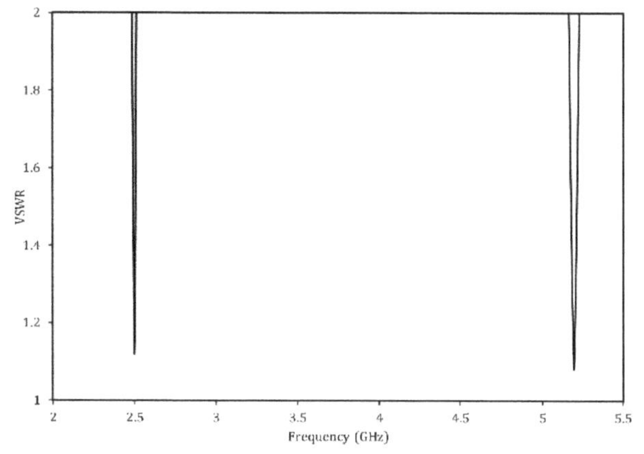

Figure 13 depicts the SAR values obtained for the conventional antenna when positioned on a flat body phantom, accounting for both resonance frequencies. These values for the lower and upper resonance bands are 0.332 W/kg and 0.234 W/kg, respectively, with a tissue sample size of 10 g. As shown in Figure 22, the most extreme bending scenario for a conventional antenna, wrapping around curved body parts like an infant's arm, results in a SAR value of 1.89 W/kg at 2.5 GHz and 0.462 W/kg at 5.2 GHz, measured across 10 g of tissue.

Figure 14. The specific absorption rate (SAR) value of the conventional antenna when placed on a child's arm (worst case scenario) with a radius of 40 mm, measured at (a) 2.5 GHz and (b) 5.2 GHz frequencies

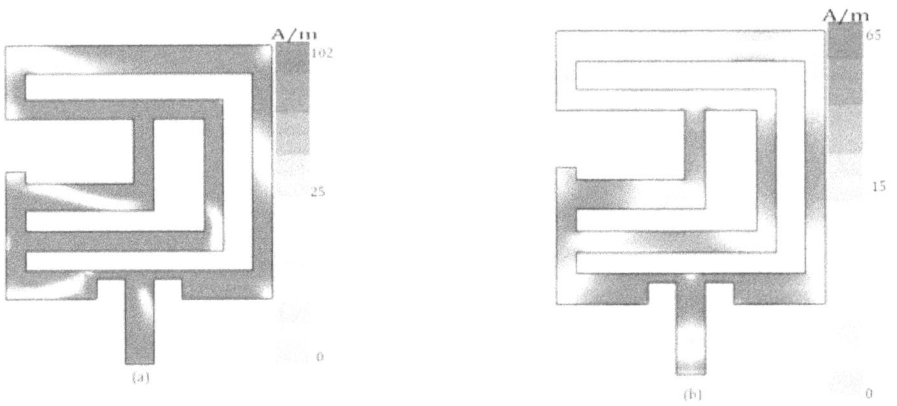

The lower SAR values obtained can be attributed to several factors. Additionally, the developed antenna emits minimal side lobes and rear lobes towards the body due to its patch antenna nature. These findings suggest that the proposed antenna could be a valuable component in numerous wearable applications, which are subject to strict safety regulations due to their critical role in maintaining human health.

CONCLUSION

In summary, this study has shown the effectiveness of integrating Electromagnetic Band Gap (EBG) structures to enhance the performance of dual-band microstrip patch antennas. The inclusion of EBG structures significantly improves key metrics such as impedance bandwidth, radiation patterns, gain, and efficiency. Comparative analysis revealed notable enhancements in antenna performance with EBG structures, including mitigation of surface wave excitation and suppression of mutual coupling between antenna elements. Additionally, examination under different bending conditions confirmed the antenna's resilience and suitability for wearable applications. SAR analysis demonstrated compliance with safety regulations, affirming the antenna's suitability for use in critical systems. Overall, this research contributes to advancing antenna design for next-generation wireless communication systems, offering insights into improved performance and safety considerations.

REFERENCES

Alam, M. S., Islam, M. T., Misran, N., & Selangor, D. E. (2011). Design analysis of an electromagnetic band gap microstrip antenna. *American Journal of Applied Sciences, 8*(12), 1374-1377.

Jaget Singh, B. S., & Badhan, , K. (2018). Slit loaded H-shaped microstrip patch antenna for 2.4 GHz. *International Journal of Applied Engineering Research,* 13(18), 13552–13554.

Mujawar, M., Gunasekaran, T., & Rashid, A. (2021). Design and Analysis of X Band Pyramidal Horn Antenna using FEKO. In *2021 International Conference on Advances in Electrical, Computing, Communication and Sustainable Technologies (ICAECT)*, Bhilai, India. 10.1109/ICAECT49130.2021.9392523

Ramesh, B., & Rajya Lakshmi, V. (2013). Design of a rectangular microstrip antenna using EBG structure. *International Journal of Engineering Research and Technology, 2*(7), 2233-2236.

Tu, Thanh, Hoc, & Yem. (2017). Double-side electromagnetic band gap structure for improving dual-band MIMO antenna performance. *REV Journal on Electronics and Communications, 7*(1).

Chapter 8
An Eight-Scallop Flower-Based Ultrathin and Bidirectional Multiple Passband FSS

Kanwar Preet Kaur
http://orcid.org/0000-0002-2030-586X
Department of Electronics and Communication Engineering, Charotar University of Science and Technology, India

Trushit Upadhyaya
http://orcid.org/0000-0001-8922-8201
Department of Electronics and Communication Engineering, Chandubhai S. Patel Institute of Technology, India

Upesh Patel
http://orcid.org/0000-0003-4872-553X
Department of Electronics and Communication Engineering, Chandubhai S. Patel Institute of Technology, India

Poonam Thanki
http://orcid.org/0000-0003-0720-0243
Department of Electronics and Communication Engineering, Chandubhai S. Patel Institute of Technology, India

ABSTRACT

A new, extremely thin, and bidirectional microwave passband FSS with multiple bands has been developed. This FSS consists of an FR4 sandwiched between three concentric metallic eight-scallop flower (ESF) rings on the front side and correspond-

DOI: 10.4018/979-8-3693-2599-5.ch008

ing complementary geometry on the back side. The thickness of the suggested FSS at the lowest transmission band is $\lambda_L/260$*. The design is insensitive to polarization and exhibits a stable response for both the TE and TM wave cases. The conformality of the design is tested for both inward and outward geometries, and further, the response of the passband FSS design is analyzed for different parameter variations. A prototype is fabricated using standard printed circuit board (PCB) techniques for practical verification. The performance of this passband FSS is validated using a circuit model and the free-space measurement technique. The outcomes are in close agreement with the simulated and circuit analysis results.*

INTRODUCTION

Frequency-selective surfaces (FSSs) (Munk, 1974; Munk, 2005) have garnered significant interest as passband and stopband filters from microwave to optical domains and have been extensively studied for various purposes, including hybrid radomes (Gao et al., 2018; Krushna Kanth & Raghavan, 2023; Shin et al., 2022; Zhou et al., 2012), dichroic sub-reflectors, and main reflectors (Daira et al., 2024; Ram Krishna & Kumar, 2015; Yuan et al., 2019), microwave absorbers (Fatima et al., 2024; Ghosh & Srivastava, 2014; Huang et al., 2022; Sharma et al., 2019), polarizers (Kiani & Dyadyuk, 2012; Sivasamy & Kanagasabai, 2020; Suganya & Natarajan, 2023), among other applications.

FSSs typically consist of periodic metallic patterns over the dielectric that control the transmission and reflection of electromagnetic (EM) waves when they interact with the structure. These two-dimensional or three-dimensional periodic arrangements of slots or patches are placed on one or both sides of a single-layer or multi-layer dielectric substrate. Recent research has shown a growing interest in the utilization of FSS for spatial filtering applications. These studies aim to demonstrate the effectiveness of FSS-based spatial filters in providing multiple independent transmission bands, making them suitable for a wide range of multiband and broadband applications. For instance, the initial bandpass FSS was proposed in Abbaspour-Tamijani et al. (2004) and Abbaspour-Tamijani et al. (2003), designed to achieve a second-order bandpass filter, with the FSS sandwiched between two antenna layers. In Bossard et al. (2006) and Govindaswamy et al. (2004), a spatial FSS-based filter was presented for near-infrared regions.

The earliest miniature first-order bandpass FSS was proposed in Sarabandi and Behdad (2007). This subwavelength design used a resonant dipole and slot patterns separated by a dielectric. Following this, numerous additional investigations have explored the utilization of subwavelength FSS structures. For example, in Behdad (2008), an FSS with second-order bandpass characteristics is presented, employing a non-resonant subwavelength periodic structure composed of dipole and slot patterns. In Behdad et al. (2009), it was shown that by increasing the number of layers from two to three, third-order filtering is achieved. A similar nonresonant structure was presented in Al-Joumayly and Behdad (2010). In Hu et al. (2009), a simple approach to designing an FSS was suggested, wherein the FSS unit cell was composed of a square loop at the top with its complement at the bottom. Furthermore, a different application of THz FSS in a remote sensing instrument was proposed in Dickie et al. (2011). In Azemi et al. (2012), a 3D bandpass FSS was developed. A multi-layered FSS for L and S band frequencies was presented in Deng et al. (2013). The X-band FSS presented in Yuan et al. (2014) was composed of the same metallic pattern on both sides of the dielectric. Rahmati and Hassani (2015) presented an FSS with a single metallic pattern comprising four small square slots, each loaded with multiple stub resonators to achieve multiple passbands in the X-band. FSS configurations for X-band shielding applications were investigated in Nauman et al. (2016), and bandpass FSS designs employing different patterns on each side of the dielectric substrate (Varuna et al., 2016). FSS structures incorporating Ferrite-SRR (Gao et al., 2017) and meandered patterns (Ghosh & Srivastava, 2017; Varuna et al., 2017) were proposed. Other works have explored various geometries and configurations, including modified plus shapes (Yadav et al., 2018), convoluted cross slots with vertical metal vias (Zhao et al., 2018), asymmetric Jerusalem cross patterns (Wang et al., 2019), and combinations of convoluted slot dipoles and slot grids (Zhao et al., 2019). An FSS based on convoluted stripe elements was proposed in Hong et al. (2019) and Sivasamy and Kanagasabai (2019), along with a dual-band pass FSS in Li et al. (2018). In Abdollahvand et al. (2020), a triple-band multilayer FSS is presented, which is transmissive at the X-band and reflective at the K-band and 30 GHz. Additionally, Gao et al. (2020) presents a triple passband FSS comprising two identical layers of three-square loops positioned on each side of a foam dielectric. Further, a multiple metallic layer FSS based on coupled SLRRs was presented in Liao et al. (2020), a dual-band single metal convoluted pattern-based FSS structure was reported in Cheng et al. (2020), and Singh et al. (2020) proposed a wideband FSS in the X-band operating region. Moreover, Mahaveer et al. (2021) introduces an FSS configuration comprising a combination of T-shaped, U-shaped, and rectangular patterns placed on one side of the dielectric, Altintaş (2021) reported a single metallic layer FSS for earth observation satellite and radar applications in the X-band, a triple-band FSS for Wi-MAX and WLAN applications was proposed in Yadav et al. (2021), Kanchana et al. (2021) presented a dual-band FSS structure, and a dual metallic layer convoluted FSS structure was presented in Li and Ne (2021). Numerous other studies have been conducted and are presented in Afzal et al. (2023), Garg et al. (2023), Han et al. (2022), Jiang et al. (2023), Kaur and Kaur (2022), Kundu (2024), Li et al. (2022), Paik and Premchand (2023), Sheng et al.

(2024), Solunke and Kothari (2024), and Xi et al. (2024), showcasing passband filter responses using a range of resonators and patterning techniques on single or both sides of dielectric substrates.

This work introduces a novel design technique to overcome limitations in traditional passband FSS. Typically, FSS structures feature metallic patterns on one side, restricting the manipulation of transmitted EM waves for multiple or broad passbands. While sandwiching a dielectric layer between identical metallic patterns offers more flexibility, it has its limitations, and employing two different metallic patterns adds complexity to the design process. To address these challenges, the proposed approach involves placing a dielectric layer between metallic patterns on the top and their complementary geometry on the bottom, creating a bidirectional multiple passband FSS. In this design, the top metallic pattern consists of three concentric eight-scallop flower (ESF) rings paired with their complementary geometry on the bottom. This configuration enables the structure to achieve six passbands without increasing thickness or adding extra layers. By increasing the number of concentric rings and optimizing resonances, a broader passband can be achieved. Additionally, the design offers bidirectional response capabilities.

Furthermore, the proposed design for a six-bidirectional passband frequency selective surface not only showcases polarization independence and angular stability but also exhibits exceptional performance in terms of conformality due to its remarkably thin profile. To ensure the practical viability and accuracy of the proposed FSS, validation procedures are carried out. This includes testing the structure's conformality to different surfaces as well as verifying its performance using both an equivalent circuit model and free-space measurement techniques. Through these rigorous validation processes, the reliability and effectiveness of the proposed FSS design are confirmed, bolstering its potential for real-world applications.

This chapter is divided into six sections for a comprehensive understanding of the proposed passband FSS design as follows. Section 2 delves into the intricacies of the proposed passband FSS unit cell design, providing detailed insights into its composition and functionality along with the insights into the bidirectional feature of the suggested FSS, exploring how it responds to EM waves from both forward and backward directions. In Section 3, the impacts of various factors such as changes in polarization angles, oblique incidence angles, geometric parameter variations, surface current, and field distributions on the performance of the proposed FSS are analyzed. Conformality analysis is presented in Section 4, where the investigation is carried out by conforming the FSS on the different radii cylindrical structures. Section 5 delves into circuit analysis, providing an in-depth examination of the electrical parameters and behavior of the FSS through an equivalent circuit model. In Section 6, the fabricated FSS is presented with details of the planar surface measurement setup used for experimental validation and the results obtained for polarization angle and oblique incidence angle variations. Finally, the chapter concludes in Section 7, summarizing the key findings of the proposed FSS design.

BIDIRECTIONAL FSS DESIGN

In this section, the design of a bidirectional FSS is discussed in detail, along with simulated results demonstrating the bidirectional characteristics of the proposed FSS and its design evolution.

Figure 1 provides a visual representation of the subwavelength unit cell, showcasing the proposed design for an ultrathin and bidirectional multiple passband frequency selective surface (FSS). The top (front) metallic pattern of the unit cell comprises three concentric rings shaped like eight-scallop flowers (ESF), while the bottom (back) metal surface mirrors this pattern with its corresponding complementary layout. Both the top and bottom patterns enclose the FR4 substrate, which is situated in between them. For a comprehensive understanding of the design, detailed dimensions for each geometric component are provided in Table 1, aiding in the replication and analysis of the proposed FSS structure.

Figure 1. Unit cell design of the proposed passband ESF FSS design with geometrical representation of (a) top (front) pattern (b) bottom (back) pattern and (c) side view, and (d) perspective view

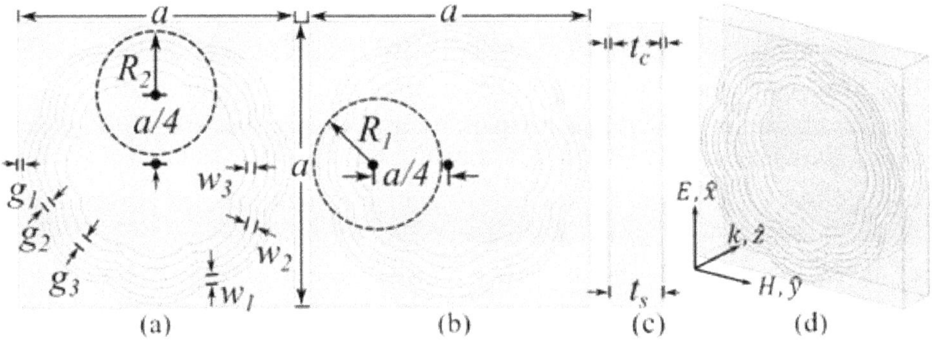

Table 1. Unit cell dimensions of the proposed passband ESF FSS design (units in mm).

a									

The construction process for the eight-scallop flower rings follows a systematic procedure outlined as follows. Initially, the outermost eight-petal flower is created by combining eight circles, each with a radius of R_1, and positioning them such that they are centered at one-quarter of the unit cell. These circles are then merged to form the outermost eight-petal flower. Subsequently, the same procedure is repeated to create another eight-petal flower, but this time with a radius of R_2. These two eight-petal flowers are then overlapped, and the overlapping region is subtracted, resulting in the outermost eight-scallop flower ring with a specified width of w_1. To obtain the additional eight-scallop flower rings, the scaling properties are utilized. By applying scaling properties iteratively, multiple eight-scallop flower rings are generated to meet the desired design requirements.

The subwavelength unit cell for the proposed passband ESF FSS design is meticulously created on an FR4 substrate, measuring 0.4mm in thickness (t_s). The substrate possesses a relative permittivity (ε_r) of 4.4 along with a loss tangent ($tan\delta$) of 0.02, which are critical parameters influencing the behavior of EM waves within the structure. Copper material is employed for both the top and bottom metallic layers, characterized by a conductivity (σ) of $5.8 \times 10^7\, S/m$ and a thickness (t_c) of 0.035mm. The dimensions of the subwavelength unit cell are carefully chosen to ensure that its size (a) remains smaller than one-fourth of the wavelength ($\lambda_L/4$) corresponding to the lowest transmission band frequency. This criterion is essential for achieving subwavelength behavior and efficient transmission characteristics.

To simulate the behavior of the FSS structure, periodic boundary conditions are enforced using Floquet port excitation within a commercial 3D full-wave HFSS simulator, which operates based on finite element methods. The simulation setup entails orienting the electric field (*E*-field) along the *X*-axis and the magnetic field (*H*-field) along the *Y*-axis, ensuring analysis of EM wave interactions within the proposed FSS unit cell.

Figure 2. Numerically simulated $S_{21}(\omega)$ and $S_{11}(\omega)$ of the proposed ESF FSS design for +Z wave propagation direction under: (a) TE-case and (b) TM-case

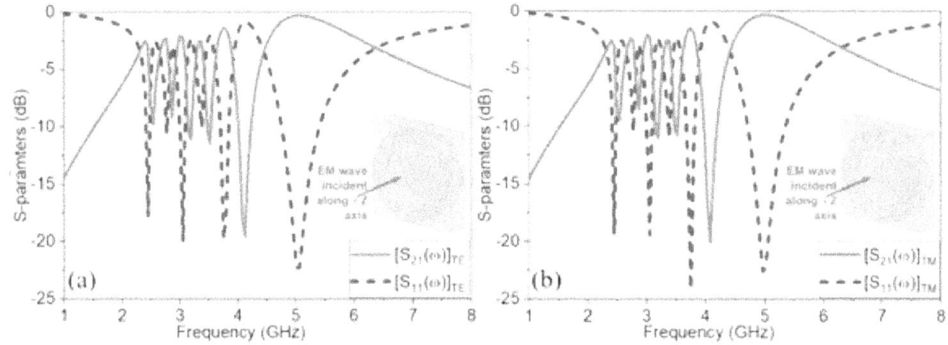

Figures 2(a) and 2(b) illustrate the S-parameters of the proposed bidirectional passband ESF FSS design for two distinct scenarios: the TE-case and TM-case. These scenarios represent the orientation of the incident EM wave when it propagates along the positive *Z*-axis, also known as the front side of the structure. In Figure 2(a), the S-parameters are depicted for the TE-case, where the *E*-field of the incident wave is transverse to the direction of propagation, while in Figure 2(b), the S-parameters represent the TM-case, where the *H*-field is transverse to the direction of propagation. These figures provide insight into the behavior of the FSS design when the EM wave impinges upon the front side of the structure.

Conversely, Figure 3 showcases the S-parameters of the proposed FSS for both the TE-case and TM-case scenarios, but this time corresponding to the incidence of the EM wave along the negative Z-axis, or the back side of the structure. This configuration allows for a comprehensive analysis of the FSS performance under different incident wave orientations, providing valuable insights into its bidirectional characteristics and transmission behavior.

Table 2 provides a comprehensive summary of the six frequencies achieved by the proposed bidirectional passband ESF FSS design, along with their corresponding -3dB bandwidths. These frequencies are analyzed for both the TE-case and TM-case scenarios, considering the forward (positive *Z*-axis) and backward (negative *Z*-axis) incidence of EM waves. The data presented in Table 2 is consistent with the observations made in Figures 2 and 3. These figures depict the S-parameters of the FSS design for different incident wave orientations and polarization states along *X-Y* directions. The similarities observed between the graphs in Figures 2 and 3, and the information presented in Table 2, highlight the bidirectional characteristics of the proposed passband ESF FSS unit cell.

REFERENCES

Abbaspour-Tamijani, A., Sarabandi, K., & Rebeiz, G. M. (2004). Antenna-filter-antenna arrays as a class of bandpass frequency-selective surfaces. *IEEE Transactions on Microwave Theory and Techniques*, 52(8), 1781–1789. 10.1109/TMTT.2004.831572

Abbaspour-Tamijani, A., Schoenlinner, B., Sarabandi, K., & Rebeiz, G. M. (2003). A new class of bandpass frequency selective structures. In *IEEE Antennas and Propagation Society International Symposium. Digest. Held in conjunction with: USNC/CNC/URSI North American Radio Sci. Meeting (Cat. No. 03CH37450)* (Vol. 2, pp. 817-820). IEEE. 10.1109/APS.2003.1219360

Abdollahvand, M., Forooraghi, K., Encinar, J. A., Atlasbaf, Z., & Martinez-de-Rioja, E. (2020). Design and demonstration of a tri-band frequency selective surface for space applications in X, K, and Ka bands. *Microwave and Optical Technology Letters*, 62(4), 1742–1751. 10.1002/mop.32225

Afzal, W., Ebrahimi, A., Robel, M. R., & Rowe, W. S. (2023). Low-profile higher-order narrowband bandpass miniaturized-element frequency-selective surface. *IEEE Transactions on Antennas and Propagation*, 71(4), 3736–3740. 10.1109/TAP.2023.3239171

Al-Joumayly, M. A., & Behdad, N. (2010). A generalized method for synthesizing low-profile, band-pass frequency selective surfaces with non-resonant constituting elements. *IEEE Transactions on Antennas and Propagation*, 58(12), 4033–4041. 10.1109/TAP.2010.2078474

Altintaş, O. (2021). A bandpass frequency selective surface filter for earth observation satellite and radar applications. *Çukurova Üniversitesi Mühendislik Fakültesi Dergisi, 36*(4), 1033-1040.

Azemi, S. N., Ghorbani, K., & Rowe, W. (2012). 3D frequency selective surfaces. *Progress in Electromagnetics Research C. Pier C*, 29, 191–203. 10.2528/PIERC12033006

Behdad, N. (2008). A second-order band-pass frequency selective surface using nonresonant subwavelength periodic structures. *Microwave and Optical Technology Letters*, 50(6), 1639–1643. 10.1002/mop.23445

Behdad, N., Al-Joumayly, M., & Salehi, M. (2009). A low-profile third-order bandpass frequency selective surface. *IEEE Transactions on Antennas and Propagation*, 57(2), 460–466. 10.1109/TAP.2008.2011202

Bossard, J. A., Werner, D. H., Mayer, T. S., Smith, J. A., Tang, Y. U., Drupp, R. P., & Li, L. (2006). The design and fabrication of planar multiband metallodielectric frequency selective surfaces for infrared applications. *IEEE Transactions on Antennas and Propagation*, 54(4), 1265–1276. 10.1109/TAP.2006.872583

Cheng, T., Jia, Z., Hong, T., Jiang, W., & Gong, S. (2020). Dual-band frequency selective surface with compact dimension and low frequency ratio. *IEEE Access : Practical Innovations, Open Solutions*, 8, 185399–185404. 10.1109/ACCESS.2020.3030131

Daira, S. E. I., Lashab, M., Berkani, H. A., Belattar, M., Gharbia, I., & Abd-Alhameed, R. A. (2024). A curved single-layer FSS design for gain improvement of a compact size CPW-fed UWB monopole antenna. *Microwave and Optical Technology Letters*, 66(1), e33943. 10.1002/mop.33943

Deng, F., Yi, X., & Wu, W. (2013). Design and performance of a double-layer miniaturized-element frequency selective surface. *IEEE Antennas and Wireless Propagation Letters*, 12, 721–724. 10.1109/LAWP.2013.2265095

Dickie, R., Cahill, R., Fusco, V., Gamble, H. S., & Mitchell, N. (2011). THz frequency selective surface filters for earth observation remote sensing instruments. *IEEE Transactions on Terahertz Science and Technology*, 1(2), 450–461. 10.1109/TTHZ.2011.2129470

Fatima, F., Akhtar, M. J., & Ramahi, O. M. (2024). Frequency Selective Surface Structures-Based RF Energy Harvesting Systems and Applications: FSS-Based RF Energy Harvesting Systems. *IEEE Microwave Magazine*, 25(3), 47–69. 10.1109/MMM.2023.3340988

Gao, B., Huang, S., Ren, Z., Chen, Y., & Wang, X. (2018). Design and verification of an integrated free-standing thick-screen FSS radome. *IEEE Antennas and Wireless Propagation Letters*, 17(9), 1630–1634. 10.1109/LAWP.2018.2859232

Gao, B., Yuen, M. M. F., & Ye, T. T. (2017). Flexible frequency selective metamaterials for microwave applications. *Scientific Reports*, 7(1), 45108. 10.1038/srep4510828322338

Gao, C., Pu, H., Gao, S., Chen, C., & Yang, Y. (2020). Design and analysis of a tri-band frequency selective surface with a second-order response. *International Journal of Microwave and Wireless Technologies*, 12(3), 205–211. 10.1017/S175907871900117X

Garg, J., Yadav, S., & Sharma, M. M. (2023). A novel miniaturized loop based angularly stable and polarization independent multiband bandpass FSS structure for Wi-Max and WLAN applications. *Sadhana*, 48(1), 14. 10.1007/s12046-022-02068-x

Garg, R., Bahl, I., & Bozzi, M. (2013). *Microstrip lines and slotlines*. Artech house.

Ghosh, S., & Srivastava, K. V. (2014). An equivalent circuit model of FSS-based metamaterial absorber using coupled line theory. *IEEE Antennas and Wireless Propagation Letters*, 14, 511–514. 10.1109/LAWP.2014.2369732

Ghosh, S., & Srivastava, K. V. (2017). An angularly stable dual-band FSS with closely spaced resonances using miniaturized unit cell. *IEEE Microwave and Wireless Components Letters*, 27(3), 218–220. 10.1109/LMWC.2017.2661683

Govindaswamy, S., East, J., Terry, F., Topsakal, E., Volakis, J. L., & Haddad, G. I. (2004). Frequency-selective surface based bandpass filters in the near-infrared region. *Microwave and Optical Technology Letters*, 41(4), 266–269. 10.1002/mop.20112

Han, Y., Feng, S., Chen, J., Chang, Y., Liao, S., & Che, W. (2022). Design of paper-based bandpass frequency selective surface using slotlines. *Microwave and Optical Technology Letters*, 64(8), 1339–1346. 10.1002/mop.33284

Hong, J. S. G., & Lancaster, M. J. (2004). *Microstrip filters for RF/microwave applications*. John Wiley & Sons.

Hong, T., Peng, K., & Wang, M. (2019). Miniaturized frequency selective surface using handshake convoluted stripe. *IEEE Antennas and Wireless Propagation Letters*, 18(10), 2026–2030. 10.1109/LAWP.2019.2936548

Hu, X. D., Zhou, X. L., Wu, L. S., Zhou, L., & Yin, W. Y. (2009). A miniaturized dual-band frequency selective surface (FSS) with closed loop and its complementary pattern. *IEEE Antennas and Wireless Propagation Letters*, 8, 1374–1377. 10.1109/LAWP.2009.2039110

Huang, H., Hua, C., & Shen, Z. (2022). Absorptive frequency-selective transmission structures based on hybrid FSS and absorber. *IEEE Transactions on Antennas and Propagation*, 70(7), 5606–5613. 10.1109/TAP.2022.3161472

Jiang, M., Du, Z., Li, Y., Sun, S., & Hu, J. (2023). Frequency Selective Surface with Quasi-Elliptic Bandpass Response for Filtering FRA Application. *IEEE Transactions on Antennas and Propagation*.

Kanchana, D., Radha, S., Sreeja, B. S., & Manikandan, E. (2021). A miniaturized flexible frequency selective surface for dual band response. *International Journal of Microwave and Wireless Technologies*, 13(8), 810–816. 10.1017/S1759078720001506

Kaur, K., & Kaur, A. (2022). Polarization independent frequency selective surface for marine and air traffic radar applications. *Sadhana*, 47(2), 81. 10.1007/s12046-022-01840-3

Kiani, G., & Dyadyuk, V. (2012, July). Low loss FSS polarizer for 70 GHz applications. In *Proceedings of the 2012 IEEE International Symposium on Antennas and Propagation* (pp. 1-2). IEEE. 10.1109/APS.2012.6348718

Krushna Kanth, V., & Raghavan, S. (2023). Metamaterial-FSS based radome using SIW technology. In *Handbook of Metamaterial-Derived Frequency Selective Surfaces* (pp. 191–223). Springer Nature Singapore.

Kundu, S. (2024). A simple planar single layered frequency selective surface with band-stop and bandpass characteristic. *Sadhana*, 49(1), 1–6. 10.1007/s12046-023-02411-w

Li, B., & Ne, R. (2021). A novel miniaturized dual-layer frequency selective surface. *AEÜ. International Journal of Electronics and Communications*, 130, 153580. 10.1016/j.aeue.2020.153580

Li, D., Shen, Z., & Li, E. P. (2018). Spurious-free dual-band bandpass frequency-selective surfaces with large band ratio. *IEEE Transactions on Antennas and Propagation*, 67(2), 1065–1072. 10.1109/TAP.2018.2882601

Li, M., Hong, M., Yang, Q., Zheng, Q., Yang, X., & Yi, Z. (2022). A design of broadband bandpass frequency selective surface. *Microwave and Optical Technology Letters*, 64(9), 1572–1578. 10.1002/mop.33331

Liao, R. X., Wong, S. W., Li, Y., Lin, J. Y., Liu, B. Y., Chen, F. C., & Quan, Z. (2020). Quasi-elliptic bandpass frequency selective surface based on coupled stubs-loaded ring resonators. *IEEE Access : Practical Innovations, Open Solutions*, 8, 113675–113682. 10.1109/ACCESS.2020.3003319

Mahaveer, U., Chandrasekaran, K. T., Mohan, M. P., Alphones, A., Siyal, M. Y., & Karim, M. F. (2021). A tri-band frequency-selective surface. *Journal of Electromagnetic Waves and Applications*, 35(7), 861–873. 10.1080/09205071.2020.1865206

Munk, B. (1974). *U.S. Patent No. 3,789,404*. Washington, DC: U.S. Patent and Trademark Office.

Munk, B. A. (2005). *Frequency selective surfaces: theory and design*. John Wiley & Sons.

Nauman, M., Saleem, R., Rashid, A. K., & Shafique, M. F. (2016). A miniaturized flexible frequency selective surface for X-band applications. *IEEE Transactions on Electromagnetic Compatibility*, 58(2), 419–428. 10.1109/TEMC.2015.2508503

Paik, H., & Premchand, K. (2023). Performance analysis of a single layer X-Band frequency selective surface based spatial filter implementing half Jerusalem cross slot. *Progress in Electromagnetics Research Letters*, 108.

Rahmati, B., & Hassani, H. R. (2015). Multiband metallic frequency selective surface with wide range of band ratio. *IEEE Transactions on Antennas and Propagation*, 63(8), 3747–3753. 10.1109/TAP.2015.2438340

Ram Krishna, R. V. S., & Kumar, R. (2015). Slotted ground microstrip antenna with FSS reflector for high-gain horizontal polarisation. *Electronics Letters*, 51(8), 599–600. 10.1049/el.2015.0339

Sarabandi, K., & Behdad, N. (2007). A frequency selective surface with miniaturized elements. *IEEE Transactions on Antennas and Propagation*, 55(5), 1239–1245. 10.1109/TAP.2007.895567

Sharma, A., Panwar, R., & Khanna, R. (2019). Experimental validation of a frequency-selective surface-loaded hybrid metamaterial absorber with wide bandwidth. *IEEE Magnetics Letters*, 10, 1–5. 10.1109/LMAG.2019.2898612

Sheng, X., Wang, H., Liu, N., & Wang, K. (2024). A Conformal Miniaturized Band-Pass Frequency Selective Surface With Stable Frequency Response for Radome Applications. *IEEE Transactions on Antennas and Propagation*, 72(3), 2423–2433. 10.1109/TAP.2024.3349677

Shin, H., Yoon, D., Na, D. Y., & Park, Y. B. (2022). Analysis of radome cross section of an aircraft equipped with a FSS radome. *IEEE Access : Practical Innovations, Open Solutions*, 10, 33704–33712. 10.1109/ACCESS.2022.3162262

Singh, C., Jha, K. R., & Sharma, S. K. (2020). Tripole type wideband bandpass frequency selective surface for X-band applications. *IET Microwaves, Antennas & Propagation*, 14(13), 1619–1625. 10.1049/iet-map.2019.1028

Sivasamy, R., & Kanagasabai, M. (2019). A novel miniaturized frequency selective surface. *International Journal of RF and Microwave Computer-Aided Engineering*, 29(6), e21691. 10.1002/mmce.21691

Sivasamy, R., & Kanagasabai, M. (2020). Design and fabrication of flexible FSS polarizer. *International Journal of RF and Microwave Computer-Aided Engineering*, 30(1), e22002. 10.1002/mmce.22002

Solunke, Y., & Kothari, A. (2024). An ultra-thin, low-RCS, dual-bandpass novel fractal-FSS for planar/conformal C&X bands applications. *AEÜ. International Journal of Electronics and Communications*, 175, 155073. 10.1016/j.aeue.2023.155073

Suganya, A., & Natarajan, R. (2023). Wideband reflecting frequency-selective surface polarizer for X-band applications. *International Journal of Communication Systems*, 36(17), e5608. 10.1002/dac.5608

Varuna, A. B., Ghosh, S., & Srivastava, K. V. (2016). An ultra thin polarization insensitive and angularly stable miniaturized frequency selective surface. *Microwave and Optical Technology Letters*, 58(11), 2713–2717. 10.1002/mop.30132

Varuna, A. B., Ghosh, S., & Srivastava, K. V. (2017). A miniaturized-element bandpass frequency selective surface using meander line geometry. *Microwave and Optical Technology Letters*, 59(10), 2484–2489. 10.1002/mop.30760

Wang, H., Yan, M., Qu, S., Zheng, L., & Wang, J. (2019). Design of a self-complementary frequency selective surface with multi-band polarization separation characteristic. *IEEE Access : Practical Innovations, Open Solutions*, 7, 36788–36799. 10.1109/ACCESS.2019.2905416

Xi, S., Xu, K. D., Yang, S., Ren, X., & Wu, W. (2024). X-band frequency selective surface with low loss and angular stability. *AEÜ. International Journal of Electronics and Communications*, 173, 154990. 10.1016/j.aeue.2023.154990

Yadav, S., Jain, C. P., & Sharma, M. M. (2018). Polarization independent dual-bandpass frequency selective surface for Wi-Max applications. *International Journal of RF and Microwave Computer-Aided Engineering*, 28(6), e21278. 10.1002/mmce.21278

Yadav, S., Sharma, M. M., & Singh, R. K. (2021). A polarization insensitive tri-band bandpass frequency selective surface for Wi-MAX and WLAN applications. *Progress in Electromagnetics Research Letters*, 101, 127–136. 10.2528/PIERL21091101

Yuan, J., Liu, S., Bian, B., Kong, X., Zhang, H., & Wang, S. (2014). A novel high-selective bandpass frequency selective surface with multiple transmission zeros. *Journal of Electromagnetic Waves and Applications*, 28(17), 2197–2209. 10.1080/09205071.2014.959620

Yuan, Y., Xi, X., & Zhao, Y. (2019). Compact UWB FSS reflector for antenna gain enhancement. *IET Microwaves, Antennas & Propagation*, 13(10), 1749–1755. 10.1049/iet-map.2019.0083

Zhao, P. C., Zong, Z. Y., Wu, W., Li, B., & Fang, D. G. (2019). Miniaturized-element bandpass FSS by loading capacitive structures. *IEEE Transactions on Antennas and Propagation*, 67(5), 3539–3544. 10.1109/TAP.2019.2902633

Zhao, Z., Li, J., Shi, H., Chen, X., & Zhang, A. (2018). A low-profile angle-insensitive bandpass frequency-selective surface based on vias. *IEEE Microwave and Wireless Components Letters*, 28(3), 200–202. 10.1109/LMWC.2018.2798166

Zhou, H., Qu, S., Lin, B., Wang, J., Ma, H., Xu, Z., Peng, W., & Bai, P. (2012). Filter-antenna consisting of conical FSS radome and monopole antenna. *IEEE Transactions on Antennas and Propagation*, 60(6), 3040–3045. 10.1109/TAP.2012.2194648

Compilation of References

Abbaspour-Tamijani, A., Sarabandi, K., & Rebeiz, G. M. (2004). Antenna-filter-antenna arrays as a class of bandpass frequency-selective surfaces. *IEEE Transactions on Microwave Theory and Techniques*, 52(8), 1781–1789. 10.1109/TMTT.2004.831572

Abbaspour-Tamijani, A., Schoenlinner, B., Sarabandi, K., & Rebeiz, G. M. (2003). A new class of bandpass frequency selective structures. In *IEEE Antennas and Propagation Society International Symposium. Digest. Held in conjunction with: USNC/CNC/URSI North American Radio Sci. Meeting (Cat. No. 03CH37450)* (Vol. 2, pp. 817-820). IEEE. 10.1109/APS.2003.1219360

Abdelwahab, H., Ebrahimi, A., Tovar-Lopez, F. J., Beziuk, G., & Ghorbani, K. (2021). Extremely sensitive microwave microfluidic dielectric sensor using a transmission line loaded with shunt LC resonators. *Sensors, 21*(20), 6811-1-6811–6813. 10.3390/s21206811

Abdollahvand, M., Forooraghi, K., Encinar, J. A., Atlasbaf, Z., & Martinez-de-Rioja, E. (2020). Design and demonstration of a tri-band frequency selective surface for space applications in X, K, and Ka bands. *Microwave and Optical Technology Letters*, 62(4), 1742–1751. 10.1002/mop.32225

Abdulkarim, Y. I., Deng, L., Karaaslan, M., Altıntaş, O., Awl, H. N., Muhammadsharif, F. F., Liao, C., Unal, E., & Luo, H. (2020). Novel metamaterials-based hypersensitized liquid sensor integrating omega-shaped resonator with microstrip transmission line. *Sensors (Switzerland), 20*(3), 943-1-943–18. 10.3390/s20030943

Abdulkarim, Y. I., Deng, L., Karaaslan, M., Dalgaç, Ş., Mahmud, R. H., Alkurt, F. O., Muhammadsharif, F. F., Awl, H. N., Huang, S., & Luo, H. (2020). The detection of chemical materials with a metamaterial-based sensor incorporating oval wing resonators. *Electronics, 9*(5), 825-1-825–15. 10.3390/electronics9050825

Abdulkarim, Y. I., Muhammadsharif, F. F., Bakır, M., Awl, H. N., Karaaslan, M., Deng, L., & Huang, S. (2021). Hypersensitized metamaterials based on a corona-shaped resonator for efficient detection of glucose. *Applied Sciences (Basel, Switzerland)*, 11(1), 1–19. 10.3390/app11010103

Adepoju, W., Bhattacharya, I., Sanyaolu, M., Bima, M. E., Banik, T., Esfahani, E. N., & Abiodun, O. (2022). Critical review of recent advancement in metamaterial design for wireless power transfer. *IEEE Access : Practical Innovations, Open Solutions*, 10, 42699–42726. 10.1109/ACCESS.2022.3167443

Afzal, W., Ebrahimi, A., Robel, M. R., & Rowe, W. S. (2023). Low-profile higher-order narrowband bandpass miniaturized-element frequency-selective surface. *IEEE Transactions on Antennas and Propagation*, 71(4), 3736–3740. 10.1109/TAP.2023.3239171

Ahmad I, Ullah S, Ullah S, Habib U, Ahmad S, Ghaffar A, Alibakhshikenari M, Khan S, Limiti E (2021) Design and analysis of a photonic crystal based planar antenna for THz applications. *Electronics 10*(16), 1941. ronic s1016 1941.10. 3390/ elect

Ai, H., Kang, Q., Wang, W., Guo, K., & Guo, Z. (2021). Multi-Beam Steering for 6G Communications Based on Graphene Metasurfaces. *Sensors (Basel)*, 21(14), 4784. 10.3390/s2114478434300521

Aïssa, B., Ali, A., & El Mellouhi, F. (2021). Oxide and Organic-Inorganic Halide Perovskites with Plasmonics for Optoelectronic and Energy Applications: A Contributive Review. *Catalysts*, 11(9), 1057. 10.3390/catal11091057

Akgol, Oguzhan, & Unal, H. (2018). Metamaterial-based multifunctional sensor design for moisture, concrete aging and ethanol density sensing applications. *Modern Physics Letters B, 32*(23). 10.1142/S0217984918502718

Akgol, O., Unal, E., Bağmancı, M., Karaaslan, M., Sevim, U. K., Öztürk, M., & Bhadauria, A. (2019). A nondestructive method for determining fiber content and fiber ratio in concretes using a metamaterial sensor based on a V-shaped resonator. *Journal of Electronic Materials*, 48(4), 2469–2481. 10.1007/s11664-019-06937-w

Akinsolu, M. O., Mistry, K. K., Liu, B., Lazaridis, P. I., & Excell, P. (2020). Machine learning-assisted antenna design optimization: A review and the state- of-the-art. *Proc. 14th Eur. Conf. Antennas Propag. (EuCAP),* (pp. 1–5). IEEE. 10.23919/EuCAP48036.2020.9135936

Alam, M. S., Islam, M. T., Misran, N., & Selangor, D. E. (2011). Design analysis of an electromagnetic band gap microstrip antenna. *American Journal of Applied Sciences, 8*(12), 1374-1377.

Ali Esmail, B., & Koziel, S. (2023). High isolation metamaterial-based dual-band MIMO antenna for 5G millimeter-wave applications. *AEÜ. International Journal of Electronics and Communications*, 158, 154470. 10.1016/j.aeue.2022.154470

Ali Kadhum, A. (2022). *Design and Analysis of Novel Reconfigurable Monopole Antenna Using Dip Switch and Covering 5G-Sub-6-GHz and C-Band Applications*. MDPI.

Ali, L., Mohammed, M. U., Khan, M., Yousuf, A. H. B., & Chowdhury, M. H. (2019). High-quality optical ring resonator-based biosensor for cancer detection. *IEEE Sensors Journal*, 20(4), 1867–1875. 10.1109/JSEN.2019.2950664

Ali, U., Jabbar, A., Yi, X., Naveed, M. A., Mehmood, M. Q., Zubair, M., & Massoud, Y. (2023). A novel fractal Hilbert curve-based low-cost and highly sensitive microwave sensor for dielectric characterization of liquid materials. *IEEE Sensors Journal*, 23(20), 23950–23957. 10.1109/JSEN.2023.3312309

Al-Joumayly, M. A., & Behdad, N. (2010). A generalized method for synthesizing low-profile, band-pass frequency selective surfaces with non-resonant constituting elements. *IEEE Transactions on Antennas and Propagation*, 58(12), 4033–4041. 10.1109/TAP.2010.2078474

Almpanis, E., Pantazopoulos, P. A., Papanikolaou, N., Yannopapas, V., & Stefanou, N. (2017). A birefringent etalon enhances the Faraday rotation of thin magneto-optical films. *Journal of Optics*, 19(7), 075102. 10.1088/2040-8986/aa7420

Al-Naib, I. (2022). Terahertz asymmetric S-shaped complementary metasurface biosensor for glucose concentration. *Biosensors (Basel)*, 12(8), 609. 10.3390/bios1208060936005005

Alonso-delPino, M., Jung-Kubiak, C., Reck, T., Llombart, N., & Chattopadhyay, G. (2019). Beam Scanning of Silicon Lens Antennas Using Integrated Piezomotors at Submillimeter Wavelengths. *IEEE Transactions on Terahertz Science and Technology*, 9(1), 47–54. 10.1109/TTHZ.2018.2881930

Alrayes, N., & Hussein, M. I. (2021). Metamaterial-based sensor design using split ring resonator and Hilbert fractal for biomedical application. *Sensing and Bio-Sensing Research*, 31, 100395-1, 100395–10. 10.1016/j.sbsr.2020.100395

Altintaş, O. (2021). A bandpass frequency selective surface filter for earth observation satellite and radar applications. *Çukurova Üniversitesi Mühendislik Fakültesi Dergisi, 36*(4), 1033-1040.

Alu, A., Engheta, N., Erentok, A., & Ziolkowski, R. W. (2007, February). Single- negative, double-negative, and low-index metamaterials and their electromagnetic applications. *IEEE Antennas & Propagation Magazine*, 49(1), 23–36. 10.1109/MAP.2007.370979

Ananda, S., Sriram Kumara, D., Wub, R. J., & Chavali, M. (2014). Graphene nanoribbon-based terahertz antenna on polyimide substrate. *Optik (Stuttgart)*, 125(19), 5546–5549. 10.1016/j.ijleo.2014.06.085

An, J., Lee, J., Lee, S. H., Park, J., & Kim, B. (2009). Separation of malignant human breast cancer epithelial cells from healthy epithelial cells using an advanced dielectrophoresis-activated cell sorter (DACS). *Analytical and Bioanalytical Chemistry*, 394(3), 801–809. 10.1007/s00216-009-2743-719308360

Arbabi, E., Arbabi, A., Kamali, S. M., Horie, Y., Faraji-Dana, M., & Faraon, A. (2018). MEMS-tunable dielectric metasurface lens. *Nature Communications*, 9(1), 812. 10.1038/s41467-018-03155-629476147

Asgari, S., Granpayeh, N., & Fabritius, T. (2020). Controllable terahertz cross-shaped three-dimensional graphene intrinsically chiral metastructure and its biosensing application. *Optics Communications*, 474, 12608. 10.1016/j.optcom.2020.126080

Azarbar, A., Masouleh, M. S., & Behbahani, A. K. (2014). A new terahertz microstrip rectangular patch array antenna. *Int J Electromagn Appl*, 4, 25–29.

Azemi, S. N., Ghorbani, K., & Rowe, W. (2012). 3D frequency selective surfaces. *Progress in Electromagnetics Research C. Pier C*, 29, 191–203. 10.2528/PIERC12033006

Babu, K. V. (2023). *Design and analysis of fractal-based THz antenna with co-axial feeding technique for wireless applications.Recent advances in graphene nanophotonics*. Springer Nature Switzerland.

Babu, K. V., Das, S., Sree, G. N. J., Madhav, B. T. P., Patel, S. K. K., & Parmar, J. (2022). Design and optimization of micro-sized wideband fractal MIMO antenna based on characteristic analysis of graphene for terahertz applications. *Optical and Quantum Electronics*, 54(5), 281. 10.1007/s11082-022-03671-2

Baena, J. D., Bonache, J., Martín, F., Sillero, R. M., Falcone, F., Lopetegi, T., Laso, M. A. G., García-García, J., Gil, I., Portillo, M. F., & Sorolla, M. (2005). Equivalent-circuit models for split-ring resonators and complementary split-ring resonators coupled to planar transmission lines. *IEEE Transactions on Microwave Theory and Techniques*, 53(4), 1451–1460. 10.1109/TMTT.2005.845211

Balanis, C. A. (2012). *Antenna theory: Analysis and design*. John Wiley & Sons., 10.1109/LMWC.2004.828009

Barron, L. D. (2004). *Molecular Light Scattering and Optical Activity* (2nd ed.). Cambridge University Press. 10.1017/CBO9780511535468

Behdad, N. (2008). A second-order band-pass frequency selective surface using nonresonant subwavelength periodic structures. *Microwave and Optical Technology Letters*, 50(6), 1639–1643. 10.1002/mop.23445

Behdad, N., Al-Joumayly, M., & Salehi, M. (2009). A low-profile third-order bandpass frequency selective surface. *IEEE Transactions on Antennas and Propagation*, 57(2), 460–466. 10.1109/TAP.2008.2011202

Bellekhiri, A., Chahboun, N., Laaziz, Y., & El, A. (2022). A New Design of 5G Planar Antenna based on metamaterials with a high gain using array antenna. *ITM Web of Conferences, 48*.

Benlakehal, M. (2022). Design and analysis of novel microstrip patch antenna array based on photonic crystal in THz. *Opt Quantum Electron, 54*(5), 1-16.2

Benlakehal, M. E., Hocini, A., & Khedrouche, D. (2022). Design and analysis of a 1 × 2 microstrip patch antenna array based on photonic crystals with a graphene load in THz. *Journal of Optics*. 10.1007/s12596-022-01006-8

Benlakehal, M. E., Hocini, A., Khedrouche, D., Temmar, M. N. E., & Denidni, T. A. (2022). Design and analysis of a 2 × 2 microstrip ratch antenna array based on periodic and non-periodic photonic crystals substrate in THz. *Optical and Quantum Electronics*, 54(3), 190. 10.1007/s11082-022-03563-5

Benlakehal, M. E., Hocini, A., Khedrouche, D., Temmar, M. N., & Denidni, T. A. (2022). Design and analysis of a 1 × 2 microstrip patch antenna array based on periodic and aperiodic photonic crystals in terahertz. *Optical and Quantum Electronics*, 54(10), 672. 10.1007/s11082-022-04076-x

Benlakehal, M. E., Hocini, A., Khedrouche, D., Temmar, M. N., & Denidni, T. A. (2022). Design and analysis of MIMO system for THz communication using terahertz patch antenna array based on photonic crystals with graphene. *Optical and Quantum Electronics*, 54(11), 693. 10.1007/s11082-022-04081-0

Bogue, R. (2017). Sensing with metamaterials: A review of recent developments. *Sensor Review*, 37(3), 305–311. 10.1108/SR-12-2016-0281

Bossard, J. A., Werner, D. H., Mayer, T. S., Smith, J. A., Tang, Y. U., Drupp, R. P., & Li, L. (2006). The design and fabrication of planar multiband metallodielectric frequency selective surfaces for infrared applications. *IEEE Transactions on Antennas and Propagation*, 54(4), 1265–1276. 10.1109/TAP.2006.872583

Caloz, C., & Itoh, T. (2006). *Electromagnetic Metamaterials Transmission Line Theory and Microwave Applications*. Wiley.

Castillo, E. S. R., Fernandez, E., Aranibar, P. C., & Vargas, D. S. (2018). Metamaterial inspired multiband planar array to detect glyphosate in water by real-time electromagnetic wave sensor. *2018 IEEE MTT-S Latin America Microwave Conference, LAMC 2018 - Proceedings*, (pp. 1–3). IEEE. 10.1109/LAMC.2018.8699077

Chandra & Dwivedi. (2020). Simulation analysis of High Gain and Low Loss Antenna with Metallic Ring and spoof SPP Transmission Line. In *7th International Conference on Signal Processing & Integrated Networks*. IEEE Explore.

Chandra & Dwivedi. (2021). Graphene based Radiation Pattern reconfigurable antenna. *International Conference for convergence in Technology*. IEEE.

Cheng, B., Cui, Z., Lu, B., Qin, Y., Liu, Q., Chen, P., He, Y., Jiang, J., He, X., Deng, X., Zhang, J., & Zhu, L. (2018). 340-GHz 3-D imaging radar with 4Tx-16Rx MIMO array. *IEEE Transactions on Terahertz Science and Technology*, 8(5), 509–519. 10.1109/TTHZ.2018.2853551

Cheng, J., & Chen, C.-L. (2011). Adaptive beam tracking and steering via electrowetting-controlled liquid prism. *Applied Physics Letters*, 99(19), 191108. 10.1063/1.3660578

Cheng, T., Jia, Z., Hong, T., Jiang, W., & Gong, S. (2020). Dual-band frequency selective surface with compact dimension and low frequency ratio. *IEEE Access : Practical Innovations, Open Solutions*, 8, 185399–185404. 10.1109/ACCESS.2020.3030131

Cheng, Y., Cao, J., & Hao, Q. (2021). Optical beam steering using liquid-based devices. *Optics and Lasers in Engineering*, 146, 106700. 10.1016/j.optlaseng.2021.106700

Chen, H.-T., Padilla, W. J., Zide, J. M., Gossard, A. C., Taylor, A. J., & Averitt, R. D. (2006). Active terahertz metamaterial devices. *Nature*, 444(7119), 597–600. 10.1038/nature0534317136089

Chen, T., Li, S., & Sun, H. (2012). Metamaterials application in sensing. *Sensors (Basel)*, 12(3), 2742–2765. 10.3390/s12030274222736975

Chen, T., Zhang, D., Huang, F., Li, Z., & Hu, F. (2020). Design of a terahertz metamaterial sensor based on split ring resonator nested square ring resonator. *Materials Research Express*, 7(9), 095802. 10.1088/2053-1591/abb496

Chi, L., Weng, Z., Qi, Y., & Drewniak, J. L. (2020). A 60 GHz PCB Wideband Antenna-in-Package for 5G/6G Applications. *IEEE Antennas and Wireless Propagation Letters*, 1225(c), 1–1. 10.1109/LAWP.2020.3006873

Chowdhury, M. Z. B., Islam, M. T., Alzamil, A., Soliman, M. S., & Samsuzzaman, M. (2024). A tunable star-shaped highly sensitive microwave sensor for solid and liquid sensing. *Alexandria Engineering Journal, 86*(July 2023), 644–662. 10.1016/j.aej.2023.12.001

Chuma, E. L., Iano, Y., Fontgalland, G., & Bravo Roger, L. L. (2018). Microwave sensor for liquid dielectric characterization based on metamaterial complementary split ring resonator. *IEEE Sensors Journal*, 18(24), 9978–9983. 10.1109/JSEN.2018.2872859

Colaco & Lohan. (2021). Metamaterial Based Multiband Microstrip Patch Antenna for 5G Wireless Technology-enabled IoT Devices and its applications. *J. Phys.: Conf. Ser.*

Costanzo, A., Augello, E., Battistini, G., Benassi, F., Masotti, D., & Paolini, G. (2023). Microwave devices for wearable sensors and IoT. *Sensors (Basel)*, 23(9), 4356. 10.3390/s2309435637177569

Costanzo, S., Venneri, F., Raffo, A., & Di Massa, G. (2018). Dual-Layer Single-Varactor Driven Reflectarray Cell for Broad-Band Beam-Steering and Frequency Tunable Applications. *IEEE Access : Practical Innovations, Open Solutions*, 6, 71793–71800. 10.1109/ACCESS.2018.2882093

Dai, J. Y., Zhao, J., Cheng, Q., & Cui, T. J. (2018). Independent control of harmonic amplitudes and phases via a time-domain digital coding metasurface. *Light, Science & Applications*, 7(1), 90. 10.1038/s41377-018-0092-z30479756

Dai, L., Zhao, X., Guo, J., Feng, S., Fu, Y., Kang, Y., & Guo, J. (2020). Microfluidics-based microwave sensor. *Sensors and Actuators. A, Physical*, 309, 111910-1, 111910–111916. 10.1016/j.sna.2020.111910

Daira, S. E. I., Lashab, M., Berkani, H. A., Belattar, M., Gharbia, I., & Abd-Alhameed, R. A. (2024). A curved single-layer FSS design for gain improvement of a compact size CPW-fed UWB monopole antenna. *Microwave and Optical Technology Letters*, 66(1), e33943. 10.1002/mop.33943

Dalgac, S., Baklr, M., Karadag, F., Karaaslan, M., Akgol, O., Unal, E., & Sabah, C. (2020). Microfluidic sensor applications by using chiral metamaterial. *Modern Physics Letters B, 34*(5), 2050031-1-2050031–14. 10.1142/S0217984920500311

Dalgaç, Ş., Karada , F., Ünal, E., Özkaner, V., Baklr, M., Akgöl, O., Sevim, U. K., Delihaclo lu, K., Öztürk, M., Karaaslan, M., & Sabah, C. (2020). Metamaterial sensor application concrete material reinforced with carbon steel fiber. *Modern Physics Letters B, 34*(10). 10.1142/S0217984920500979

Dalgac, S., Akdogan, V., Kiris, S., Incesu, A., Akgol, O., Unal, E., Basar, M. T., & Karaaslan, M. (2021). Investigation of methanol contaminated local spirit using metamaterial based transmission line sensor. *Measurement*, 178(April), 109360-1, 109360–109369. 10.1016/j.measurement.2021.109360

Dalgaç, Ş., Furat, M., Karaaslan, M., Akgöl, O., Karadağ, F., Zile, M., & Bakir, M. (2020). Grease oil humidity sensor by using metamaterial. *Journal of Electromagnetic Waves and Applications*, 34(18), 2488–2498. 10.1080/09205071.2020.1824690

Das, P., Singh, A. K., & Mandal, K. (2022). Metamaterial loaded highly isolated tunable polarisation diversity MIMO antennas for THz applications. *Optical and Quantum Electronics*, 54(4), 250. 10.1007/s11082-022-03641-8

Das, S., Anveshkumar, N., Dutta, J., & Biswas, A. (Eds.). (2021). *Advances in terahertz technology and its applications*. Springer. 10.1007/978-981-16-5731-3

De Galarreta, C. R., Alexeev, A. M., Au, Y., Lopez-Garcia, M., Klemm, M., Cryan, M., Bertolotti, J., & Wright, C. D. (2018). Nonvolatile Reconfigurable Phase-Change Metadevices for Beam Steering in the Near Infrared. *Advanced Functional Materials*, 28(10), 1704993. 10.1002/adfm.201704993

De Souza, F. A. L., Amorim, R. G., Scopel, W. L., & Scheicher, R. H. (2017). Electrical detection of nucleotides via nanopores in a hybrid graphene/h-BN sheet. *Nanoscale*, 9(6), 2207–2212. 10.1039/C6NR07154F28120993

Dell'Olio, F., & Passaro, V. M. (2007). Optical sensing by optimized silicon slot waveguides. *Optics Express*, 15(8), 4977–4993. 10.1364/OE.15.00497719532747

Deng, F., Yi, X., & Wu, W. (2013). Design and performance of a double-layer miniaturized-element frequency selective surface. *IEEE Antennas and Wireless Propagation Letters*, 12, 721–724. 10.1109/LAWP.2013.2265095

Devapriya, A. T., & Robinson, S. Investigation on metamaterial antenna for terahertz applications. J. Microw.Optoelectron. Electromagn. Appl. (2019). 10. 1590/ 2179- 10742 019v1 8i315 77

Dewan, R., Rahim, M. K. A., Hamid, M. R., Yusoff, M. F. M., Samsuri, N. A., Murad, N. A., & Kamardin, K. (2017). Artificial magnetic conductor for various antenna applications: An overview. *International Journal of RF and Microwave Computer-Aided Engineering*, 27(6), 1–18. 10.1002/mmce.21105

Dickie, R., Cahill, R., Fusco, V., Gamble, H. S., & Mitchell, N. (2011). THz frequency selective surface filters for earth observation remote sensing instruments. *IEEE Transactions on Terahertz Science and Technology*, 1(2), 450–461. 10.1109/TTHZ.2011.2129470

Ding, C., Sun, H.-H., Ziolkowski, R. W., & Jay Guo, Y. (2018, December). A dual layered loop array antenna for base stations with enhanced crosspolarization discrimination. *IEEE Transactions on Antennas and Propagation*, 66(12), 6975–6985. 10.1109/TAP.2018.2869216

Duan, Z., Tang, X., Wang, Z., Zhang, Y., Chen, X., Chen, M., & Gong, Y. (2017). Observation of the reversed Cherenkov radiation. *Nature Communications*, 8(1), 14901. 10.1038/ncomms1490128332487

Duma, V.-F., & Rolland, J. P. (2010). Mechanical Constraints and Design Considerations for Polygon Scanners. In Pisla, D., Ceccarelli, M., Husty, M., & Corves, B. (Eds.), *New Trends in Mechanism Science* (pp. 475–483). Springer Netherlands. 10.1007/978-90-481-9689-0_55

Dussopt, L., & Rebeiz, G. M. (2003, April). Intermodulation distortion and power handling in RF MEMS switches, varactors, and tunable filters. *IEEE Transactions on Antennas and Propagation*, 51(4), 1247–1256.

Dwivedi, S. (2023). Proposed Design for Beyond 5G Antenna for Upgraded Applications with review. *2023 14th International Conference on Computing Communication and Networking Technologies (ICCCNT)*. IEEE. 10.1109/ICCCNT56998.2023.10307737

Dwivedi. (2022). *Design and Analysis of Metamaterial Waveguide Antenna for Broadband Applications*. Springer. 10.1007/978-981-19-5224-1_42

Dwivedi. (2023). *Antenna Array for reconfiguration*. IGI Global. 10.4018/978-1-6684-5955-3

Dwivedi, S. (2022). *ReconfigurableArray antenna*. Intech Open., 10.5772/intechopen.106343

Dwivedi, S. (2023). *Metamaterials-Based Antenna for 5G and Beyond*. IGI Global. 10.4018/978-1-6684-8287-2.ch001

Ebrahimi, A., Scott, J., & Ghorbani, K. (2018). Differential sensors using microstrip lines loaded with two split-ring resonators. *IEEE Sensors Journal*, 18(14), 5786–5793. 10.1109/JSEN.2018.2840691

Eduardo, V. P. (2022). FORMAT: A Reconfigurable Tile-Based Antenna Array System for 5G and 6G Millimeter-Wave Testbeds. *IEEE Systems Journal*, 16(3), 4489–4500. 10.1109/JSYST.2022.3146360

Ejaz, J. W., Anpalagan, A., Imran, M. A., Jo, M., Naeem, M., Bin Qaisar, S., & Wang, W. (2016). *2016 Internet of Things (IoT) in 5G Wireless Communications*. IEEE Access.

El-Kenawy, E. S. M., Ibrahim, A., Mirjalili, S., Zhang, Y. D., Elnazer, S., & Zaki, R. M. (2022). Optimized ensemblealgorithm for predicting metamaterial antenna parameters. *Computers, Materials & Continua*, 2022, 023884. 10. 32604/ cmc

Elwi, T. A. (2018). Metamaterial based a printed monopole antenna for sensing applications. *International Journal of RF and Microwave Computer-Aided Engineering*, 28(7), 1–10. 10.1002/mmce.21470

Emara, M. K., Stuhec-Leonard, S. K., Tomura, T., Hirokawa, J., & Gupta, S. (2020). Laser-Drilled All- Dielectric Huygens' Transmit-Arrays as 120 GHz Band Beamformers. *IEEE Access : Practical Innovations, Open Solutions*, 8, 153815–153825. 10.1109/ACCESS.2020.3018297

Esmail, B. A. F., Koziel, S., & Szczepanski, S. (2022). Overview of planar antenna loading metamaterials for gain performance enhancement: The two decades of progress. *IEEE Access : Practical Innovations, Open Solutions*, 10, 27381–27403. 10.1109/ACCESS.2022.3157634

Fallahi, A., & Perruisseau-Carrier, J. (2012). Design of tunable biperiodic graphene metasurfaces. *hysical Review B*, *86* (19), p. 195408.

Faraji-Dana, M., Arbabi, E., Kwon, H., Kamali, S., Arbabi, A., Bartholomew, J., & Faraon, A. (2019). Hyperspectral imager with folded metasurface optics. *ACS Photonics*, 6(8), 2161–2167. 10.1021/acsphotonics.9b00744

Fatima, F., Akhtar, M. J., & Ramahi, O. M. (2024). Frequency Selective Surface Structures-Based RF Energy Harvesting Systems and Applications: FSS-Based RF Energy Harvesting Systems. *IEEE Microwave Magazine*, 25(3), 47–69. 10.1109/MMM.2023.3340988

Fitri, I., & Akbart, A. A. (2018). A new gridded parasitic patch stacked microstrip antenna for enhanced wide bandwidth in 60 GHz Band. *2017 International Conference on Broadband Communication, Wireless Sensors and Powering, BCWSP 2017*. IEEE. 10.1109/BCWSP.2017.8272571

Fowler, C., Silva, S., Thapa, G., & Zhou, J. (2022). High efficiency ambient RF energy harvesting by a metamaterial perfect absorber. *Optical Materials Express*, 12(3), 1242–1250. 10.1364/OME.449494

Fu, Y., Fu, X., Yang, S., Peng, S., Wang, P., Liu, Y., Yang, J., Wu, J., & Cui, T. J. (2023). Two-dimensional terahertz beam manipulations based on liquid-crystal-assisted programmable metasurface. *Applied Physics Letters*, 123(11), 111703. 10.1063/5.0167812

Galindo-Romera, G., Javier Herraiz-Martínez, F., Gil, M., Martínez-Martínez, J. J., & Segovia-Vargas, D. (2016). Submersible printed split-ring resonator-based sensor for thin-film detection and permittivity characterization. *IEEE Sensors Journal*, 16(10), 3587–3596. 10.1109/JSEN.2016.2538086

Gan, H. Y., Zhao, W. S., Liu, Q., Wang, D. W., Dong, L., Wang, G., & Yin, W. Y. (2020). Differential microwave microfluidic sensor based on microstrip complementary split-ring resonator (MCSRR) structure. *IEEE Sensors Journal*, 20(11), 5876–5884. 10.1109/JSEN.2020.2973196

Gansel, J. K., Thiel, M., Rill, M. S., Decker, M., Bade, K., Saile, V., Freymann, G., Linden, S., & Wegener, M. (2009). Gold helix photonic metamateri-al as broadband circular polarizer. *Science*, 325(5947), 1513–1515. 10.1126/science.117703119696310

Gao, B., Huang, S., Ren, Z., Chen, Y., & Wang, X. (2018). Design and verification of an integrated free-standing thick-screen FSS radome. *IEEE Antennas and Wireless Propagation Letters*, 17(9), 1630–1634. 10.1109/LAWP.2018.2859232

Gao, B., Yuen, M. M. F., & Ye, T. T. (2017). Flexible frequency selective metamaterials for microwave applications. *Scientific Reports*, 7(1), 45108. 10.1038/srep4510828322338

Gao, C., Pu, H., Gao, S., Chen, C., & Yang, Y. (2020). Design and analysis of a tri-band frequency selective surface with a second-order response. *International Journal of Microwave and Wireless Technologies*, 12(3), 205–211. 10.1017/S175907871900117X

Garg, J., Yadav, S., & Sharma, M. M. (2023). A novel miniaturized loop based angularly stable and polarization independent multiband bandpass FSS structure for Wi-Max and WLAN applications. *Sadhana*, 48(1), 14. 10.1007/s12046-022-02068-x

Garg, P., & Jain, P. (2023). A review of metamaterial absorbers and their application in sensors and radar cross-section reduction. *Microwave and Optical Technology Letters*, 65(2), 387–411. 10.1002/mop.33496

Garg, R., Bahl, I., & Bozzi, M. (2013). *Microstrip lines and slotlines*. Artech house.

Geng, Z., Zhang, X., Fan, Z., Lv, X., & Chen, H. (2017). A route to terahertz metamaterial biosensor integrated with microfluidics for liver cancer biomarker testing in early stage. *Scientific Reports*, 7(1), 16378. 10.1038/s41598-017-16762-y29180650

Ghodrati, M., Farmani, A., & Mir, A. (2019). Nanoscale Sensor-Based Tunneling Carbon Nanotube Transistor for Toxic Gases Detection: A First-Principle Study. *IEEE Sensors Journal*, 19(17), 7373–7377. 10.1109/JSEN.2019.2916850

Ghodrati, M., & Mir, A. (2022). Improving the Performance of a Doping-Less Carbon Nanotube FET with Dual Junction Source and Drain Regions: Numerical Studies. *Journal of Circuits, Systems, and Computers*, 31(10), 2250182. 10.1142/S0218126622501821

Ghodrati, M., Mir, A., & Farmani, A. (2020). Carbon nanotube field effect transistors-based gas sensors. In *Nanosensors for Smart Cities* (pp. 171–183). Elsevier. 10.1016/B978-0-12-819870-4.00036-0

Ghodrati, M., Mir, A., & Farmani, A. (2021). Non-destructive label-free biomaterials detection using tunneling carbon nanotube-based biosensor. *IEEE Sensors Journal*, 21(7), 8847–8854. 10.1109/JSEN.2021.3054120

Ghodrati, M., Mir, A., & Farmani, A. (2022). Proposing of SPR biosensor based on 2D $Ti_3C_2T_x$ MXene for uric acid detection ımmobilized by uricase enzyme. *Journal of Computational Electronics*, 22, 560–569. 10.1007/s10825-022-01959-w

Ghodrati, M., Mir, A., & Farmani, A. (2022). Sensitivity-Enhanced Surface Plasmon Resonance Sensor with Bimetal/ Tungsten Disulfide (WS_2)/MXene ($Ti_3C_2T_x$) Hybrid Structure. *Plasmonics*, 17(5), 1973–1984. 10.1007/s11468-022-01685-w

Ghodrati, M., Mir, A., & Farmani, A. (2023). Numerical analysis of a surface plasmon resonance based biosensor using molybdenum disulfide, molybdenum trioxide, and MXene for the diagnosis of diabetes. *Diamond and Related Materials*, 132, 109633. 10.1016/j.diamond.2022.109633

Ghodrati, M., Mir, A., & Farmani, A. (2024). *2D Materials/Heterostructures/Metasurfaces in Plasmonic Sensing and Biosensing*.

Ghodrati, M., Mir, A., & Naderi, A. (2020). New structure of tunneling carbon nanotube FET with electrical junction in part of drain region and step impurity distribution pattern. *AEÜ. International Journal of Electronics and Communications*, 117, 153102. 10.1016/j.aeue.2020.153102

Ghodrati, M., Mir, A., & Naderi, A. (2021). Proposal of a doping-less tunneling carbon nanotube field-effect transistor. *Materials Science and Engineering B*, 265, 115016. 10.1016/j.mseb.2020.115016

Ghosh, S., & Srivastava, K. V. (2014). An equivalent circuit model of FSS-based metamaterial absorber using coupled line theory. *IEEE Antennas and Wireless Propagation Letters*, 14, 511–514. 10.1109/LAWP.2014.2369732

Ghosh, S., & Srivastava, K. V. (2017). An angularly stable dual-band FSS with closely spaced resonances using miniaturized unit cell. *IEEE Microwave and Wireless Components Letters*, 27(3), 218–220. 10.1109/LMWC.2017.2661683

Gneiding, N., Zhuromskyy, O., Shamonina, E., & Peschel, U. (2014, October). Circuit model optimization of a nano split ring resonator dimer antenna operating in infrared spectral range. *Journal of Applied Physics*, 116(16), 164311. 10.1063/1.4900479

Goswami, S., Sarmah, K., Sarma, A., Sarma, K. K., & Baruah, S. (2019). Design considerations pertaining to the application of complementary split ring resonators in microstrip antennas. In *Emerging Innovations in Microwave and Antenna Engineering* (pp. 25–56). Springer. 10.4018/978-1-5225-7539-9.ch002

Govindaswamy, S., East, J., Terry, F., Topsakal, E., Volakis, J. L., & Haddad, G. I. (2004). Frequency-selective surface based bandpass filters in the near-infrared region. *Microwave and Optical Technology Letters*, 41(4), 266–269. 10.1002/mop.20112

Govind, G., & Akhtar, M. J. (2019). Metamaterial-inspired microwave microfluidic sensor for glucose monitoring in aqueous solutions. *IEEE Sensors Journal*, 19(24), 11900–11907. 10.1109/JSEN.2019.2938853

Goyal, R., & Vishwakarma, D. K. (2018). Design of a graphene-based patch antenna on glass substrate for high-speed terahertz communications. *Microwave and Optical Technology Letters*, 60(7), 1594–1600. 10.1002/mop.31216

GSMA. (2019). *5G the internet of things and wearable devices: Radiofrequency exposure*. GSMA Public Policy.

Gulsu, M. S., Bagci, F., Can, S., Yilmaz, A. E., & Akaoglu, B. (2021). Minkowski-like fractal resonator-based dielectric sensor for estimating the complex permittivity of binary mixtures of ethanol, methanol and water. *Sensors and Actuators. A, Physical*, 330, 112841-1, 112841–10. 10.1016/j.sna.2021.112841

Guo, L., Huang, F., & Tang, X. (2014). A novel integrated MEMS helix antenna for terahertz applications. *Optik (Stuttgart)*, 125(1), 101–103. 10.1016/j.ijleo.2013.06.016

Gupta, U., & Dwivedi, S. (2021). Frequency Reconfigurable Antenna Using Metamaterial Split Ring Resonators for Smart Applications", *2021 International Conference on Electrical, Communication, and Computer Engineering (ICECCE)*. IEEE. 10.1109/ICECCE52056.2021.9514067

Han, Y., Feng, S., Chen, J., Chang, Y., Liao, S., & Che, W. (2022). Design of paper-based bandpass frequency selective surface using slotlines. *Microwave and Optical Technology Letters*, 64(8), 1339–1346. 10.1002/mop.33284

Haq, T., Ruan, C., Zhang, X., Ullah, S., Fahad, A. K., & He, W. (2020). Extremely sensitive microwave sensor for evaluation of dielectric characteristics of low-permittivity materials. *Sensors, 20*(7), 1916-1-1916–1917. https://doi.org/10.3390/s2007191

Haq, T., & Koziel, S. (2023). New complementary resonator for permittivity- and thickness-based dielectric characterization. *Sensors (Basel)*, 23(22), 1–14. 10.3390/s2322913838005525

Harnsoongnoen, S. (2021). Metamaterial-inspired microwave sensor for detecting the concentration of mixed phosphate and nitrate in water. *IEEE Transactions on Instrumentation and Measurement*, 70, 9509906. 10.1109/TIM.2021.3086901

Harrsion, L., Ravan, M., Tandel, D., Zhang, K., Patel, T., & Amineh, R. K. (2020). Material identification using a microwave sensor array and machine learning. *Electronics (Switzerland)*, 9(2). Advance online publication. 10.3390/electronics9020288

Hashemi, M. R. M., Yang, S.-H., Wang, T., Sepúlveda, N., & Jarrahi, M. (2016). Electronically-Controlled Beam-Steering through Vanadium Dioxide Metasurfaces. *Scientific Reports*, 6(1), 35439. 10.1038/srep3543927739471

Helena. (2020). *IEEE Access : Practical Innovations, Open Solutions*, 8, 177064–177083.

Hernandez-Cardoso, G. G., Rojas-Landeros, S. C., Alfaro-Gomez, M., Hernandez-Serrano, A. I., Salas-Gutierrez, I., Lemus-Bedolla, E., Castillo-Guzman, A. R., Lopez-Lemus, H. L., & Castro-Camus, E. (2017). Terahertz imaging for early screening of diabetic foot syndrome: A proof of concept. *Scientific Reports*, 7(1), 42124. 10.1038/srep4212428165050

He, Z., Li, L., Ma, H., Pu, L., Xu, H., Yi, Z., Cao, X., & Cui, W. (2021). Graphene-based metasurface sensing applications in terahertz band. *Results in Physics*, 21, 103795. 10.1016/j.rinp.2020.103795

Hiep, L. T. H., Pham, T. S., Khuyen, B. X., Tung, B. S., Ngo, Q. M., Hien, N. T., Minh, N. T., & Lam, V. D. (2022). Enhanced transmission efficiency of magneto-inductive wave propagating in non-homogeneous 2-D magnetic metamaterial array. *Physica Scripta*, 97(2), 025504. 10.1088/1402-4896/ac4a3a

Hlali, A., Oueslati, A., & Zairi, H. (2021). Numerical simulation of tunable terahertz graphene-based sensor for breast tumor detection. *IEEE Sensors Journal*, 21(8), 9844–9851. 10.1109/JSEN.2021.3060326

Hocini, A., Temmar, M. N., Khedrouche, D., & Zamani, M. (2019). Novel approach for the design and analysis of a terahertz microstrip antenna based on photonic crystals. *Photonics and Nanostructures*, 36, 100723. 10.1016/j.photonics.2019.100723

Hofmann, U., Janes, J., & Quenzer, H.-J. (2012). High-Q MEMS Resonators for Laser Beam Scanning Displays. *Micromachines*, 3(2), 509–528. 10.3390/mi3020509

Holloway, D., Dienstfrey, A., Kuester, E. F., O'Hara, J. F., Azad, A. K., & Taylor, A. J. (2009). A Discussion on the Interpretation and Characterization of Metafilms/Metasurfaces: The Two Dimensional Equivalent of Metamaterials. *Metamaterials (Amsterdam)*, 3(2), 100–112. 10.1016/j.metmat.2009.08.001

Holloway, K., Kuester, E. F., Baker-Jarvis, J., & Kabos, P. (2003). A Double Negative (DNG) Composite Medium Composed of Magneto-Dielectric Spherical Particles Embedded in a Matrix. *IEEE Transactions on Antennas and Propagation*, 51(10), 2596–2603. 10.1109/TAP.2003.817563

Hong, J. S. G., & Lancaster, M. J. (2004). *Microstrip filters for RF/microwave applications.* John Wiley & Sons.

Hong, T., Peng, K., & Wang, M. (2019). Miniaturized frequency selective surface using hand-shake convoluted stripe. *IEEE Antennas and Wireless Propagation Letters*, 18(10), 2026–2030. 10.1109/LAWP.2019.2936548

Hong, W., Jiang, Z. H., Yu, C., Zhou, J., Chen, P., Yu, Z., Zhang, H., Yang, B., Pang, X., Jiang, M., Cheng, Y., Al-Nuaimi, M. K. T., Zhang, Y., Chen, J., & He, S. (2017, December). Multibeam antenna technologies for 5G wireless communications. *IEEE Transactions on Antennas and Propagation*, 65(12), 6231–6249. 10.1109/TAP.2017.2712819

Hosseininejad, S. E., Rouhi, K., Neshat, M., Faraji-Dana, R., Cabellos-Aparicio, A., Abadal, S., & Alarcón, E. (2019). Reprogrammable Graphene-based Metasurface Mirror with Adaptive Focal Point for THz Imaging. *Scientific Reports*, 9(1), 2868. 10.1038/s41598-019-39266-330814570

Huang, H., Hua, C., & Shen, Z. (2022). Absorptive frequency-selective transmission structures based on hybrid FSS and absorber. *IEEE Transactions on Antennas and Propagation*, 70(7), 5606–5613. 10.1109/TAP.2022.3161472

Huang, H., Li, X., & Liu, Y. (2018, June). A novel vector synthetic dipole antenna and its common aperture array. *IEEE Transactions on Antennas and Propagation*, 66(6), 3183–3188. 10.1109/TAP.2018.2819894

Huff, G. H., Bahukudumbi, P. B., Everett, W. N., Beskok, A., Bevan, M. A., Lagoudas, D., & Ounaies, Z. (2007). Microfluidic reconfiguration of antennas. *Proc. Antenna Appl. Symp.*, (pp. 241-258). IEEE.

Hui, S. Y. R., Zhong, W., & Lee, C. K. (2014). A critical review of recent progress in mid-range wireless power transfer. *IEEE Transactions on Power Electronics*, 29(9), 4500–4511. 10.1109/TPEL.2013.2249670

Hu, X. D., Zhou, X. L., Wu, L. S., Zhou, L., & Yin, W. Y. (2009). A miniaturized dual-band frequency selective surface (FSS) with closed loop and its complementary pattern. *IEEE Antennas and Wireless Propagation Letters*, 8, 1374–1377. 10.1109/LAWP.2009.2039110

Hu, Y., Luo, X., Chen, Y., Liu, Q., Li, X., Wang, Y., Liu, N., & Duan, H. (2019). 3D-integrated metasurfaces for full-colour holography. *Light, Science & Applications*, 8(1), 86. 10.1038/s41377-019-0198-y31645930

Iftimie, N., Faktorová, D., Fabo, P., Savin, A., & Steigmann, R. (2018). Evaluation of dielectric materials properties using microwave enhanced metamaterials sensor. *IOP Conference Series: Materials Science and Engineering, 444*(2), 022007-1-022007–022009. 10.1088/1757-899X/444/2/022007

Im, H., Shao, H., Park, Y. I., Peterson, V. M., Castro, C. M., Weissleder, R., & Lee, H. (2014). Label-free detection and molecular profiling of exosomes with a nano-plasmonic sensor. *Nature Biotechnology*, 32(5), 490–495. 10.1038/nbt.288624752081

Islam, M. T., Samsuzzaman, M., Islam, M. T., & Kibria, S. (2018). Experimental breast phantom imaging with metamaterial-inspired nine-antenna sensor array. *Sensors, 18*(12), 4427-1-4427–19. 10.3390/s18124427

Islam, M. R., Islam, M. T., M, M. S., Bais, B., Almalki, S. H. A., Alsaif, H., & Islam, M. S. (2022). Metamaterial sensor based on rectangular enclosed adjacent triple circle split ring resonator with good quality factor for microwave sensing application. *Scientific Reports*, 12(1), 6792. 10.1038/s41598-022-10729-435474227

Jafari, F., & Rad Malekshahi, M. (2023). A low-cost microwave microfluidic sensor based on planar ring resonator. *IEEE Sensors Journal*, 23(18), 21070–21077. 10.1109/JSEN.2023.3301813

Jaget Singh, B. S., & Badhan, , K. (2018). Slit loaded H-shaped microstrip patch antenna for 2.4 GHz. *International Journal of Applied Engineering Research*, 13(18), 13552–13554.

Jain, M. C., Nadaraja, A. V., Vizcaino, B. M., Roberts, D. J., & Zarifi, M. H. (2020). Differential microwave resonator sensor reveals glucose-dependent growth profile of *E. coli* on solid agar. *IEEE Microwave and Wireless Components Letters*, 30(5), 531–534. 10.1109/LMWC.2020.2980756

Javed, A., Arif, A., Zubair, M., Mehmood, M. Q., & Riaz, K. (2020). A low-cost multiple complementary split-ring resonator-based microwave sensor for contactless dielectric characterization of liquids. *IEEE Sensors Journal*, 20(19), 11326–11334. 10.1109/JSEN.2020.2998004

Jiang, H., Choudhury, S., Kudyshev, Z. A., Wang, D., Prokopeva, L. J., Xiao, P., Jiang, Y., & Kildishev, A. V. (2019). Enhancing sensitivity to ambient refractive index with tunable few-layer graphene/hBN nanoribbons. *Photonics Research*, 7(7), 815–822. 10.1364/PRJ.7.000815

Jiang, L., Zeng, S., Quyang, Q., Dinh, X. Q., Coquet, P., Qu, J., He, S., & Yong, K. T. (2017). Graphene-TMDC-Graphene Hybrid Plasmonic Metasurface for Enhanced Biosensing: A Theoretical Analysis. *Physica Status Solidi. A, Applications and Materials Science*, 214(12), 1700563. 10.1002/pssa.201700563

Jiang, L., Zeng, S., Xu, Z., Quyang, Q., Zhang, D. H., Chong, P. H. J., Coquet, P., He, S., & Yong, K. T. (2017). Multifunctional hyperbolic nanogroove metasurface for submolecular detection. *Small*, 13(30), 1700600. 10.1002/smll.20170060028597602

Jiang, M., Du, Z., Li, Y., Sun, S., & Hu, J. (2023). Frequency Selective Surface with Quasi-Elliptic Bandpass Response for Filtering FRA Application. *IEEE Transactions on Antennas and Propagation.*

Jiang, X., Liang, B., Zou, X., Yin, L., & Cheng, J. (2014). Broadband field rotator based on acoustic metamaterials. *Applied Physics Letters*, 104(8), 083510. 10.1063/1.4866333

Jiang, Z. H., Yun, S., Lin, L., Bossard, J. A., Werner, D. H., & Mayer, T. S. (2013). Tailoring dispersion for broadband low-loss optical metamaterials using deep-subwavelength inclusions. *Scientific Reports*, 3(1), 1571. 10.1038/srep0157123535875

Kanchana, D., Radha, S., Sreeja, B. S., & Manikandan, E. (2021). A miniaturized flexible frequency selective surface for dual band response. *International Journal of Microwave and Wireless Technologies*, 13(8), 810–816. 10.1017/S1759078720001506

Karl, N., Reichel, K., Chen, H.-T., Taylor, A. J., Brener, I., Benz, A., Reno, J. L., Mendis, R., & Mittleman, D. M. (2014). An electrically driven terahertz metamaterial diffractive modulator with more than 20 dB of dynamic range. *Applied Physics Letters*, 104(9), 091115. 10.1063/1.4867276

Kaur, K., & Kaur, A. (2022). Polarization independent frequency selective surface for marine and air traffic radar applications. *Sadhana*, 47(2), 81. 10.1007/s12046-022-01840-3

Kayal, S., Shaw, T., & Mitra, D. (2020). Design of metamaterial-based compact and highly sensitive microwave liquid sensor. *Applied Physics. A, Materials Science & Processing*, 126(1), 1–9. 10.1007/s00339-019-3186-4

Kazemi, N., Abdolrazzaghi, M., & Musilek, P. (2021). Comparative analysis of machine learning techniques for temperature compensation in microwave sensors. *IEEE Transactions on Microwave Theory and Techniques*, 69(9), 4223–4236. 10.1109/TMTT.2021.3081119

Keerthi, R.S., Dhabliya, D., Elangovan, P., Borodin, K., Parmar, J., & Patel, S.K. (2021). Tunable high-gain and multiband microstrip antenna based on liquid/copper split-ring resonator superstrates for C/X band communication. *Phys. B Condens. Matter.* . physb. 2021. 413203.10. 1016/j

Ke, J. C., Dai, J. Y., Zhang, J. W., Chen, Z., Chen, M. Z., Lu, Y., Zhang, L., Wang, L., Zhou, Q. Y., Li, L., Ding, J. S., Cheng, Q., & Cui, T. J. (2022). Frequency-modulated continuous waves controlled by space-time-coding metasurface with nonlinearly periodic phases. *Light, Science & Applications*, 11(1), 1. 10.1038/s41377-022-00973-836104318

Khanikev, A. B., Wu, C., & Shevets, G. (2013). Fano-resonant metamaterials and their applications. *Nanophotonics*, 2(4), 247–264. 10.1515/nanoph-2013-0009

Khezzar, D., Khedrouche, D., & Denidni, T. A. (2021). New design of a broadband PBG-based antenna for THz band applications. *Photonics and Nanostructures*, 46, 100947. 10.1016/j.photonics.2021.100947

Kiani, G., & Dyadyuk, V. (2012, July). Low loss FSS polarizer for 70 GHz applications. In *Proceedings of the 2012 IEEE International Symposium on Antennas and Propagation* (pp. 1-2). IEEE. 10.1109/APS.2012.6348718

Kiani, S., Rezaei, P., & Navaei, M. (2020). Dual-sensing and dual-frequency microwave SRR sensor for liquid samples permittivity detection. *Measurement*, 160, 107805-1, 107805–107808. 10.1016/j.measurement.2020.107805

Kodera, T., Sounas, D. L., & Caloz, C. (2011). Artificial Faraday rotation using a ring metamaterial structure without static magnetic field. *Applied Physics Letters*, 99(3), 31114. 10.1063/1.3615688

Komar, A., Paniagua-Domínguez, R., Miroshnichenko, A., Yu, Y. F., Kivshar, Y. S., Kuznetsov, A. I., & Neshev, D. (2018). Dynamic Beam Switching by Liquid Crystal Tunable Dielectric Metasurfaces. *ACS Photonics*, 5(5), 1742–1748. 10.1021/acsphotonics.7b01343

Koutsoupidou, M., Karanasiou, I. S., & Uzunoglu, N. (2014). Substrate constructed by an array of split ring resonators for a THz planar antenna. *Journal of Computational Electronics*, 13(3), 593–598. 10.1007/s10825-014-0575-y

Kowerdziej, R., Wróbel, J., & Kula, P. (2019). Ultrafast electrical switching of nanostructured metadevice with dual-frequency liquid crystal. *Scientific Reports*, 9(1), 20367. 10.1038/s41598-019-55656-z31889047

Kraus, J. D., & Marhefka, R. J. (2003). *Antennas for all applications* (3rd ed.). McGraw-Hill.

Kretschmann. E, H. Raether, & Notizen, (1965). Radiative decay of non radiative surface plasmons excited by light. *Z. Naturforsch. A,23*, 2135–2136.

Krishna, D., Maheshwari, T., Prakash, K. B., Saheb, S. G., & Muhhidin, S. (2022). *Design a Frequency Reconfigurable Antenna for 5G\6G Applications. IJIRSE, 8*(6).

Krishna, C. M., Das, S., Nella, A., Lakrit, S., & Madhav, B. T. P. (2021). A micro-sized rhombus shaped THz antenna for high-speed short-range wireless communication applications. *Plasmonics*, 16(6), 2167–2177. 10.1007/s11468-021-01472-z

Krishna, ChM., Das, S., Lakrit, S., Lavadiya, S., Madhav, B. T. P., & Sorathiya, V. (2021). Design and analysis of a super wideband (0.09–30.14 THz) graphene-based log periodic dipole array antenna for terahertz applications. *Optik (Stuttgart)*, 247, 167991. 10.1016/j.ijleo.2021.167991

Krushna Kanth, V., & Raghavan, S. (2023). Metamaterial-FSS based radome using SIW technology. In *Handbook of Metamaterial-Derived Frequency Selective Surfaces* (pp. 191–223). Springer Nature Singapore.

Kumar, A., Rajawat, M. S., Mahto, S. K., & Sinha, R. (2021). Metamaterial-inspired complementary split ring resonator sensor and second-order approximation for dielectric characterization of fluid. *Journal of Electronic Materials*, 50(10), 5925–5932. 10.1007/s11664-021-09099-w

Kumari, R., Tomar, V. K., & Sharma, A. (2022). Miniaturization and performance enhancement of super wide band four element MIMO antenna using DNG metamaterial for THz applications. *Optical and Quantum Electronics*, 54(9), 577. 10.1007/s11082-022-04011-0

Kumari, R., Yadav, A., Sharma, S., Gupta, T. D., Varshney, S. K., & Lahiri, B. (2021). Tunable Van der Waal's optical metasurfaces (VOMs) for biosensing of multiple analytes. *Optics Express*, 29(16), 25800–25811. 10.1364/OE.43228434614900

Kundu, S. (2024). A simple planar single layered frequency selective surface with band-stop and bandpass characteristic. *Sadhana*, 49(1), 1–6. 10.1007/s12046-023-02411-w

Kurland, Z., Goyette, T., & Gatesman, A. (2023). A novel technique for ultrathin inhomogeneous dielectric powder layer sensing using a W-Band metasurface. *Sensors (Basel)*, 23(2), 842. Advance online publication. 10.3390/s2302084236679638

Kurosawa, H., Choi, B., Sugimoto, Y., & Iwanaga, M. (2017). High-performance metasurface polarizers with extinction ratios exceeding 12000. *Optics Express*, 25(4), 4446–4455. 10.1364/OE.25.00444628241647

Kurs, A., Karalis, A., Moffatt, R., Joannopoulos, J. D., Fisher, P., & Soljačić, M. (2007). Wireless power transfer via strongly coupled magnetic resonances. *Science*, 317(5834), 83–86. 10.1126/science.114325417556549

Kushwaha, R. K., & Karuppanan, P. (2021). Investigation and design of microstrip patch antenna employed on PCs substrates in THz regime. *Aust J Electr Electron Eng*, 18(2), 118–125. 10.1080/1448837X.2021.1936779

Kushwaha, R. K., Karuppanan, P., & Malviya, L. D. (2018). Design and analysis of novel microstrip patch antenna on photonic crystal in THz. *Physica B, Condensed Matter*, 545, 107–112. 10.1016/j.physb.2018.05.045

Kushwaha, R. K., Karuppanan, P., & Srivastava, Y. (2018). Proximity feed multiband patch antenna array with SRR and PBG for THz applications. *Optik (Stuttgart)*, 175, 78–86. 10.1016/j.ijleo.2018.08.139

La Spada, L., Tarparelli, R., & Vegni, L. (2014). Spectral Green's function for SPR meta-structures. *Mater. Sci,Forum 792*, 110–114. 10.

La Spada, L., McManus, T. M., Dyke, A., Haq, S., Zhang, L., Cheng, Q., & Hao, Y. (2016). Surface wave cloak from graded refractive index nanocomposites. *Scientific Reports*, 6(1), 22045–22322. 10.1038/srep2936327416815

Lan, F., Wang, L., Zeng, H., Liang, S., Song, T., Liu, W., Mazumder, P., Yang, Z., Zhang, Y., & Mittleman, D. M. (2023). Real-time programmable metasurface for terahertz multifunctional wave front engineering. *Light, Science & Applications*, 12(1), 191. 10.1038/s41377-023-01228-w37550383

Lapine, M., & Tretyakov, S. (2007). Contemporary notes on metamaterials. *IET Microwaves, Antennas & Propagation*, 1(1), 3–11. 10.1049/iet-map:20050307

Larsson, E. M., Alegret, J., Käll, M., & Sutherland, D. S. (2007). Sensing characteristics of nir localized surface plasmon resonances in gold nanorings for application as ultrasensitive biosensors. *Nano Letters*, 7(5), 1256–1263. 10.1021/nl070161217430004

Lee, W., & Yoon, Y.-K. (2020). Wireless power transfer systems using metamaterials: A review. *IEEE Access : Practical Innovations, Open Solutions*, 8, 147930–147947. 10.1109/ACCESS.2020.3015176

Lee, W., & Yoon, Y.-K. (2023). High-efficiency wireless-power-transfer system using fully rollable Tx/Rx coils and metasurface screen. *Sensors (Basel)*, 23(4), 4. 10.3390/s2304197236850570

Lee, Y., Kim, S.-J., Park, H., & Lee, B. (2017). Metamaterials and Metasurfaces for Sensor applications. *Sensors (Basel)*, 17(8), 1726. 10.3390/s1708172628749422

Leitis, A., Tittl, A., Liu, M., Lee, B. H., Gu, M. B., Kivshar, Y. S., & Altug, H. (2019). Angle-multiplexed all-dielectric metasurfaces for broadband molecular fingerprint retrieval. *Science Advances*, 5(5), eaaw2871. 10.1126/sciadv.aaw287131123705

Leonhardt, U. (2007). Invisibility cup. *Nature Photonics*, 1(4), 207–208. 10.1038/nphoton.2007.38

Letaief, K. B., Chen, W., Shi, Y., Zhang, J., & Zhang, Y. J. A. (2019, August). The Roadmap to 6G: AI Empowered Wireless Networks. *IEEE Communications Magazine*, 57(8), 84–90. 10.1109/MCOM.2019.1900271

Li, A., Sun, W., Yi, W., & Zuo, Q. (2016). Investigation of beam steering performances in rotation Risley-prism scanner. *Optics Express*, 24(12), 12840. 10.1364/OE.24.01284027410303

Liao, R. X., Wong, S. W., Li, Y., Lin, J. Y., Liu, B. Y., Chen, F. C., & Quan, Z. (2020). Quasi-elliptic bandpass frequency selective surface based on coupled stubs-loaded ring resonators. *IEEE Access : Practical Innovations, Open Solutions*, 8, 113675–113682. 10.1109/ACCESS.2020.3003319

Li, B., Lin, Q., & Li, M. (2023). Frequency–angular resolving LiDAR using chip-scale acousto-optic beam steering. *Nature*, 620(7973), 316–322. 10.1038/s41586-023-06201-637380781

Li, B., & Ne, R. (2021). A novel miniaturized dual-layer frequency selective surface. *AEÜ. International Journal of Electronics and Communications*, 130, 153580. 10.1016/j.aeue.2020.153580

Li, C., Chen, C., Yu, C., Jau, H., Lv, J., Qing, X., Lin, C., Cheng, C., Wang, C., Wei, J., Yu, Y., & Lin, T. (2017). Arbitrary Beam Steering Enabled by Photomechanically Bendable Cholesteric Liquid Crystal Polymers. *Advanced Optical Materials*, 5(4), 1600824. 10.1002/adom.201600824

Li, C., Yu, P., Huang, Y., Zhou, Q., Wu, J., Li, Z., Tong, X., Wen, Q., Kuo, H.-C., & Wang, Z. M. (2020). Dielectric metasurfaces: From wavefront shaping to quantum platforms. *Progress in Surface Science*, 95(2), 100584. 10.1016/j.progsurf.2020.100584

Li, D., Shen, Z., & Li, E. P. (2018). Spurious-free dual-band bandpass frequency-selective surfaces with large band ratio. *IEEE Transactions on Antennas and Propagation*, 67(2), 1065–1072. 10.1109/TAP.2018.2882601

Li, F., Zhou, P., Wang, T., He, J., Yu, H., & Shen, W. (2017). A Large-Size MEMS Scanning Mirror for Speckle Reduction Application. *Micromachines*, 8(5), 140. 10.3390/mi8050140

Li, L., Wen, J., Wang, Y., Jin, Y., Wen, Y., Sun, J., Zhao, Q., Li, B., & Zhou, J. (2023). A Transparent broadband all-dielectric water-based metamaterial absorber based on laser cutting. *Physica Scripta*, 98(5), 055516. 10.1088/1402-4896/accc15

Li, L., Zong, X., & Liu, Y. (2020). All-metallic metasurfaces towards high-performance magneto-plasmonic sensing devices. *Photonics Research*, 8(11), 1742. 10.1364/PRJ.399926

Li, M., Hong, M., Yang, Q., Zheng, Q., Yang, X., & Yi, Z. (2022). A design of broadband bandpass frequency selective surface. *Microwave and Optical Technology Letters*, 64(9), 1572–1578. 10.1002/mop.33331

Lin, Y.-J., Chen, K.-M., & Wu, S.-T. (2009). Broadband and polarization-independent beam steering using dielectrophoresis-tilted prism. *Optics Express*, 17(10), 8651. 10.1364/OE.17.00865119434198

Li, Q., Bao, W., Nie, Z., Xia, Y., Xue, Y., Wang, Y., Yang, S., & Zhang, X. (2021). A non-unitary metasurface enables continuous control of quantum photon-photon interactions from bosonic to fermionic. *Nature Photonics*, 15(4), 267–271. 10.1038/s41566-021-00762-6

Liu, G., Yu, M., Liu, Z., Pan, P., Liu, X., Huang, S., & Wang, Y. (2016). Multi-Band High Refractive Index Susceptibility of Plasmonic Structures with Network-Type Metasurface. *Plasmonics*, 11(2), 677–682. 10.1007/s11468-015-0101-5

Li, Y., Wang, Y., Qi, S., Ren, Q., Kang, L., Campbell, S. D., Werner, P. L., & Werner, D. H. (2020). Predicting Scattering From Complex Nano-Structures via Deep Learning. *IEEE Access : Practical Innovations, Open Solutions*, 8, 139983–139993. 10.1109/ACCESS.2020.3012132

Lu, F., Tan, Q., Ji, Y., Guo, Q., Guo, Y., & Xiong, J. (2018). A novel metamaterial inspired high-temperature microwave sensor in harsh environments. *Sensors, 18*(9), 2879-1-2879–12. 10.3390/s18092879

Lu, C., Huang, X., Tao, X., Rong, C., & Liu, M. (2020). Comprehensive analysis of side-placed metamaterials in wireless power transfer system. *IEEE Access : Practical Innovations, Open Solutions*, 8, 152900–152908. 10.1109/ACCESS.2020.3017492

Lu, C., Hu, X., Shi, K., Hu, Q., Zhu, R., Yang, H., & Gong, Q. (2015). An actively ultrafast tunable giant slow-light effect in ultrathin nonlinear metasurfaces. *Light, Science & Applications*, 4(6), e302. 10.1038/lsa.2015.75

Maci, S., Minatti, G., Casaletti, M., & Bosiljevac, M. (2011). Metasurfing: Addressing waves on impenetrable metasurfaces. *IEEE Antennas and Wireless Propagation Letters*, 10, 1499–1502. 10.1109/LAWP.2012.2183631

Mahaveer, U., Chandrasekaran, K. T., Mohan, M. P., Alphones, A., Siyal, M. Y., & Karim, M. F. (2021). A tri-band frequency-selective surface. *Journal of Electromagnetic Waves and Applications*, 35(7), 861–873. 10.1080/09205071.2020.1865206

Mahmud, R.H. (2020). Terahertz microstrip patch antennas for the surveillance applications. *Kurd J Appl Res (KJAR), 5*, 17–27. ce. 2020.1.2.10. 24017/ scien

Ma, Q., Bai, G. D., Jing, H. B., Yang, C., Li, L., & Cui, T. J. (2019). Smart metasurface with self-adaptively reprogrammable functions. *Light, Science & Applications*, 8(1), 98. 10.1038/s41377-019-0205-331700618

Ma, Q., Hong, Q., Gao, X., Jing, H., Liu, C., Bai, G., Cheng, Q., & Cui, T. (2020). Smart sensing metasurface with self-defined functions in dual polarizations. *Nanophotonics*, 9(10), 3271–3278. 10.1515/nanoph-2020-0052

Marqués, R., Mesa, F., Martel, J., & Medina, F. (2003). Comparative analysis of edge- and broadside-coupled split ring resonators for metamaterial design - theory and experiments. *IEEE Transactions on Antennas and Propagation*, 51(10), 2572–2581. 10.1109/TAP.2003.817562

Ma, S., Zhang, P., Mi, X., & Zhao, H. (2022). Highly sensitive terahertz sensor based on graphene metamaterial absorber. *Optics Communications*, 528, 129021. 10.1016/j.optcom.2022.129021

Mata-Contreras, J., Su, L., & Martín, F. (2017). Microwave sensors based on symmetry properties and metamaterial concepts: a review of some recent developments (invited paper). *2017 IEEE 18th Wireless and Microwave Technology Conference, WAMICON 2017*, (pp. 1–6). IEEE. 10.1109/WAMICON.2017.7930278

Mayani, M. G., Herraiz-Martinez, F. J., Domingo, J. M., & Giannetti, R. (2021). Resonator-based microwave metamaterial sensors for instrumentation: Survey, classification, and performance comparison. *IEEE Transactions on Instrumentation and Measurement*, 70, 1–14. 10.1109/TIM.2020.3040484

Megens, M., Yoo, B.-W., Chan, T., Yang, W., Sun, T., Chang-Hasnain, C. J., Wu, M. C., & Horsley, D. A. (2014). *High-speed 32×32 MEMS optical phased array* (W. Piyawattanametha & Y.-H. Park, Eds.; p. 89770H). 10.1117/12.2044197

Memon, M. U., Salim, A., Jeong, H., & Lim, S. (2020). Metamaterial inspired radio frequency-based touchpad sensor system. *IEEE Transactions on Instrumentation and Measurement*, 69(4), 1344–1352. 10.1109/TIM.2019.2908507

Meng, C., Thrane, P. C. V., Ding, F., Gjessing, J., Thomaschewski, M., Wu, C., Dirdal, C., & Bozhevolnyi, S. I. (2021). Dynamic piezoelectric MEMS-based optical metasurfaces. *Science Advances*, 7(26), eabg5639. 10.1126/sciadv.abg563934162551

Meng, X., Depauw, V., Gomard, G., El Daif, O., Trompoukis, C., Drouard, E., Jamois, C., Fave, A., Dross, F., Gordon, I., & Seassal, C. (2012). Design, fabrication and optical characterization of photonic crystal assisted thin film monocrystalline-silicon solar cells. *Optics Express*, 20(S4), A465–A475. 10.1364/OE.20.00A46522828615

Miao, Z.-W., Hao, Z.-C., Wang, Y., Jin, B.-B., Wu, J.-B., & Hong, W. (2018). A 400-GHz High-Gain Quartz-Based Single Layered Folded Reflectarray Antenna for Terahertz Applications. *IEEE Transactions on Terahertz Science and Technology*, 9(1), 78–88. 10.1109/TTHZ.2018.2883215

Milius, C., Andersen, R. B., Lazaridis, P. I., Zaharis, Z. D., Muhammad, B., & Jes, T. B. (2021). Metamaterial-Inspired Antennas: A Review of the State of the Art and Future Design Challenges. *IEEE Access : Practical Innovations, Open Solutions*, 9, 89846–89865. 10.1109/ACCESS.2021.3091479

Mohammadi, P., Mohammadi, A., & Kara, A. (2023). Dual-frequency microwave resonator for noninvasive detection of aqueous glucose. *IEEE Sensors Journal*, 23(18), 21246–21253. 10.1109/JSEN.2023.3303170

Mohammadi, S., & Zarifi, M. H. (2021). Differential microwave resonator sensor for real-time monitoring of volatile organic compounds. *IEEE Sensors Journal*, 21(5), 6105–6114. 10.1109/JSEN.2020.3041810

Mujawar, M., Gunasekaran, T., & Rashid, A. (2021). Design and Analysis of X Band Pyramidal Horn Antenna using FEKO. In *2021 International Conference on Advances in Electrical, Computing, Communication and Sustainable Technologies (ICAECT)*, Bhilai, India. 10.1109/ICAECT49130.2021.9392523

Mun, J., & Rho, J. (2019). Importance of higher-order multipole transitions on chiral nearfield interactions. *Nanophotonics*, 8(5), 941–948. 10.1515/nanoph-2019-0046

Munk, B. (1974). *U.S. Patent No. 3,789,404*. Washington, DC: U.S. Patent and Trademark Office.

Munk, B. A. (2005). *Frequency selective surfaces: theory and design*. John Wiley & Sons.

Muñoz-Enano, J., Vélez, P., Gil, M., & Martín, F. (2020). Planar microwave resonant sensors: A review and recent developments. *Applied Sciences, 10*(7), 2615-1-2615–2630. 10.3390/app10072615

Naderi, A., & Ghodrati, M. (2017). Improving band-to-band tunneling in a tunneling carbon nanotube field effect transistor by multi-level development of impurities in the drain region. *Eur. Phys. J. Plus,132*.

Naderi, A., & Ghodrati, M. (2018). An efficient structure for T-CNTFETs with intrinsic-n-doped impurity distribution pattern in drain region. *Turk J. Electr. Eng*, 26(5), 2335–2346. 10.3906/elk-1709-180

Naderi, A., & Ghodrati, M. (2018). Cut Off Frequency Variation by Ambient Heating in Tunneling p-i-n CNTFETs. *ECS Journal of Solid State Science and Technology : JSS*, 7(2), M6–M10. 10.1149/2.0241802jss

Naderi, A., Ghodrati, M., & Baniardalani, S. (2020). The use of a Gaussian doping distribution in the channel region to improve the performance of a tunneling carbon nanotube field-effect transistor. *Journal of Computational Electronics*, 19(1), 283–290. 10.1007/s10825-020-01445-1

Najumunnisa, M., Sastry, A. S. C., Madhav, B. T. P., Das, S., Hussain, N., Ali, S. S., & Aslam, M. (2022). A metamaterial inspired AMC backed dual band antenna for ISM and RFID applications. *Sensors (Basel)*, 22(20), 8065. 10.3390/s2220806536298414

Nauman, M., Saleem, R., Rashid, A. K., & Shafique, M. F. (2016). A miniaturized flexible frequency selective surface for X-band applications. *IEEE Transactions on Electromagnetic Compatibility*, 58(2), 419–428. 10.1109/TEMC.2015.2508503

Nejati, A., Sadeghzadeh, R. A., & Geran, F. (2014). Effect of photonic crystal and frequency selective surface implementation on gain enhancement in the microstrip patch antenna at terahertz frequency. *Physica B, Condensed Matter*, 449, 113–120. 10.1016/j.physb.2014.05.014

Nemati, A., Yuan, G., Deng, J., Huang, A., Wang, W., Toh, Y. T., Teng, J., & Wang, Q. (2022). Controllable Polarization-Insensitive and Large-Angle Beam Switching with Phase-Change Metasurfaces. *Advanced Optical Materials*, 10(5), 2101847. 10.1002/adom.202101847

Niksan, O., Jain, M. C., Shah, A., & Zarifi, M. H. (2022). A nonintrusive flow rate sensor based on microwave split-ring resonators and thermal modulation. *IEEE Transactions on Microwave Theory and Techniques*, 70(3), 1954–1963. 10.1109/TMTT.2022.3142038

Nocentini, S., Martella, D., Wiersma, D. S., & Parmeggiani, C. (2017). Beam steering by liquid crystal elastomer fibres. *Soft Matter*, 13(45), 8590–8596. 10.1039/C7SM02063E29105720

Okan, T. (2021). High efficiency unslotted ultra-wideband microstrip antenna for sub-terahertz short range wireless communication systems. *Optik (Stuttgart)*, 242, 166859. 10.1016/j.ijleo.2021.166859

Otto, A. (1968). Excitation of nonradiative surface plasma waves in silver by the method of frustrated total reflection. *Zeitschrift für Physik*, 216(4), 398–410. 10.1007/BF01391532

Paik, H., & Premchand, K. (2023). Performance analysis of a single layer X-Band frequency selective surface based spatial filter implementing half Jerusalem cross slot. *Progress in Electromagnetics Research Letters*, 108.

Pandya, A., Upadhyay, T. K., & Pandya, K. (2021). Design of metamaterial based multilayer antenna for Navigation/Wifi/ Satellite applications. *Progress in Electromagnetics Research M. Pier M*, 99, 103–113. 10.2528/PIERM20100105

Park, S., Cha, S., Shin, G., & Ahn, Y. (2017). Sensing viruses using terahertz nano-gap metamaterials. *Biomedical Optics Express*, 8(8), 3551–3558. 10.1364/BOE.8.00355128856034

Parvin, T., Ahmed, K., Alatwi, A. M., & Rashed, A. N. Z. (2021). Differential optical absorption spectroscopy-based refractive index sensor for cancer cell detection. *Optical Review*, 28(1), 134–143. 10.1007/s10043-021-00644-w

Patel, S. K., & Kosta, Y. P. (2012). Meandered multiband metamaterial square microstrip patch antenna design. Waves Rand. *Complex Media*, 2012, 723837. 10. 1080/ 17455 030

Pendry, J. B. (2000). Negative Refraction Makes a Perfect Lens. *Physical Review Letters*, 85(18), 3966–3969. 10.1103/PhysRevLett.85.396611041972

Pendry, J. B., Holden, A. J., Robbins, D. J., & Stewart, W. (1999). Magnetism from conductors and enhanced nonlinear phenomena. *IEEE Transactions on Microwave Theory and Techniques*, 47(11), 2075–2084. 10.1109/22.798002

Pham, T. S., Bui, H. N., & Lee, J.-W. (2019). Wave propagation control and switching for wireless power transfer using tunable 2-D magnetic metamaterials. *Journal of Magnetism and Magnetic Materials*, 485, 126–135. 10.1016/j.jmmm.2019.04.034

Pham, T. S., Ranaweera, A. K., Ngo, D. V., & Lee, J. W. (2017). Analysis and experiments on Fano interference using a 2D metamaterial cavity for field localized wireless power transfer. *Journal of Physics. D, Applied Physics*, 50(305102), 1–10. 10.1088/1361-6463/aa7988

Pozar, D. M., Targonski, S. D., & Syrigos, H. D. (1997). Design of millimeter wave microstrip reflectarrays. *IEEE Transactions on Antennas and Propagation*, 45(2), 287–296. 10.1109/8.560348

Prakash, D., & Gupta, N. (2022). Applications of metamaterial sensors: A review. *International Journal of Microwave and Wireless Technologies*, 14(1), 19–33. 10.1017/S1759078721000039

Prakash, D., & Gupta, N. (2023). CSRR based metamaterial inspired sensor for liquid concentration detection using machine learning. *Progress in Electromagnetics Research C. Pier C*, 130(February), 255–267. 10.2528/PIERC22110101

Puentes Vargas, M. (2014). Fundamentals of metamaterial structures. In *Planar Metamaterial Based Microwave Sensor Arrays for Biomedical Analysis and Treatment* (pp. 7–31). Springer. 10.1007/978-3-319-06041-5_2

Qi, S., Wang, Y., Li, Y., Wu, X., Ren, Q., & Ren, Y. (2020). Two-Dimensional Electromagnetic Solver Based on Deep Learning Technique. *IEEE Journal on Multiscale and Multiphysics Computational Techniques*, 5, 83–88. 10.1109/JMMCT.2020.2995811

Qiu, G., Gai, Z., Saleh, L., Tang, J., Gui, T., Kullak-Ublick, G. A., & Wang, J. (2021). Thermoplasmonic-assisted cyclic cleavage amplification for self-validating plasmonic detection of SARS-CoV-2. *ACS Nano*, 15(4), 7536–7546. 10.1021/acsnano.1c0095733724796

Qiu, G., Gai, Z., Tao, Y., Schmitt, J., Kullak-Ublick, G. A., & Wang, J. (2020). Dual-functional plasmonic photothermal biosensors for highly accurate severe acute respiratory syndrome coronavirus 2 detection. *ACS Nano*, 14(5), 5268–5277. 10.1021/acsnano.0c0243932281785

Qi, Z.-M., Matsuda, N., Itoh, K., Murabayashi, M., & Lavers, C. (2002). A design for improving the sensitivity of a mach–zehnder interferometer to chemical and biological measurands. *Sensors and Actuators. B, Chemical*, 81(2-3), 254–258. 10.1016/S0925-4005(01)00960-1

Rabbani, M. G., Islam, M. T., Hoque, A., Bais, B., Albadran, S., Islam, M. S., & Soliman, M. S. (2024). Orthogonal centre ring field optimization triple-band metamaterial absorber with sensing application. *Engineering Science and Technology, an International Journal, 49*(June 2023), 101588. 10.1016/j.jestch.2023.101588

Rahman, N. A., Zakaria, Z., Rahim, R. A., Dasril, Y., & Mohd Bahar, A. A. (2017). Planar microwave sensors for accurate measurement of material characterization: A review. [Telecommunication Computing Electronics and Control]. *Telkomnika*, 15(3), 1108–1118. 10.12928/telkomnika.v15i3.6684

Rahmati, B., & Hassani, H. R. (2015). Multiband metallic frequency selective surface with wide range of band ratio. *IEEE Transactions on Antennas and Propagation*, 63(8), 3747–3753. 10.1109/TAP.2015.2438340

Rainville, P. J., & Harackewiez, F. J. (1992, December). Magnetic tuning of a microstrip antenna fabricated on a ferrite film. *IEEE Microw. Guided Wave*, 2(12), 483–485.

Ram Krishna, R. V. S., & Kumar, R. (2015). Slotted ground microstrip antenna with FSS reflector for high-gain horizontal polarisation. *Electronics Letters*, 51(8), 599–600. 10.1049/el.2015.0339

Ramesh, B., & Rajya Lakshmi, V. (2013). Design of a rectangular microstrip antenna using EBG structure. *International Journal of Engineering Research and Technology, 2*(7), 2233-2236.

Ranaweera, A. L. A. K., Duong, T. P., & Lee, J.-W. (2014). Experimental investigation of compact metamaterial for high efficiency mid-range wireless power transfer applications. *Journal of Applied Physics*, 116(4), 043914. 10.1063/1.4891715

Ranaweera, A. L. A. K., Pham, T. S., Bui, H. N., Ngo, V., & Lee, J.-W. (2019). An active metasurface for field-localizing wireless power transfer using dynamically reconfigurable cavities. *Scientific Reports*, 9(1), 1. 10.1038/s41598-019-48253-731409834

Ren, Q., Wang, Y., Li, Y., & Qi, S. (2022). *Sophisticated Electromagnetic Forward Scattering Solver via Deep Learning*. Springer Singapore. 10.1007/978-981-16-6261-4

Reyes-Vera, E., Acevedo-Osorio, G., Arias-Correa, M., & Senior, D. E. (2019). A submersible printed sensor based on a monopole-coupled split ring resonator for permittivity characterization. *Sensors (Switzerland), 19*(8), 1936-1-1936–12. 10.3390/s19081936

Rifat, A., Rahmani, M., Xu, L., & Miroshnichenko, A. (2018). Hybrid Metasurface Based Tunable Near-Perfect Absorber and Plasmonic Sensor. *Materials (Basel)*, 11(7), 1091. 10.3390/ma1107109129954060

Rong, C., Lu, C., Zeng, Y., Tao, X., Liu, X., Liu, R., He, X., & Liu, M. (2021). A critical review of metamaterial in wireless power transfer system. *IET Power Electronics*, 14(9), 1541–1559. 10.1049/pel2.12099

Roslan, N. H., Awang, A. H., & Hizan, H. M. (2008). *The effect of photonic crystal parameters on the terahertz photonic crystal cavities microstrip antenna performances*. In: *Proceedings of the 2018 IEEE International RF and Microwave Conference (RFM)*, Penang, Malaysia.

Rotman, W. (1962, January). Plasma simulation by artificial dielectrics and parallelplate media. *IRE Transactions on Antennas and Propagation*, 10(1), 82–95. 10.1109/TAP.1962.1137809

RoyChoudhury, S., Rawat, V., Jalal, A. H., Kale, S. N., & Bhansali, S.RoyChoudhury. (2016). Recent advances in metamaterial split-ring-resonator circuits as biosensors and therapeutic agents. *Biosensors & Bioelectronics*, 86, 595–608. 10.1016/j.bios.2016.07.02027453988

Saadat-Safa, M., Nayyeri, V., Ghadimi, A., Soleimani, M., & Ramahi, O. M. (2019). A pixelated microwave near-field sensor for precise characterization of dielectric materials. *Scientific Reports*, 9(1), 1–12. 10.1038/s41598-019-49767-w31527610

Sabapathy, T., Jamlos, M. F. B., Ahmad, R. B., Jusoh, M., Jais, M. I., & Kamarudin, M. R. (2013). ELECTRONICALLY RECONFIGURABLE BEAM STEERING ANTENNA USING EMBEDDED RF PIN BASED PARASITIC ARRAYS (ERPPA). *Electromagnetic Waves*, 140, 241–261. 10.2528/PIER13042906

Saha, C., & Siddiqui, J. Y. (2012, July). Theoretical model for estimation of resonance frequency of rotational circular split-ring resonators. *Electromagnetics*, 32(6), 345–355. 10.1080/02726343.2012.701540

Salim, A., & Lim, S. (2018). Review of recent metamaterial microfluidic sensors. *Sensors, 18*(1), 232-1-232–25. 10.3390/s18010232

Sarabandi, K., & Behdad, N. (2007). A frequency selective surface with miniaturized elements. *IEEE Transactions on Antennas and Propagation*, 55(5), 1239–1245. 10.1109/TAP.2007.895567

Saraswat, R. K., & Kumar, M. (2020, October). A quad band metamaterial miniaturized antenna for wireless applications with gain enhancement. *Wireless Personal Communications*, 114(4), 3595–3612. 10.1007/s11277-020-07548-z

Sehrai, D. A., Muhammad, F., Kiani, S. H., & Kim, S. (2020). Gain-Enhanced Metamaterial Based Antenna for 5GCommunication Standards. *Computers, Materials & Continua*, 64(3), 1587–1599. 10.32604/cmc.2020.011057

Seng, S. M., You, K. Y., Esa, F., & Mayzan, M. Z. H. (2020). Dielectric and magnetic properties of epoxy with dispersed iron phosphate glass particles by microwave measurement. *Journal of Microwaves, Optoelectronics and Electromagnetic Applications*, 19(2), 165–1676. 10.1590/2179-10742020v19i2824

Shahzad, W., Hu, W., Ali, Q., Raza, H., Abbas, S. M., & Ligthart, L. P. (2022). A low-cost metamaterial sensor based on DS-CSRR for material characterization applications. *Sensors (Basel)*, 22(5), 1–11. 10.3390/s2205200035271147

Shalini, M., & Madhan, M. (2019). Design and analysis of a dualpolarized graphene based microstrip patch antenna for terahertz applications. *Optik (Stuttgart)*, 194, 163050. 10.1016/j.ijleo.2019.163050

Shamim, S. M., Das, S., Hossain, M. A., & Madhav, B. T. P. (2021). Investigations on Graphene-Based Ultra-Wideband (UWB) microstrip patch antennas for Terahertz (THz) Applications. *Plasmonics*, 16(5), 1623–1631. 10.1007/s11468-021-01423-8

Shamim, S. M., Uddin, M. S., Hasan, M., & Samad, M. (2021). Design and implementation of miniaturized wideband microstrip patch antenna for high-speed terahertz applications. *Journal of Computational Electronics*, 20(1), 604–610. 10.1007/s10825-020-01587-2

Shamonina, E., & Solymar, L. (2007). Metamaterials: How the Subject Started. *Metamaterials (Amsterdam)*, 1(1), 12–18. 10.1016/j.metmat.2007.02.001

Shangguan, M., Xia, H., Wang, C., Qiu, J., Lin, S., Dou, X., Zhang, Q., & Pan, J.-W. (2017). Dual-frequency Doppler lidar for wind detection with a superconducting nanowire single-photon detector. *Optics Letters*, 42(18), 3541. 10.1364/OL.42.00354128914897

Sharma, A., Panwar, R., & Khanna, R. (2019). Experimental validation of a frequency-selective surface-loaded hybrid metamaterial absorber with wide bandwidth. *IEEE Magnetics Letters*, 10, 1–5. 10.1109/LMAG.2019.2898612

Shaw, D. (2007). Design and analysis of an asymmetrical liquid-filled lens. *Optical Engineering (Redondo Beach, Calif.)*, 46(12), 123002. 10.1117/1.2821426

Sheng, X., Wang, H., Liu, N., & Wang, K. (2024). A Conformal Miniaturized Band-Pass Frequency Selective Surface With Stable Frequency Response for Radome Applications. *IEEE Transactions on Antennas and Propagation*, 72(3), 2423–2433. 10.1109/TAP.2024.3349677

Shen, Z., Zhang, H., & Zhang, J. (2024). A high Q-factor metamaterial sensor based on electromagnetically induced transparency-like. *Materials Research Express*, 11(1), 015801. 10.1088/2053-1591/ad1a61

Shin, H., Yoon, D., Na, D. Y., & Park, Y. B. (2022). Analysis of radome cross section of an aircraft equipped with a FSS radome. *IEEE Access : Practical Innovations, Open Solutions*, 10, 33704–33712. 10.1109/ACCESS.2022.3162262

Shrekenhamer, D., Chen, W.-C., & Padilla, W. J. (2013). Liquid Crystal Tunable Metamaterial Absorber. *Physical Review Letters*, 110(17), 177403. 10.1103/PhysRevLett.110.17740323679774

Sihvola, A. (2007). Metamaterials in electromagnetics. *Metamaterials (Amsterdam)*, 1(1), 2–11. 10.1016/j.metmat.2007.02.003

Silin, R. A., & Chepurnykh, I. P. (2001). On Media with Negative Dispersion. *Commun. Technol. Electron.*, 46, 1121–1125.

Sim, M. S., You, K. Y., & Esa, F. (2020). Electromagnetic metamaterials in microwave regime. In *Handbook of Research on Recent Developments in Electrical and Mechanical Engineering* (pp. 64–86). Springer. 10.4018/978-1-7998-0117-7.ch002

Sim, M. S., You, K. Y., Dewan, R., Esa, F., Salim, M. R., Kew, S. Y. N., & Hamid, F. (2023). Dual-band metamaterial microwave absorber using ring and circular patch with slits. *Advanced Electromagnetics*, 12(4), 36–44. 10.7716/aem.v12i4.2324

Sim, M. S., You, K. Y., Esa, F., & Chan, Y. L. (2021). Nanostructured electromagnetic metamaterials for sensing applications. In *Applications of Nanomaterials in Agriculture* (pp. 141–164). Food Science, and Medicine. 10.4018/978-1-7998-5563-7.ch009

Sim, M. S., You, K. Y., Esa, F., Dimon, M. N., & Khamis, N. H. (2018). Multiband metamaterial microwave absorbers using split ring and multiwidth slot structure. *International Journal of RF and Microwave Computer-Aided Engineering*, 28(7), 1–13. 10.1002/mmce.21473

Singh, M., Singh, S., & Islam, M. T. (2021). Highly efficient ultra-wideband MIMO patch antenna array for short range THz applications, emerging trends in terahertz engineering and system technologies devices, materials, imaging, data acquisition and processing. Springer. 10.1007/978-981-15-9766-4

Singh, A., & Singh, S. (2015). A trapezoidal microstrip patch antenna on photonic crystal substrate for high-speed THz applications. *Photonics and Nanostructures*, 14, 52–62. 10.1016/j.photonics.2015.01.003

Singhal, S. (2019). Elliptical ring terahertz fractal antenna. *Optik (Stuttgart)*, 194, 163129. 10.1016/j.ijleo.2019.163129

Singh, C., Jha, K. R., & Sharma, S. K. (2020). Tripole type wideband bandpass frequency selective surface for X-band applications. *IET Microwaves, Antennas & Propagation*, 14(13), 1619–1625. 10.1049/iet-map.2019.1028

Sivasamy, R., & Kanagasabai, M. (2019). A novel miniaturized frequency selective surface. *International Journal of RF and Microwave Computer-Aided Engineering*, 29(6), e21691. 10.1002/mmce.21691

Sivasamy, R., & Kanagasabai, M. (2020). Design and fabrication of flexible FSS polarizer. *International Journal of RF and Microwave Computer-Aided Engineering*, 30(1), e22002. 10.1002/mmce.22002

Sivukhin, D. V. (1957). The Energy of Electromagnetic Fields in Dispersive Media. *Opt. Spektrosk.*, 3, 308–312.

Slimi, M., Jmai, B., Dinis, H., Gharsallah, A., & Mendes, P. M. (2022). Metamaterial Vivaldi antenna array for breast cancer detection. *Sensors (Basel)*, 22(10), 3945. 10.3390/s2210394535632355

Smith, D. R., Padilla, W. J., Vier, D. C., Nemat-Nasser, S. C., & Schultz, S. (2000). Composite Medium with Simultaneously Negative Permeability and Permittivity. *Physical Review Letters*, 84(18), 4184–4186. 10.1103/PhysRevLett.84.418410990641

Smith, N. R., Abeysinghe, D. C., Haus, J. W., & Heikenfeld, J. (2006). Agile wide-angle beam steering with electrowetting microprisms. *Optics Express*, 14(14), 6557. 10.1364/OE.14.00655719516833

Solntsev, A. S., Agarwal, G. S., & Kivshar, Y. S. (2021). Metasurfaces for quantum photonics. *Nature Photonics*, 15(5), 327–336. 10.1038/s41566-021-00793-z

Solunke, Y., & Kothari, A. (2024). An ultra-thin, low-RCS, dual-bandpass novel fractal-FSS for planar/conformal C&X bands applications. *AEÜ. International Journal of Electronics and Communications*, 175, 155073. 10.1016/j.aeue.2023.155073

Song, J., & Huang, J. (2023). A microfluidic antenna-sensor for contactless characterization of complex permittivity of liquids. *IEEE Sensors Journal*, 23(22), 27251–27261. 10.1109/JSEN.2023.3318213

Song, R., Deng, Q., Zhou, S., & Pu, M. (2021). Catenary-based phase change metasurfaces for mid-infrared switchable wavefront control. *Optics Express*, 29(15), 23006. 10.1364/OE.43484434614576

Son, P. T., Khuyen, B. X., Tung, B. S., Hiep, L. T. H., & Lam, V. D. (2022). A critical review on wireless power transfer systems using metamaterials. *Vietnam Journal of Science and Technology*, 60(4), 4. 10.15625/2525-2518/16954

Stevens, C. J. (2015). Magnetoinductive waves and wireless power transfer. *IEEE Transactions on Power Electronics*, 30(11), 6182–6190. 10.1109/TPEL.2014.2369811

Suganya, A., & Natarajan, R. (2023). Wideband reflecting frequency-selective surface polarizer for X-band applications. *International Journal of Communication Systems*, 36(17), e5608. 10.1002/dac.5608

Su, L., Mata-contreras, J., Vélez, P., & Martín, F. (2017). A review of sensing strategies for microwave sensors based on metamaterial-inspired resonators: Dielectric characterization, displacement, and angular velocity measurements for health diagnosis, telecommunication, and space applications. *International Journal of Antennas and Propagation*, 2017, 1–13. 10.1155/2017/5619728

Sumit Kumar, A. S. (2020). Fifth Generation Antennas: A Comprehensive Review of Design and Performance Enhancement Techniques. *IEEE Access, Digital Object Identifier.* IEEE. .10.1109/ACCESS.2020.3020952

Sun, Y., Zhang, L., Shi, H., Cao, S., Yang, S., & Wu, Y. (2021). Near-infrared plasma cavity metasurface with independently tunable double Fano resonances. *Results in Physics*, 25, 104204. 10.1016/j.rinp.2021.104204

Tabassum, S., Nayemuzzaman, S. K., Kala, M., Kumar Mishra, A., & Mishra, S. K. (2022). Metasurfaces for sensing applications: Gas, bio and chemical. *Sensors, 22*(18), 6896-1-6896–29. 10.3390/s22186896

Takase, H., & Takahara, J. (2021). Switchable wavefront control using an all-dielectric metasurface mediated by VO_2. *Applied Physics Express*, 14(3), 032007. 10.35848/1882-0786/abdd13

Tamayama, Y., Yasui, K., Nakanishi, T., & Kitano, M. (2014). A linear-to-circular polarization converter with half transmission and half reflection using a single-layered metamaterial. *Applied Physics Letters*, 105(2), 021110. 10.1063/1.4890623

Tan, C., Wang, S., Li, S., Liu, X., Wei, J., Zhang, G., & Ye, H. (2022). Cancer Diagnosis Using Terahertz-Graphene-Metasurface-Based Biosensor with Dual-Resonance Response. *Nanomaterials (Basel, Switzerland)*, 12(21), 3889. 10.3390/nano1221388936364665

Tang, Z., Liu, J., Cai, Y.-M., Wang, J., & Yin, Y. (2018, April). A wideband differentially fed dual-polarized stacked patch antenna with tuned slot excitations. *IEEE Transactions on Antennas and Propagation*, 66(4), 2055–2060. 10.1109/TAP.2018.2800764

Tan, L. R., Wu, R. X., Wang, C. Y., & Poo, Y. (2013, March). Magnetically Tunable Ferrite Loaded SIW Antenna. *IEEE Antennas and Wireless Propagation Letters*, 12, 273–275. 10.1109/LAWP.2013.2248113

Tatartschuk, E., Gneiding, N., Hesmer, F., Radkovskaya, A., & Shamonina, E. (2012, May). Mapping inter-element coupling in metamaterials: Scaling down to infrared. *Journal of Applied Physics*, 111(9), 094904. 10.1063/1.4711092

Tefek, U., Sari, B., Alhmoud, H. Z., & Hanay, M. S. (2023). Permittivity-based microparticle classification by the integration of impedance cytometry and microwave resonators. *Advanced Materials*, 35(46), 2304072. 10.1002/adma.20230407237498158

Temmar, M. N. E., Hocini, A., Khedrouche, D., & Denidni, T. A. (2021). Analysis and design of MIMO indoor communication system using terahertz patch antenna based on photonic crystal with graphene. *Photonics and Nanostructures*, 43, 100867. 10.1016/j.photonics.2020.100867

Temmar, M. N. E., Hocini, A., Khedrouche, D., & Zamani, M. (2019). Analysis and design of a terahertz microstrip antenna based on a synthesized photonic bandgap substrate using BPSO. *Journal of Computational Electronics*, 18(1), 231–240. 10.1007/s10825-019-01301-x

Temmar, M. N., Hocini, A., Khedrouche, D., & Denidni, T. A. (2020). Enhanced flexible terahertz microstrip antenna based on modified silicon-air photonic crystal. *Optik (Stuttgart)*, 217, 164897. 10.1016/j.ijleo.2020.164897

Tesla, N. (1914). *Apparatus for transmitting electrical energy* (United States Patent US1119732A). https://patents.google.com/patent/US1119732A/en

Thippeswamy, M. C., Kuchibhatla, S. A. R., & Rajagopal, P. (2021). Concentric shell gradient index metamaterials for focusing ultrasound in bulk media. *Ultrasonics*, 114, 106424. 10.1016/j.ultras.2021.10642433819870

Tiwari, N. K., Singh, S. P., & Akhtar, M. J. (2020). Adulteration detection in petroleum products using directly loaded coupled-line-based metamaterial-inspired submersible microwave sensor. *IET Science, Measurement & Technology*, 14(3), 376–385. 10.1049/iet-smt.2018.5687

Tretyakov, S. (2003). *Analytical Modeling in Applied Electromagnetics*. Artech House. google-Books-ID: MZ3tpGtadhcC.

Tretyakov, S. A. (2017). A personal view on the origins and developments of the metamaterial concept. *Journal of Optics*, 19(1), 013002. 10.1088/2040-8986/19/1/013002

Tripathi, S. K., & Kumar, A. (2019). High gain miniaturised photonic band gap terahertz antenna for size-limited applications. *Aust J Electr Electron Eng*, 16(2), 74–80. 10.1080/1448837X.2019.1602944

Troudi, Z., Machac, J., & Osman, L. (2019, October). Miniaturised planar band-pass filter based on interdigital arm SRR. *IET Microwaves, Antennas & Propagation*, 13(12), 2081–2086. 10.1049/iet-map.2018.5708

Tu, Thanh, Hoc, & Yem. (2017). Double-side electromagnetic band gap structure for improving dual-band MIMO antenna performance. *REV Journal on Electronics and Communications, 7*(1).

Upadhyaya, T. K., Kosta, S. P., Jyoti, R., & Palandoken, M. (2014, October 07). Negative refractive index material-inspired 90-deg electrically tilted ultra wideband resonator. *Optical Engineering (Redondo Beach, Calif.)*, 53(10), 107104–107104. 10.1117/1.OE.53.10.107104

Urja Sudhir Ingle. (2022). *Reconfigurable Antenna for 5G Applicationsx.* ACCAI.

Varuna, A. B., Ghosh, S., & Srivastava, K. V. (2016). An ultra thin polarization insensitive and angularly stable miniaturized frequency selective surface. *Microwave and Optical Technology Letters*, 58(11), 2713–2717. 10.1002/mop.30132

Varuna, A. B., Ghosh, S., & Srivastava, K. V. (2017). A miniaturized-element bandpass frequency selective surface using meander line geometry. *Microwave and Optical Technology Letters*, 59(10), 2484–2489. 10.1002/mop.30760

Vasic, B., Isic, G. & Gajic, R. (2013). Localized surface plasmon resonances in graphene ribbon arrays for sensing of dielectric environment at infrared frequencies. *Journal of Applied Physics*, 113, 013110-013110-013117.

Vasu Babu, K., Das, S., Varshney, G., Sree, G. N. J., & Madhav, B. T. P. (2022). A micro-scaled graphene-based treeshaped wideband printed MIMO antenna for terahertz applications. *Journal of Computational Electronics*, 21(1), 289–303. 10.1007/s10825-021-01831-3

Vasu Babu, K., Sree, G. N. J., Kumari, S. V., & Das, S. (2022). Design and analysis of a CPW-fed fractal MIMO THz antenna using an array of parasitic elements. In Das, S., Nella, A., & Patel, S. K. (Eds.), *Terahertz devices, circuits and systems*. Springer. 10.1007/978-981-19-4105-4_4

Velez, P., Paredes, F., Casacuberta, P., Elgeziry, M., Su, L., Munoz-Enano, J., Costa, F., Genovesi, S., & Martin, F. (2023). Portable reflective-mode phase-variation microwave sensor based on a rat-race coupler pair and gain/phase detector for dielectric characterization. *IEEE Sensors Journal*, 23(6), 5745–5756. 10.1109/JSEN.2023.3240771

Velez, P., Su, L., Grenier, K., Mata-Contreras, J., Dubuc, D., & Martin, F. (2017). Microwave microfluidic sensor based on a microstrip splitter/combiner configuration and split ring resonators (SRRs) for dielectric characterization of liquids. *IEEE Sensors Journal*, 17(20), 6589–6598. 10.1109/JSEN.2017.2747764

Venkatesh, S., Lu, X., Saeidi, H., & Sengupta, K. (2020). A high-speed programmable and scalable terahertz holographic metasurface based on tiled CMOS chips. *Nature Electronics*, 3(12), 785–793. 10.1038/s41928-020-00497-2

Veselago, V. G. (1968). The Electrodynamics of Substances with Simultaneously Negative Values of ε and μ. *Physics Uspekhi*, 10, 509–514. 10.1070/PU1968v010n04ABEH003699

Viswanathan, A. P., Moolat, R., Mani, M., Shameena, V. A., & Pezholil, M. (2020). A simple electrically small microwave sensor based on complementary asymmetric single split resonator for dielectric characterization of solids and liquids. *International Journal of RF and Microwave Computer-Aided Engineering*, 30(12), 1–13. 10.1002/mmce.22462

Vivek, A., Shambavi, K., & Alex, Z. C. (2019). A review: Metamaterial sensors for material characterization. *Sensor Review*, 39(3), 417–432. 10.1108/SR-06-2018-0152

Walser, R. (1999). Metamaterials: What are they? What are they good for? *Acta Obstetricia et Gynecologica Scandinavica*, 97, 388–393.

Wang, B. X., Zhao, W. S., Wang, D. W., Wang, J., Li, W., & Liu, J. (2021). Optimal design of planar microwave microfluidic sensors based on deep reinforcement learning. *IEEE Sensors Journal*, 21(24), 27441–27449. 10.1109/JSEN.2021.3124294

Wang, B., Teo, K. H., Nishino, T., Yerazunis, W., Barnwell, J., & Zhang, J. (2011). Experiments on wireless power transfer with metamaterials. *Applied Physics Letters*, 98(25), 254101. 10.1063/1.3601927

Wang, B.-X., Xu, C., Duan, G., Xu, W., & Pi, F. (2023). Review of broadband metamaterial absorbers: From principles, design strategies, and tunable properties to functional applications. *Advanced Functional Materials*, 33(14), 2213818. 10.1002/adfm.202213818

Wang, C., Chen, Y., & Yang, S. (2018, November). Dual-band dual-polarized antenna array with flat-top and sharp cutoff radiation patterns for 2G/3G/LTE cellular bands. *IEEE Transactions on Antennas and Propagation*, 66(11), 5907–5917. 10.1109/TAP.2018.2866596

Wang, G., Zhu, F., Lang, T., Liu, J., Hong, Z., & Qin, J. (2021). All-metal terahertz metamaterial biosensor for protein detection. *Nanoscale Research Letters*, 16(1), 109. 10.1186/s11671-021-03566-334191133

Wang, H., Yan, M., Qu, S., Zheng, L., & Wang, J. (2019). Design of a self-complementary frequency selective surface with multi-band polarization separation characteristic. *IEEE Access : Practical Innovations, Open Solutions*, 7, 36788–36799. 10.1109/ACCESS.2019.2905416

Wang, H., Zhou, L., Zhang, X., & Xie, H. (2018). Thermal Reliability Study of an Electrothermal MEMS Mirror. *IEEE Transactions on Device and Materials Reliability*, 18(3), 422–428. 10.1109/TDMR.2018.2860286

Wang, M., Guo, J., Shi, Y., Wang, M., Song, G., & Yin, R. (2023). A metamaterial-incorporated wireless power transmission system for efficiency enhancement. *International Journal of Circuit Theory and Applications*, 51(7), 3051–3065. 10.1002/cta.3587

Wang, Q., Chen, Y., Mao, J., Yang, F., & Wang, N. (2023). Metasurface-assisted terahertz sensing. *Sensors (Basel)*, 23(13), 1–17. 10.3390/s2313590237447747

Wang, S. R., Chen, M. Z., Ke, J. C., Cheng, Q., & Cui, T. J. (2022). Asynchronous Space-Time-Coding Digital Metasurface. *Advancement of Science*, 9(24), 2200106. 10.1002/advs.20220010635751468

Wang, S. R., Dai, J. Y., Zhou, Q. Y., Ke, J. C., Cheng, Q., & Cui, T. J. (2023). Manipulations of multi-frequency waves and signals via multi-partition asynchronous space-time-coding digital metasurface. *Nature Communications*, 14(1), 1. 10.1038/s41467-023-41031-037666804

Wang, Y., Gao, H., & Ren, Q. (2022). Differential Operator Approximation Based Tightly Coupled Multiphysics Solver Using Cascaded Fourier Network. *Advanced Theory and Simulations*, 5(11), 2200409. 10.1002/adts.202200409

Wang, Y., & Ren, Q. (2022). A versatile inversion approach for space/temperature/time-related thermal conductivity via deep learning. *International Journal of Heat and Mass Transfer*, 186, 122444. 10.1016/j.ijheatmasstransfer.2021.122444

Wang, Y., & Ren, Q. (2023). *Deep Learning-Based Forward Modeling and Inversion Techniques for Computational Physics Problems* (1st ed.). CRC Press., 10.1201/9781003397830

Wang, Y., Zhou, J., Ren, Q., Li, Y., & Su, D. (2021). 3-D Steady Heat Conduction Solver via Deep Learning. *IEEE Journal on Multiscale and Multiphysics Computational Techniques*, 6, 100–108. 10.1109/JMMCT.2021.3106539

Wang, Z., Chen, J., Khan, S. A., Li, F., Shen, J., Duan, Q., Liu, X., & Zhu, J. (2022). Plasmonic Metasurfaces for Medical Diagnosis Applications: A Review. *Sensors (Basel)*, 22(1), 133. 10.3390/s2201013335009676

Wei, Z., Cao, Y., Su, X., Gong, Z., Long, Y., & Li, H. (2013). Highly efficient beam steering with a transparent metasurface. *Optics Express*, 21(9), 10739–10745. 10.1364/OE.21.01073923669930

Wen, J., Ren, Q., Peng, R., & Zhao, Q. (2022a). Multi-functional tunable ultra-broadband water-based metasurface absorber with high reconfigurability. *Journal of Physics. D, Applied Physics*, 55(28), 285103. 10.1088/1361-6463/ac683e

Wen, J., Ren, Q., Peng, R., & Zhao, Q. (2022b). Ultrabroadband Saline-Based Metamaterial Absorber With Near Theoretical Absorption Bandwidth Limit. *IEEE Antennas and Wireless Propagation Letters*, 21(7), 1388–1392. 10.1109/LAWP.2022.3169467

Wen, L.-H., Gao, S., Mao, C.-X., Luo, Q., Hu, W., Yin, Y., & Yang, X. (2018). A wideband dual-polarized antenna using shorted dipoles. *IEEE Access : Practical Innovations, Open Solutions*, 6, 39725–39733. 10.1109/ACCESS.2018.2855425

Wu, G.-B., Dai, J. Y., Cheng, Q., Cui, T. J., & Chan, C. H. (2022). Sideband-free space–time-coding metasurface antennas. *Nature Electronics*, 5(11), 808–819. 10.1038/s41928-022-00857-0

Wu, J., Shen, Z., Ge, S., Chen, B., Shen, Z., Wang, T., Zhang, C., Hu, W., Fan, K., Padilla, W., Lu, Y., Jin, B., Chen, J., & Wu, P. (2020). Liquid crystal programmable metasurface for terahertz beam steering. *Applied Physics Letters*, 116(13), 131104. 10.1063/1.5144858

Wu, J., Yang, D., Huang, X., Li, Y., & Xia, Y. (2021). The design and experiment of a novel microwave gas sensor loaded with metamaterials. *Physics Letters, Section A: General. Physics Letters. [Part A]*, 389, 127080-1, 127080–127086. 10.1016/j.physleta.2020.127080

Xie, B., Gao, Z., Wang, C., Ali, L., Muhammad, A., Meng, F., Qian, C., Ding, X., Adhikari, K. K., & Wu, Q. (2023). High-sensitivity liquid dielectric characterization differential sensor by 1-bit coding DGS. *Sensors, 23*(1), 372-1-372–12. 10.3390/s23010372

Xi, S., Xu, K. D., Yang, S., Ren, X., & Wu, W. (2024). X-band frequency selective surface with low loss and angular stability. *AEÜ. International Journal of Electronics and Communications*, 173, 154990. 10.1016/j.aeue.2023.154990

Xu, H.-X., Hu, G., Wang, Y., Wang, C., Wang, M., Wang, S., Huang, Y., Genevet, P., Huang, W., & Qiu, C.-W. (2021). Polarization-insensitive 3D conformal-skin metasurface cloak. *Light, Science & Applications*, 10(1), 75. 10.1038/s41377-021-00507-833833215

Xu, T., Xu, X., & Lin, Y. S. (2021). Tunable terahertz free spectra range using electric split-ring metamaterial. *Journal of Microelectromechanical Systems*, 30(2), 309–314. 10.1109/JMEMS.2021.3057354

Yadav, S., Jain, C. P., & Sharma, M. M. (2018). Polarization independent dual-bandpass frequency selective surface for Wi-Max applications. *International Journal of RF and Microwave Computer-Aided Engineering*, 28(6), e21278. 10.1002/mmce.21278

Yadav, S., Sharma, M. M., & Singh, R. K. (2021). A polarization insensitive tri-band bandpass frequency selective surface for Wi-MAX and WLAN applications. *Progress in Electromagnetics Research Letters*, 101, 127–136. 10.2528/PIERL21091101

Yalcin, A., Popat, K. C., Aldridge, J. C., Desai, T. A., Hryniewicz, J., Chbouki, N., Little, B. E., King, O., Van, V., & Chu, S. (2006). Optical sensing of biomolecules using microring resonators. *IEEE Journal of Selected Topics in Quantum Electronics*, 12(1), 148–155. 10.1109/JSTQE.2005.863003

Yang, D., Liu, S., & Zhao, Z. (2017, May). A broadband dual-polarized printed dipole antenna with low cross-polarization and high isolation for base station applications. *Microwave and Optical Technology Letters*, 59(5), 1107–1111. 10.1002/mop.30467

Yang, D., Wang, W., Lv, E., Wang, H., Liu, B., Hou, Y., & Chen, J. (2022). Programmable VO2 metasurface for terahertz wave beam steering. *iScience*, 25(8), 104824. 10.1016/j.isci.2022.10482435992076

Yang, P. (2020). *Reconfigurable 3-D Slot Antenna Design for 4G and Sub-6G Smartphones with Metallic Casing*. MDPI. 10.3390/electronics9020216

Yao, J., Ou, J. Y., Savinov, V., Chen, M. K., Kuo, H. Y., Zheludev, N. I., & Tsai, D. P. (2022). Plasmonic anapole metamaterial for refractive index sensing. *PhotoniX*, 3(1), 23. 10.1186/s43074-022-00069-x

Yasin, A., Gogosh, N., Sohail, S. I., Abbas, S. M., Shafique, M. F., & Mahmoud, A. (2023). Relative permittivity measurement of microliter volume liquid samples through microwave filters. *Sensors, 23*(6), 2884-1-2884–13. 10.3390/s23062884

Ye, L., Zhang, G., & You, Z. (2017). 5 V Compatible Two-Axis PZT Driven MEMS Scanning Mirror with Mechanical Leverage Structure for Miniature LiDAR Application. *Sensors (Basel)*, 17(3), 521. 10.3390/s1703052128273880

Ye, Q., Wang, J., Liu, Z., Deng, Z. Ch., Kong, X. T., Xing, F., Chen, X. D., Zhou, W. Y., Zhang, C. P., & Tian, J. G. (2013). Polarization-dependent optical absorption of graphene under total internal reflection. *Applied Physics Letters*, 102(2), 021912. 10.1063/1.4776694

Yoo, S., & Park, Q.-H. (2019). Metamaterials and chiral sensing: A review of fundamentals and applications. *Nanophotonics*, 8(2), 249–261. 10.1515/nanoph-2018-0167

Yotter, R. A., & Wilson, D. M. (2004). Sensor technologies for monitoring metabolic activity in single cells-part ii: Nonoptical methods and applications. *IEEE Sensors Journal*, 4(4), 412–429. 10.1109/JSEN.2004.830954

You, K. Y. (2017). Materials characterization using microwave waveguide system. In *Microwave Systems and Applications* (pp. 341–358). 10.5772/66230

You, K. Y., & Abbas, Z. (2011). Waveguide techniques for determination of moisture content in materials. In *Antenna and Applied Electromagnetic Application* (pp. 113–130).

You, K. Y., & Esa, F. Bin, & Abbas, Z. (2017). Macroscopic characterization of materials using microwave measurement methods - A survey. *2017 Progress in Electromagnetics Research Symposium-Fall (PIERS-FALL)*, pp. 194–204. 10.1109/PIERS-FALL.2017.8293135

You, K. Y., & Sim, M. S. (2018). Precision permittivity measurement for low-loss thin planar materials using large coaxial probe from 1 to 400 MHz. *Journal of Manufacturing and Materials Processing, 2*(4), 81-1-81–15. 10.3390/jmmp2040081

You, K. Y., Sim, M. S., & Abdullah, S. N. (2020). Emerging microwave technologies for agricultural and food processing. In *Precision Agriculture Technologies for Food Security and Sustainability* (pp. 94–148). 10.4018/978-1-7998-5000-7.ch005

You, K. Y., Sim, M. S., Mutadza, H., Esa, F., & Chan, Y. L. (2017). Free-space measurement using explicit, reference-plane and thickness-invariant method for permittivity determination of planar materials. *2017 Progress in Electromagnetics Research Symposium-Fall (PIERS-FALL)*, (pp. 222–228). IEEE. 10.1109/PIERS-FALL.2017.8293139

Younssi, M., Jaoujal, A., Yaccoub, M. D., El Moussaoui, A., & Aknin, N. (2013). Study of a microstrip antenna with and without superstrate for terahertz frequency. *International Journal of Innovation and Applied Studies*, 2(4), 369–371.

Yuan, Y., Yu, X., Ouyang, Q., Shao, Y., Song, J., Qu, J. & Yong, K.T. (2018). Highly anisotropic black phosphorous-graphene hybrid architecture for ultrassensitive plasmonic biosensing: Theoretical insight. *2D Materials*, 5, 025015.

Yuan, J., Liu, S., Bian, B., Kong, X., Zhang, H., & Wang, S. (2014). A novel high-selective band-pass frequency selective surface with multiple transmission zeros. *Journal of Electromagnetic Waves and Applications*, 28(17), 2197–2209. 10.1080/09205071.2014.959620

Yuan, Y., Xi, X., & Zhao, Y. (2019). Compact UWB FSS reflector for antenna gain enhancement. *IET Microwaves, Antennas & Propagation*, 13(10), 1749–1755. 10.1049/iet-map.2019.0083

Yu, N., Genevet, P., Kats, M. A., Aieta, F., Tetienne, J.-P., Capasso, F., & Gaburro, Z. (2011). Light propagation with phase discontinuities: Generalized laws of reflection and refraction. *Science*, 1210713. Sun,S., He, Q., Xiao, S., Xu, Q., Li, X., & Zhou, L. (2012). Gradient-index meta-surfaces as a bridge linking propagating waves and surface waves. *Nature Materials*, 11, 426.

Yu, N., Genevet, P., Kats, M. A., Aieta, F., Tetienne, J.-P., Capasso, F., & Gaburro, Z. (2011). Light Propagation with Phase Discontinuities: Generalized Laws of Reflection and Refraction. *Science*, 334(6054), 333–337. 10.1126/science.121071321885733

Zainud- Deen., S. H., Gaber, Shaymaa. M., & Awadalla, K. H. (2012). Beam steering reflectarray using varactor diodes. *2012 Japan-Egypt Conference on Electronics, Communications and Computers*, (pp. 178–181). IEEE. 10.1109/JEC-ECC.2012.6186979

Zhang, Y., Hong, W., Yu, C., Kuai, Z. Q., Don, Y. D., & Zhou, J. Y. (2008). Planar ultrawideband antennas with multiple notched bands based on etched slots on the patch and or split ring resonators on the feed line. *IEEE Trans. Antennas Propag., 56*(9), 3063–3068. doi:.928815 doi:10.1109/tap.2008

Zhang, C., Liu, Q., Peng, X., Ouyang, Z., & Shen, S. (2021). Sensitive THz sensing based on Fano resonance in all-polymeric Bloch surface wave structure. *Nanophotonics*, 10(15), 3879–3888. 10.1515/nanoph-2021-0339

Zhang, L., Chen, X. Q., Liu, S., Zhang, Q., Zhao, J., Dai, J. Y., Bai, G. D., Wan, X., Cheng, Q., Castaldi, G., Galdi, V., & Cui, T. J. (2018). Space-time-coding digital metasurfaces. *Nature Communications*, 9(1), 4334. 10.1038/s41467-018-06802-030337522

Zhang, R., Wang, K., Wang, X., Luo, X., & Zhao, C. (2023). Broadband and precise reconfiguration of megahertz electromagnetic metamaterials for wireless power transfer. *Physica Scripta*, 98(11), 115529. 10.1088/1402-4896/acfea9

Zhang, S., Chen, X., & Pedersen, G. F. (2019). Mutual coupling suppression with decoupling ground for massive MIMOantenna arrays. *IEEE Transactions on Vehicular Technology*, 68(8), 7273–7282. 10.1109/TVT.2019.2923338

Zhang, W., Song, Q., Zhu, W., Shen, Z., Chong, P., Tsai, D. P., Qiu, C., & Liu, A. Q. (2018). Metafluidic metamaterial: A review. *Advances in Physics: X*, 3(1), 165–184. 10.1080/23746149.2017.1417055

Zhang, Z., Yang, J., He, X., Zhang, J., Huang, J., Chen, D., & Han, Y. (2018). Plasmonic Refractive Index Sensor with High Figure of Merit Based on Concentric-Rings Resonator. *Sensors (Basel)*, 18(2), 116. 10.3390/s1801011629300331

Zhao, P. C., Zong, Z. Y., Wu, W., Li, B., & Fang, D. G. (2019). Miniaturized-element bandpass FSS by loading capacitive structures. *IEEE Transactions on Antennas and Propagation*, 67(5), 3539–3544. 10.1109/TAP.2019.2902633

Zhao, W. S., Wang, B. X., Wang, D. W., You, B., Liu, Q., & Wang, G. (2022). Swarm intelligence algorithm-based optimal design of microwave microfluidic sensors. *IEEE Transactions on Industrial Electronics*, 69(2), 2077–2087. 10.1109/TIE.2021.3063873

Zhao, Z., Li, J., Shi, H., Chen, X., & Zhang, A. (2018). A low-profile angle-insensitive bandpass frequency-selective surface based on vias. *IEEE Microwave and Wireless Components Letters*, 28(3), 200–202. 10.1109/LMWC.2018.2798166

Zheludev, N. I., & Kivshar, Y. S. (2012). From metamaterials to metadevices. *Nature Materials*, 11(11), 11. Advance online publication. 10.1038/nmat343123089997

Zheng, B., Rao, X., Shan, Y., Yu, C., Zhang, J., & Li, N. (2023). Multiple-Beam Steering Using Graphene-Based Coding Metasurfaces. *Micromachines*, 14(5), 1018. 10.3390/mi1405101837241641

Zheng, H., Pham, T. S., Chen, L., & Lee, Y. (2024). Metamaterial Perfect Absorbers for Controlling Bandwidth: Single-Peak/Multiple-Peaks/Tailored-Band/Broadband. *Crystals*, 14(1), 1. 10.3390/cryst14010019

Zhong, J., Xu, X., & Lin, Y. S. (2021). Tunable terahertz metamaterial with electromagnetically induced transparency characteristic for sensing application. *Nanomaterials (Basel, Switzerland)*, 11(9), 2175. 10.3390/nano1109217534578491

Zhou, H., Hu, D., Yang, C., Chen, C., Ji, J., Chen, M., Chen, Y., Yang, Y., & Mu, X. (2018). Multi-band sensing for dielectric property of chemicals using metamaterial integrated microfluidic sensor. *Scientific Reports*, 8(1), 1–11. 10.1038/s41598-018-32827-y30287826

Zhou, H., Qu, S., Lin, B., Wang, J., Ma, H., Xu, Z., Peng, W., & Bai, P. (2012). Filter-antenna consisting of conical FSS radome and monopole antenna. *IEEE Transactions on Antennas and Propagation*, 60(6), 3040–3045. 10.1109/TAP.2012.2194648

Zhuang, X., Zhang, W., Wang, K., Gu, Y., An, Y., Zhang, X., Gu, J., Luo, D., Han, J., & Zhang, W. (2023). Active terahertz beam steering based on mechanical deformation of liquid crystal elastomer metasurface. *Light, Science & Applications*, 12(1), 14. 10.1038/s41377-022-01046-636596761

Zhu, H. L., Cheung, S. W., Chung, K. L., & Yuk, T. I. (2013). Linear-to-circular polarization conversion using metasurface. *IEEE Transactions on Antennas and Propagation*, 61(9), 4615–4623. 10.1109/TAP.2013.2267712

Zhu, J., Wang, Z., Lin, S., Jiang, S., Liu, X., & Guo, S. (2020). Low-cost flexible plasmonic nanobump metasurfaces for label-free sensing of serum tumor marker. *Biosensors & Bioelectronics*, 150, 111905. 10.1016/j.bios.2019.11190531791874

About the Contributors

Jingda Wen received the Master degree in Beihang University in 2022, where he also got a Bachelor degree in 2020. His research interest including metamaterial and metasurface. He has authored more than 8 academic articles including SCI journals such as Applied Physics Letters, IEEE AWPL, and so on. Moreover, he serves as a reviewer and editorial board member for SCI journals. In addition, he recevied more than 15 kinds of awards and scholarships from Beihang University and the Chinese Ministry of Education.

***

Raimi Dewan received his Bachelor Degree in Electrical and Telecommunication engineering, Master Degree (Electrical) and Ph.D. (Electrical Engineering) from Universiti Teknologi Malaysia (UTM), in 2010, 2013, and 2019, respectively. His research area is in Wireless Communication Engineering, RF & Microwave and Biomedical Application. He is currently a senior lecturer at the Department of Biomedical Engineering and Health Sciences, Faculty of Electrical Engineering, Universiti Teknologi Malaysia (UTM). Prior to joining UTM, he was a UHF RFID Product Coordinator at HID Global. From 2019 to 2020, He joined the Institute of Electronics and Telecommunications of Rennes, University of Rennes 1, France as a Research Engineer Cum Postdoctoral Researcher. He was actively working with Naval Group (the Project Lead) and Seribase for the confidential industrial projects under the Ministry of Arms (DGA) France. In 2018, he was a Senior Manufacturing Engineer with Keysight Technologies. He was also an RF Design Engineer I with Smartrac Technology. He also serves as the active Reviewer for various International journals, including (but not limited to) IEEE ACCESS, IEEE MWCL, IEEE AWPL, IEEE Journal of Radio Frequency Identification, Progress in Electromagnetic Research, the International Journal of Electronics and Communications, Elsevier-Renewable and Sustainable Energy Reviews, Plasmonics by Springer, Wireless Personal Communications, IET Electronic Letters, IET Microwaves, Antennas and Propagation, and the International Journal of RF and Microwave Computer-Aided Engineering. He has published more than 80 articles related in international journals, conferences and book chapters.

Smrity Dwivedi is working as an Assistant Professor in the department of Electronics Engg., IIT (BHU), Varanasi. She is Senior Member of IEEE, Fellow of IETE, Member of IET, LMISC, Guest Editor in Elsevier, Life member of VEDA society. She completed her doctoral work from IIT (BHU), VNS, INDIA in RF and Microwaves. She worked as a research fellow (postdoctoral type work) from IIT-BHU, INDIA. She has published many research papers in international journals, international conferences and national conferences. She is also published many chapters in reputed books.

Fahmiruddin Esa was born in 1984. He graduated from Universiti Putra Malaysia (UPM) with a Bachelor's degree in Physics in 2006. Two years later, he pursued his studies at Universiti Teknologi Malaysia (UTM) and obtained a Master of Science (Physics) degree through coursework in 2010. In 2011, he began his Ph.D. studies in Microwave at the Faculty of Science, UPM, and completed it in 2015. Currently, he holds a position as a lecturer at the Department of Physics and Chemistry, Faculty of Applied Sciences and Technology, Pagoh Higher Education Hub, Universiti Tun Hussein Onn Malaysia (UTHM), 84600 Panchor, Johor, Malaysia. His research interest lies in experimental and simulation techniques for the characterization of microwave-absorbing materials.

Maryam Ghodrati research interests include nanoelectronics, Optical sensors, and biosensors. She has authored over 10 academic articles including ISI journals such as IEEE Sensor, The European Physical Journal Plus, Materials Science and Engineering: B, etc.

Fandi Hamid is working as a military Signal Officer in the Malaysian Armed Forces. He was born in Sarawak, Malaysia, in 1981. He received his Bachelor of Engineering (Electrical – Electronics) from the Universiti Teknologi Malaysia, Johor, Malaysia, in 2003 and Master of Engineering (Telecommunications) from the Universiti Malaya, Kuala Lumpur, Malaysia, in 2016. He is currently pursuing his Doctor of Philosophy (PhD) at the Faculty of Electrical Engineering, Universiti Teknologi Malaysia, Johor, Malaysia. His research interest includes electromagnetic pulse (EMP), metamaterials, microwave antennas, and antenna theory.

Manh Kha Hoang received the B.E and M.E degree in Electronics and Telecommunications Engineering both from Hanoi University of Science and Technology, in 2002 and 2004, respectively. He obtained his Dr.- Eng. (PhD) degree in Communications Engineering from the University of Paderborn, Germany, in 2016 with specialization in parameter estimation of missing data and application to indoor positioning. He is now working as a dean of Faculty of Electronics, Hanoi University of Industry. He has served as the CoTPC Chairs of RICE (2019), TPC member of ATC (2018, 2019), reviewer of REV-ECIT, ATC, Journal of Science and Technology (ISSN 1859-3585). His research interests include digital signal processing, wireless communication, indoor positioning, machine learning and pattern classification, and nature-inspired algorithm application.

Kanwar Preet Kaur (Member, IEEE) completed her Ph.D. in Microwave Metamaterial Absorber at Charotar University of Science and Technology, Changa, Anand, Gujarat, India, in 2020 and her M. E. degree in Communication Systems from Jabalpur Engineering College, Jabalpur, M. P., India in 2008. She is currently an Assistant Professor with the Electronics and Communication Engineering Department, Charotar University of Science and Technology (CHARUSAT). She has published more than 20 research articles, mostly in SCI/Scopus journals, international conferences, and book chapters. Her expertise extends to high-frequency techniques, with a particular emphasis on ultrathin multiband/broadband absorbers, the design of energy harvesting systems, bandpass/bandstop FSS, and patch antennas.

Stephanie Yen Nee Kew received her Bachelor of Science (Honours) from Quest International University (QIU). Her journey into research exploration has been marked by one of her achievements in the International Engineering Innovation Challenge (EIC) in 2021 and 2022, organized by the Institute of Engineers (IES) in Singapore. In 2023, she attained the CHST UTAR Research Centre Excellence Award at Lee Kong Chian Faculty of Engineering and Science, Universiti Tunku Abdul Rahman (UTAR). Moving forward, she joined an industrial training program at the Faculty of Electrical Engineering, Universiti Teknologi Malaysia (UTM) and is actively involved in research on microwave measurements and material characterization. She also became a distinguished member of Engineering for Change (E4C), United States in 2024.

About the Contributors

Mehaboob Mujawar, Associate Professor and HOD, Department of Artificial Intelligence and Data Science, Bearys Institute of Technology, Mangalore. He completed his B.E degree in Electronics and Telecommunication Engineering in 2015 from Don Bosco College of Engineering, Fatorda and had secured 2nd Rank in Electronics Telecommunication Department. He has completed M.E in Industrial Automation and Radio Frequency Engineering from Goa College of Engineering in 2017 and had secured 3rd Rank to Goa University. He also holds M.Tech in AI. He has completed his PHD in the year 2023. He has published four books. He has Published/Presented 36 papers in the National/International journals and Conferences. He has published 12 book chapters with reputed publishers like CRC Press, Springer and Nova Publishers USA. He is reviewer of reputed journals and has worked as reviewer for international conferences. He is a member of IETE, ISTE, IAENG and INSC. He has 2 Indian Patent and one South African Patent to his credit. He has received Seven Awards for Excellent Performance in Academics and Research. He has 5 years teaching experience, currently working as Associate Professor and HOD, Department of Artificial Intelligence and Data Science at Bearys Institute of Technology, Mangalore. His area of interest is Antenna Design, Printed Circuit Board Design, Microwave and RF Engineering, Artificial Intelligence and Robotics, Data Science, Design of Algorithms

DiviyaDevi Paramasivam received a Bachelor of Engineering in Bio-Medical in 2022 from Universiti Teknologi Malaysia (UTM). Currently, she is a first-year Ph.D. student pursuing her studies in Wireless Communication at UTM. She is working on designing, simulating, fabricating, and testing the Radio Frequency Identification (RFID) tag for on-body applications. She has been a Graduate Research Assistant for 3 months in 2023 to complete a funded project that focuses on the incorporation of metamaterial with RFID tags to enhance the on-body read range. Recently, she was appointed as a Graduate Research Assistant for a project that focuses on implantable antenna. Her areas of interest include Smart Healthcare, On-body Wireless Devices, RFID, and Implantable Antenna. Her long-term career goal is to pursue a researcher position in the Research and Development department.

Upesh Patel received his M.Tech. and Ph.D. in Communication Systems from the Charotar University of Science and Technology (CHARUSAT), Changa, in 2011 and 2020 respectively. He is currently working as an Associate Professor and the Head of Department with the Department of Electronics and Communication Engineering, Faculty of Technology and Engineering, CHARUSAT. He has published several research articles and his current research interests include commercial antenna design, RF, and microwave engineering.

Thanh Son Pham was born in Hanoi, Vietnam. He received the B.Sc., and M. Sc. degrees in engineering physics from Vietnam National University, Hanoi, Vietnam, in 2011 and 2013, respectively. He received the Ph.D. degree in electronics engineering from Kyung Hee University, Seoul, Korea, in 2019. From 2011 to now, he was a researcher at Vietnam Academy of Science and Technology, Hanoi, Vietnam. His current research interests include metamaterial for wireless power transfer, metasurface and photonic structure. He has published more than 30 SCIE papers in these research fields.

Xuan Thanh Pham received the Ph.D. degree from the School of Electronics and Information Engineering, Kyung Hee University, Seoul, Korea, in 2021; the M.S. degree in Electronics and Telecommunications Engineering from Ha Noi University of Science and Technology, Ha Noi, Viet Nam, in 2013. Currently, he works as a lecturer in Faculty of Electronics Engineering, Hanoi University of Industry. His research interests and area are CMOS analog, mix-signal integrated circuits and sensors for biomedical applications. He serves as a reviewer for the AEÜ - International Journal of Electronics and Communications, Elsevier; Journal of Military Science and Technology, Journal of Science and Technology (ISSN 1859-3585); TPC member of ATC (2023), Chair session Integrated Circuit (ATC 2023).

Subuh Pramono was born in Indonesia. He received the B. Eng degree in Electrical Engineering from Telkom University, Indonesia in 2003, and the M. Eng degree in Electrical Engineering from Institut Teknologi Bandung (ITB), Indonesia, in 2009. He got his Ph.D from Graduate School of Science and Engineering, Chiba University, Japan in 2024. He has published three books, four book chapters, and three Indonesia patents. He has published 84 papers in the international journals and conferences. He is currently a lecturer at Department of Electrical Engineering, Faculty of Engineering, Universitas Sebelas Maret, Surakarta-Indonesia, and he is serving as reviewer in some reputable journals. He is a member of IEEE, IEEE Comsoc, and IEEE Antennas & Propagation Society. His main areas of research interests are wireless and mobile communications, antenna and propagation, wireless sensor networks, and synthetic aperture radar.

Man Seng Sim was born in Johor, Malaysia, in 1993. He obtained his Bachelor of Science in Applied Physics with Honors from Universiti Tun Hussein Onn Malaysia (UTHM) in 2017. In 2019, he completed his Master of Philosophy in Electrical Engineering from Universiti Teknologi Malaysia (UTM). He is currently a Ph.D. research student in the Faculty of Electrical Engineering at UTM. His research interests include microwave metamaterials, microwave absorbers, microwave sensors, and material characterization.

Poonam Thanki received her M. Tech. degree in Electronics and Communication Engineering from PIET, GTU, Gujarat in 2012, and the Ph.D. degree in Electronics and Communication Engineering from the Charotar University of Science and Technology, (CHARUSAT), Changa, in February 2022 in the field of Reconfigurable antennas. Currently, she is an Assistant Professor at the Department of Electronics and Communication Engineering, Faculty of Technology and Engineering, CHARUSAT. She has authored several research papers/book chapters in reputed journals and conferences. Her research interests include Reconfigurable antennas, MIMO antennas, and Wireless Communication.

Trushit Upadhyaya (Senior Member, IEEE) received an M.E. degree from the Institute of Telecommunication Research, University of South Australia, Adelaide, in 2007, and a Ph.D. degree in satellite antennas from the Charotar University of Science and Technology (CHARUSAT), Changa, in 2014. He is currently working as a Professor with the Department of Electronics and Communication Engineering and the Principal of Chandubhai S. Patel Institute of Technology (CSPIT), Faculty of Technology and Engineering, CHARUSAT. His research interests include antenna system design and applied electromagnetics and he has carried out several research projects and published several papers/articles/book chapters.

Yinpeng Wang received the B.S. degree in Electronic and Information Engineering and the M.S. degree in Electronic Science and Technology both at Beihang University in 2020 and 2023, respectively. He is now a PhD student at National University of Singapore. From 2017 to 2018, he was a researcher at the Physical Experiment Center, Beihang University. In 2018, he worked as a research assistant at the Spintronics Interdisciplinary Center. Since 2018, he has been a member of the Institute of EMC Technology. Since 2018, Mr. Wang has published 2 academic monographs and more than 20 peer-reviewed technical papers in international journals and conferences. He serves as a reviewer for Springer, IOP, Elsevier, and IEEE journals.

Kok Yeow You was born in 1977. He obtained his B.Sc. Physics (Honours) degree from Universiti Kebangsaan Malaysia (UKM) in 2001. He pursued his M.Sc. in Microwave at the Faculty of Science in 2003 and his Ph.D. in Wave Propagation at the Institute for Mathematical Research in 2006 at Universiti Putra Malaysia (UPM), Serdang, Selangor, Malaysia. He is currently working as a senior lecturer at the Faculty of Electrical Engineering, Faculty of Engineering, Universiti Teknologi Malaysia (UTM), Skudai, Johor, Malaysia. His main research interest lies in the theory, simulation, and instrumentation of electromagnetic wave propagation at microwave frequencies, with a focus on the development of microwave passive devices and sensors for medical and agricultural applications.

Index

O

P

R

S

U

W

Milton Keynes UK
Ingram Content Group UK Ltd.
UKHW010228300724
446304UK00005B/102

9 798369 325995